Antonie van Leeuwenhoek **60**: 131, 1991.

Editorial

The study of quantitative aspects of growth and metabolism in microorganisms started in 1950 with the classical work of Monod. In his paper the yield factor was defined, and further it contained the equation which gives the relation between the rate of substrate uptake and substrate concentration. In 1960 the relation was laid between the yield factor and the ATP production during substrate breakdown. Subsequently rapid progress was made because of the introduction of continuous cultivation in the chemostat. This led to the realization that the yield factor was dependent on the specific growth rate and to the introduction of the maintenance concept. In the early seventies theoretical calculations were performed on the amount of ATP required for the formation of cellular material and of the major macromolecules in the microbial cell. This also led to the realization that Y_{ATP} is not a constant and is dependent on the nature of the growth substrate, the pathway of its breakdown and the specific growth rate. An important new development was then the introduction of the method for material balancing during bacterial growth. At the same time calculations on the thermodynamic efficiency of microbial growth started to be a matter of study. Afterwards non-equilibrium thermodynamics, first utilized to describe the bioenergetics of mitochondria, were applied to microbial growth and this led to a new method for the description of microbial growth.

Finally, in biochemistry the metabolic control theory was developed in the early seventies. According to this theory, in a sequence of biochemical reactions a certain control coefficient can be ascribed to each reaction step. Each of the steps exerts a certain control over the rate of the overall reaction. This theory was applied to microbial growth in the late eighties.

From this enumeration it becomes clear that in the field of the quantitative description of microbial growth and metabolism a number of important new developments have occurred successively. These innovations suggested many new experimental approaches, yielding a wealth of experimental data. The data have been used to develop mathematical models for the description of various aspects of microbial growth, e.g. the production of certain metabolites with micro-organisms. With these models it has become possible to direct the metabolism of a micro-organism in such a way that more of a certain desired product is made.

From the preceding section the reader could get the impression that a full quantitative description of all aspects of microbial growth and metabolism can now be given. This is not at all true. We are still far from a full understanding of all aspects of microbial growth and metabolism. A large number of intriguing questions remain.

In this Special Issue the state of the art in this field is given. A number of papers are reviews on certain aspects of the field. A number of other papers give new experimental material. A new development can be recognized: a number of papers deal with individual cells in a population. Furthermore, a number of papers deal with quantitative aspects of product formation and its optimization. I would like to thank all the contributors to this Special Issue for their enthusiasm to participate. All contributors were able to meet the very tight deadline by which their contribution had to be completed. I hope that the readers will find the contributed papers interesting and stimulating. I hope the papers will convince the readers that quantitative aspects of microbial growth and metabolism is a very dynamic field of study, which has great promise for our understanding of the functioning of the microbial cell.

Vrije Universiteit, Amsterdam   A.H. Stouthamer

Antonie van Leeuwenhoek **60**: 133–143, 1991.

Quantitation of microbial metabolism

B. Sonnleitner
Institute for Biotechnology, ETH Zürich Hönggerberg, CH 8093 Zürich, Switzerland

Key words: bioprocess analysis: (on line, *in situ, in vivo,* continuous), bioprocess automation, density of data, (mathemetical) models, relaxation time, validation of data

Abstract

Quantitation is a characteristic property of natural sciences and technologies and is the background for all kinetic and dynamic studies of microbial life. This presentation concentrates therefore on materials and methods as tools necessary to accomplish a sound, quantitative and mechanistic understanding of metabolism. Mathematical models are the software, bioreactors, actuators and analytical equipment are the hardware used. Experiments must be designed and performed in accordance with the relaxation times of the biosystem investigated; some of the respective consequences are discussed and commented in detail. Special emphasis is given to the required density, accuracy and reproducibility of data as well as their validation.

Introduction

The quantitative knowledge of biological reactions, of concentrations or activities of the reactands, of type and extent of regulation, or of the driving forces of proliferation and product formation is of paramount interest for the elucidation of the mechanisms of life as well as for the technical exploitation of living organisms or parts therefrom. With the emergence of modern disciplines with a molecular framework, (micro)biology is being shifted from mysticism to a honest, true science. Quantitation is a decisive characteristic property of natural sciences. Understanding of mechanisms and driving forces distinguishes a technology from a mere technique. Kinetic knowledge is good enough to describe (biochemical) processes but there is a need to develop the perception of dynamics in order to understand and predict them.

Historically, there are two extreme, different approaches to investigate microbial metabolism. They have so far been restricted to distinct disciplines and no unifying view was visible in the past. But there is an urgency for more efficient and prag-matic approaches to promote progress of both scientific research and technological applications. Figure 1 tries to give a principal synopsis of the current state of the art.

The older approach has been successfully exploited in the engineering disciplines and is a black box approach: cells are a catalyst with the property to convert a starting material (the substrate) to a product (either biomass or a product of metabolism). The specific rate of conversion is an internal property of the catalyst and described by simple kinetic equations.

Biochemists and molecular biologists, however, have revealed plenty of intracellular details and characterized those subsystems (enzymes, organelles, genes, etc). Yet, the knowledge is incomplete due to the complexity of a microbial cell's structure; only fragments of the biosystem – the cell – but not the structured entity can be described quantitatively.

A reasonable compromise is the exploitation of mathematical models with a rough internal structure for biological systems. They identify important key elements within the cells and quantify

134

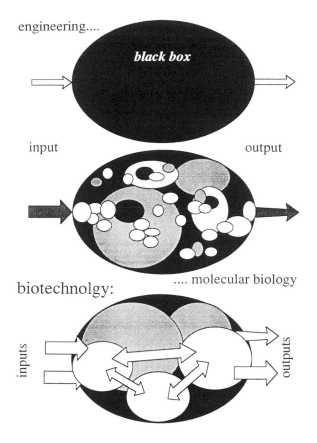

engineering....

black box

input   output

biotechnolgy: .... molecular biology

inputs   outputs

Fig. 1. Schematic comparison of different types of models. *Top*: black box model widely used in engineering sciences: the biological systems has 2 connections to the environment: input and output. The characteristic property of the system is to convert the input vector according to rate equations into the output vector. The rate equations are of pure descriptive nature and must not be used to predict values in the space outside the original data base used to evaluate the model parameters. *Middle*: collection of single, detailed informations about elements or subsystems of cells, typically the compiled knowledge accomplished in molecular biological investigations. Many interconnections (as rate equations) are normally ill defined or not known. Some structural elements are well understood and quantified (indicated by white areas), others less (grey zones) and the rest not yet (black). *Bottom*: a compromise approach yielding an appropriate tool to predict (= calculate) reaction rates and yields. Biological knowledge is condensed to several relevant key elements. Their interconnection is defined by a relatively small but complete set of rate equations. Outputs are calculated from inputs but the algorithms are based on mechanistic assumptions; the mechanisms need not necessarily be molecular, even hypothetical mechanisms are helpful. Such models may – with necessary criticism and care – be used to extrapolate outside the data space which served for construction and validation of the model.

them by balance and rate equations. These keys need not necessarily reflect a physical structure, they may well be hypothetical or lumped structural elements, e.g. a 'respiratory bottleneck', or 'residual biomass', or 'active part of the biomass'. This intermediate approach is suited to explain and understand physiological properties or peculiarities. This way of reasoning offers the chance to perform predicting calculations, i.e. design better, well directed experiments, to calculate metabolic fluxes which cannot be measured directly, or to predict the regulatory behavior of populations although concrete molecular mechanisms may not be addressed.

Materials and methods are most decisive tools

The objective of this contribution is not to present individual, single results of quantitation, it is rather designed to comment and critically evaluate the tools which are necessary for reliable and correct quantitation. These tools are of different nature: materials are the hardware and methods are the software and both types are required. The quality of the results and conclusions produced is determined by the least reliable tool used.

Models

Quantitation is not just measurement of (single, isolated) concentration values or of their time courses. It requires more, namely evaluation of these raw data and interpretation using mathematical models although many scientists seem not to be actively aware of this fact. An inherent property of models is that they always simplify reality.

The Monod model, for instance, which is widely used to characterize microbial growth, considers only two dependent state variables: the mass concentration of cells and the concentration of a single limiting substrate. This is most likely an over-simplification of any real biological system but some important objectives can be achieved:

– An efficient data reduction because both the

number of state variables and the number of model parameters are low.

- In spite of the simple structure the model is generally useful because it describes most bio-processes provided they are autocatalytic or constant.
- The model is worldwide known and therefore an appropriate vehicle to exchange or compare the data characterizing a distinct cultivation system.

The few parameters are unequivocal and have a clear biological meaning. This simplification is reasonable but it is decisive to identify the correct key elements and the range of validity. The key elements of the Monod model are the following:

- Only growth of a great number of homogeneous, vital cells is described but neither product formation, nor maintenance, nor inhibition, nor lag nor cell death.
- Growth is an autocatalytic reaction as long as all essential medium components are in excess.
- The specific growth rate adapts to the concentration of the unique nutrient which becomes limiting first.
- There is a constant relation between growth and substrate consumption.

The last key element is the reason why this mechanistically improper formulation works. If the model assumptions and constraints do not hold true, e.g. when maintenance can no longer be neglected or when more than one substrate can be cometabolized in parallel or alternative paths can be exploited for one substrate at the same time, it becomes obvious that the realistic origin $\rightarrow$ effect mechanism must be: substrate consumption is the origin and growth is the effect. It is plausible to choose formulations in the following sequence: $q_S = f(s_i)$, and only thereafter calculate growth as $\mu = f_1(q_{s_i})$ or product formation as $q_P = f_2(q_{s_i})$. The general formulation for n possible parallel substrates and m possible alternative ways of utilization of a substrate would be:

$$\mu = \sum_{i=1}^{n} \sum_{j=1}^{m} (f_j(q_{s_i}))$$

Another important aspect in quantification of microbial metabolism is the fact that cells (i.e.: biomass) are normally regarded neither structured nor segregated: they just have a mass concentration – x, for the great unknown – as if they were a homogeneously dissolved single substance. Of course, the unstructured and unsegregated models are much simpler to handle mathematically and to verify experimentally. But they prohibit the possibility to formulate the knowledge about intracellular key elements in a quantitative way. A structure would easily allow to formulate certain activities and temporal developments of microbial cells independently. To segregate biomass would permit to account for the individual or genealogical age of cells in a population or to consider progression through their cell cycle (eukaryotes only, not prokaroytes), i.e. population dynamics. Structured and/or segregated models are more complicated to handle and currently not easy to verify experimentally by direct measurements. But the complexity of such models is not necessarily overwhelming: Palsson and his group have shown several times that a few – but correctly chosen – lumped structural key elements of cells permit a sufficiently accurate and precise description of metabolic behavior when compared with the respective, significantly more detailed model (see, e.g. Palsson & Joshi 1987; Lee & Palsson 1990; Domach et al. 1984).

A distinction between batch kinetics and chemostat kinetics can sometimes be found in the literature. Although this is incorrect in the usual context – not the kinetics are different, just the balance equations contain different terms – it indicates strongly that a different type of experimentation may well result in different behavior and, hence, other models and/or parameters. The dramatic difference is that steady state investigations (classical chemostat studies) cannot reveal the information which is necessary to predict transient behavior (e.g. in a batch, fed batch, disturbed chemostat). The simpler models currently in use contain static rate equations only, e.g. $q_S = f(s)$. Neither the physiological state of the population nor the historical development of the state vector is used in those expressions although some information to accomplish this goal were experimentally accessible and

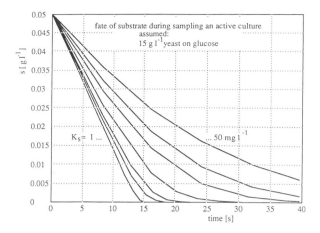

Fig. 2. Example to demonstrate the potentially very short relaxation times of realistic biological systems: an intermediately dense culture of a microbe grown under limited substrate supply (fed batch or continuous culture) is assumed to be withdrawn (i.e. sampled) from the ideal bioreactor at time 0. The substrate concentration falls rapidly $\rightarrow 0$ because the sample is no longer further supplied with fresh substrate. Depending on the characteristic properties of the organism – here is the K_s value varied – the transition to complete limitation is a matter of a few seconds. Very important are these implications with respect to accuracy and representativity, i.e. reliability, of chemical or biological analyses of a culture. Parameters used for simulation of simple Monod model: $x = 15\,g\,l^{-1}$, $\mu_{max} = 0.5\,h^{-1}$, $Y_{X/S} = 0.5\,g\,g^{-1}$, K_s varied: 1, 3, 5, 10, 20, 30, 50 mg l^{-1}.

available (batch, pulse, shift technique). This fact identifies urgent needs for further developments: obviously, well structured models must be promoted which are capable of explaining transients such as immediate responses, delays or dead times.

Appropriate experimentation hardware: bioreactors

Biology is an experimental science. Besides the theoretical background (e.g. models), it requires proper hardware (equipment for cultivation and analyses) and software tools (methods and techniques). Quantitative characterization of microbial activities is the objective in the present context. It is, under these auspicies, wise to use reactors that approximate the ideal types as close as possible. Laboratory scale bioreactors are believed to be

ideal reactors. This is true when compared with production scale but cannot be accepted as such. Depending on the smallest relevant relaxation time of the biological system investigated, even the lab scale reactors' performance may significantly contribute to the data measured; these are always a superposition of shares that originate from the biological system (kinetics) and from the mechanical system (mass and energy transfer, mixing, shear). Some metabolites have a low relaxation time in the order of seconds or less (e.g. ATP, NAD(P)H), some others in the order of minutes (e.g. RNA) and others an order or more greater (e.g. DNA, cell envelope) (Klingenberg 1974; Harder & Roels 1982). We in Zürich have therefore decided to preferentially use and promote high performance bioreactors with a mixing time $\ll 1\,s$ and an oxygen transfer capacity of $> 10\,g\,l^{-1}\,h^{-1}$ for physiological studies with dilute cultures (Sonnleitner & Fiechter 1988). This allows most probably to study biological reactions and regulations exclusively and reliably in the time domaine of a few seconds and, of course, greater in vivo. Such situations are much more frequent than generally assumed and expected; see the always underestimated example in Fig. 2.

Relaxation times of microbial cultures

The awareness of the actual relaxation times of biological systems or parts therefrom must have consequences: the analytical methods and tools used must be able to cope with this situation. Sampling and sample preparation are both integral factors of chemical analysis.

A good approach is the consequent application of on line in situ sensors. They are generally noninvasive and do, therefore, not disturb the bioreaction. *On line* means fully automatic or that there is not any manual interaction of personnel. But it does not rule out that these sensors have a significant delay such as the membrane covered electrodes or others with bridged electrolytes. *In situ* means that the sensor is built in or mounted in (hopefully a well mixed space of) the reactor. These sensors have no dead time because distances for transport are virtually nil. Contrarily, sensors mounted in a bypass always suffer from the time required for physical transport and the reactions taking place in the culture aliquot during its finite mean residence time in the bypass (outside the ideal reactor).

Only a few reports exist where the importance of short sample removal times and appropriate inactivation of the sample is explicitly mentioned. Gschwend-Petrik (1983) has withdrawn her samples in 3 to 4 s and then cooled on ice. Postma et al. (1988) have transferred their samples within 3 s from the reactor into liquid nitrogen. Wehnert (1989) claims to sample and inactivate cells in far less than 1 s. Theobald et al. (1991) described a rapid sampling method using precooled ($-25°C$) equipment. Unfortunately, there is no commercially available, automatic and complete sampling system of this type on the market. Holst et al. (1988) have inactivated cells during sampling by addition of KCN; they used a catheter probe which was originally designed to sample blood from humans (heparin instead of KCN) and is commercialized. Of course, only KCN sensitive metabolic paths are inactivated by this system, a considerable fraction of metabolism still remains operating. A suitable method to remove cell free supernatant from a bioreactor is the use of cross flow filtration membranes mounted either in situ or in a bypass with short mean residence time of the culture; several systems are commercially available (see Schügerl 1991). Such filtration devices usually prepare samples to be collected for manual analysis or for on line analytical instruments which work in batch mode, e.g., GC, HPLC or FIA. The data produced in this way are time discrete (not continuous) and delayed by a sometimes considerable dead time; correct synchronization with other data may no longer be trivial but is important, especially in rapid transient experiments. These delays do not obstacle scientific investigations but could limit the application for process control.

Reproducibility and reliability of experiments

Reproducibility of biological experiments is a major aspect for both scientific research and process

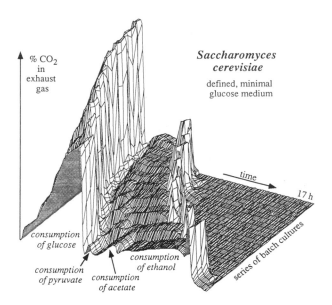

Fig. 3. Series of repetitive batch cultures of *Saccharomyces cerevisiae* with an increasing amount of inoculum left in the reactor. From left to right: relative cultivation time, longest experiments: 17 h, time step between plotted data: 5 min. From front to back: first batch culture inoculated from shake flask, repetition of experiment after harvesting of 95% of the culture and refilling with sterile medium, second and third repetition, next series with only 90% being replaced, and so on until only 50 % have been replaced (last series). From bottom to top: CO_2 in the exhaust gas (scale: 0 to 5%).

development. It can be dramatically increased with potent equipment and methods. Highly accurate measurements of as many state variables as possible and precise control of environmental and operational conditions is a prerequisite. A 'culture parameter' is not constant per se, just by intention, it must be actively made constant by closed loop control. Our experience shows further that the reduction of manual operations by consequent automation and extensive use of on line measurement devices effectively increases reproducibility. An example is given in Fig. 3 for a completely automated, i.e. unattended, series of repetitive batch cultures of yeast. The hardware has become quite robust in recent years and, in turn, cultivations and quantitative information derived therefrom have become more reliable.

It is also important to decide a priori which of the environmental and operating conditions should be made parameters, i.e. held constant by closed loop control. They should be as many as possible. However, we found it crucial to treat all data from a bioprocess primarily as variables and to decide on line whether they actually remain constant, i.e. behave as a parameter, or fluctuate more than inside a predefined window, i.e. behave as a variable. Then, the resulting documentation is really complete and correct, exactly the necessary data are stored. Moreover, the quality of constancy must be good enough. Table 1 gives an example for realistic values which characterize the present potential of our equipment at ETH Zürich. This means, of course, that the investments for the appropriate experimentation equipment is relatively high but we know this fact from other disciplines (medicine, physics, etc) sufficiently well.

Density of quantitative information

The independent variable in most biological experiments is time. An important exception are the steady state experiments (e.g. in chemostat cultures) where the dependent variables do theoretically not change with time; there, the independent variable is the dilution rate or its inverse, the mean residence time. It is nowadays established that most experimental studies of physiology are based on the investigation of the trajectory of a dependent variable with respect to the independent variable (time course, x-D-diagram) because, obviously, much more information can be derived from these data than from single analytical values (e.g. yield or conversion at the end of a culture). Such characteristic values have been shown many times to be not independent of growth rate and conditions (see, e.g. Stouthamer & Bettenhaussen 1975; Pirt 1965).

An important determinant in physiological studies is the necessary frequency of analyses (if they are not continuous signals), of data storage and evaluation. Considerable information about dynamics may be lost or overseen when this aspect is underestimated. Figure 4 gives an illustrative example for batch cultures, and Fig. 5 for spontaneously synchronized chemostat cultures of *Saccharomyces cerevisiae*. Although the duration of an

Table 1. Characteristic ranges, relative accuracy of typical biotechnological state variable measurements or precision of cultivation parameters (i.e. control variables) listed according to type: physical, chemical and biological variables.

Variable	Type/Range	Units	Accuracy/Precision
Physical variables			
Temperature	0– 150	°C	0.01%
rpm	0–3000	min^{-1}	0.2%
Pressure	0– 2	bar	0.1%
Weight	90– 100	kg	0.1%
	0– 1	kg	0.01%
Liquid flux	0– 8	$m^3 h^{-1}$	1%
	0– 2	$kg h^{-1}$	0.5%
Dilution rate	0– 1	h^{-1}	<0.5%
Gas flux	0– 2	vvm	0.1%
Foam	on/off	–	–
Bubbles	on/off	–	–
Level	on/off	–	–
Chemical variables			
pH	2 – 12	units	0.1%
pO_2	0 –100	% sat	1%
pCO_2	0 –100	mbar	1%
Exhaust-O_2	16 – 21	%	1%
Exhaust-CO_2	0 – 5	%	1%
Fluorescence	0 – 5	V	–
Redox	– 0.6– 0.3	V	0.2%
MS: volatiles			membrane
methanol, ethanol	0– 10		
acetone	0– 10	$g l^{-1}$	1–5%
butanols	0– 10		
On line FIA			= f (dilution)
glucose ($<100 h^{-1}$)	0–100		<2%
NH_4^+ ($<20 h^{-1}$)	0– 10	gl^{-1}	1%
PO_4''' ($<15 h^{-1}$)	0– 10		1–4%
GLU ($<20 h^{-1}$)	0– 10		<2%
On line HPLC			
phenols ($<5 h^{-1}$)	0–100	$mg l^{-1}$	2–5%
phthalates	0–100	$mg l^{-1}$	2–5%
organic acids	0– 1	$g l^{-1}$	1–4%
erythromycins	0– 20	$g l^{-1}$	<8%
by products	0– 5	$g l^{-1}$	2–5%
On line GC			capillary
acetic acid ($<10 h^{-1}$)	0– 5		2–7%
acetoin	0–10		<2%
butanediol (R, S)	0–10	$g l^{-1}$	<8%
meso-butanediol	0–10		<8%
ethanol	0– 5		2%
glycerol	0– 1		<9%
Biological variables			
RQ	0.5– 20	M M^{-1}	largely depends on error propagation
x: OD-sensors	0 –100	AU	highly variable
x: βμgmeter	–		
Physiological state	software	–	reference patterns
μ (= D)	0 – 1	h^{-1}	<0.5%

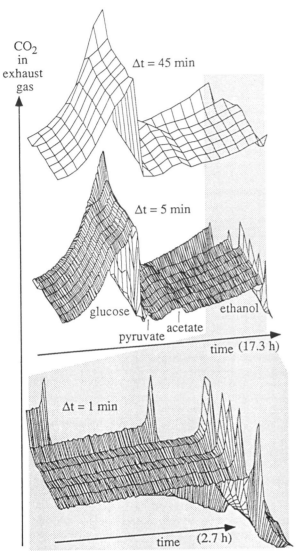

CO₂ in exhaust gas

Δt = 45 min

Δt = 5 min

glucose   ethanol

pyruvate   acetate

time (17.3 h)

Δt = 1 min

time   (2.7 h)

Fig. 4. Importance of high data density for the quantitative characterization of microbial cultures visualized for a series of batch cultures with *Saccharomyces cerevisiae* on chemically defined minimal medium. All 3 subplots show the CO_2 concentration in the exhaust gas, a continuous measurement, but plotted in different time resolution (Δt; as if it were either an off line or a discontinuous on line analysis). Time increases from left to right; first batch is the series in the back, following batches are added towards the front. The first batch had been inoculated with 5% YEPD preculture, the second batch was conducted by replacing 95% of the culture with defined medium, and so on. Unknown components from the complex medium do obviously influence the growth characteristics in very low concentrations: the exponential character of growth throughout the glucose phase and the growth rate thereafter. *Top*: data available every 45 min: the growth phase on glucose can well be distinguished from the subsequent phase on ethanol. In the intermediate diauxic lag phase is, if at all, only little activity obvious. The exponential character of growth cannot be confirmed unequivocally. *Middle*: data available every 5 min: additionally, activities in the diauxic lag phase become obvious: after depletion of glucose, 2 other substrates are metabolized with CO_2 as a by-product formed [other analyses reveal that the intermediately excreted pyruvic and acetic acids are reconsumed as indicated]. Growth on glucose is not exponential over the entire range (shoulder before maximal CO_2 production). At the end of the culture, some non reproducible activities can be deduced (peaks). *Bottom*: zoomed view of the last part of the growth phase on ethanol (indicated by shaded area) with data available every 1 min: the last activity turns out to be reproducibly of very short duration [we have indications that residual acetate is consumed after ethanol depletion].

entire batch is somewhat below 1 day and the single growth phases extend over several hours each, there are characteristic phases which are reproducibly found considerably shorter. A reasonable frequency of analyses employing manual techniques is in the range of tens of minutes but is, in this case, improperly low and would obscure most interesting phases. It is, therefore, necessary to use more frequent – or better continuous – on line (= automated) techniques which do neither disturb the system nor waste too much of the samples.

Some of the signals that are available from on line sensors are difficult to calibrate or to interpret per se, i.e. when isolated from other information. Such signals are, for instance, the redox potential or optical density. If, however, these signals are used in combination with others, they provide some further, useful information to characterize a cultivation. Exploitation of all three dimensions (1: value of signal, 2: time, 3: individual variables, i.e. signals produced by different sensors and analytical equipment) of the on line and off line information, i.e. working with patterns of (time) trajectories, yields more information than just the sum of single, isolated trajectories. A prototype variable to characterize the physiological state is the respiratory quotient; but this can only be calculated from raw data (CO_2 and O_2 balances) which do per se, i.e. individually, *not* describe the physiological state. This information gain also becomes obvious when, for instance, pattern recognition algorithms are fed with a varying number of parallel sensor signals: the more signals the easier and

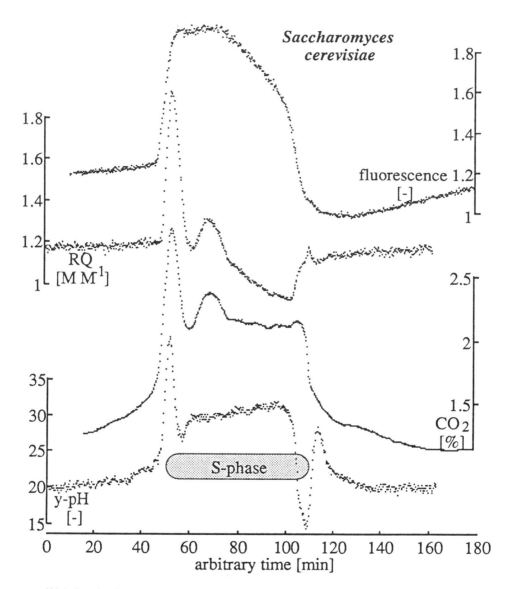

Fig. 5. Importance of high data density to reveal population dynamics and fast metabolic regulation: trajectories of 2 directly measured variables and 2 software sensors (calculated data) during 1 oscillation of a spontaneously synchronized chemostat culture of *Saccharomyces cerevisiae* on defined glucose medium at dilution rate D = 0.13 h⁻¹. The fine structure of the oscillations is highly reproducible and resolved here with a time distance of 15 s (between consecutive dots). Unfortunately, mechanistic explanations for all the individual fine structures cannot be given at the moment. [Flow cytometric measurements, however, confirm that one subpopulation undergoes the cell cycle S-phase, as indicated. These cells excrete ethanol which is consumed by the other subpopulation; data not shown.]

safer is the recognition of a cultivations state (Locher et al. 1990).

The latter technique even allows to localize drifting or defective sensors when reliable reference (or prototype) patterns are available. It is one method for on line validation of signals which is of considerable importance in scientific research. Well working alternatives are the multiple Kalman filter approach, neural networks or fuzzy control. King (1991) applied several different models to describe foreseen process states, e.g. normal operation, infection, or sensor drift. Test of the model probabilities allowed to discriminate on line and to identify the most probable state of a culture.

Accuracy of generated data

The values for accuracy/precision of measurement and control given in Table 1 are optimal values that can be accomplished with high performance equipment. But it is a fact that the accuracy of measurement is often less than the precision of control. Temperature, for instance in our equipment, can be determined to $\pm 0.2°$ accuracy but is routinely controlled with a precision of $\pm 0.01°$ (the better value is entered in Table 1). The reasons are manifold but most probably, there are not sufficiently accurate references available for calibration. In other cases, the analytical instrument (or sensor) may not be sufficiently sensitive, linear or is too noisy. A third category refers to errors introduced during sampling and sample preparation. However, there is an absolute method for testing the accuracy of sets of analytical measurements: the carbon balance must yield 100 % since (bio)chemical reactions do not convert elements. The carbon recovery can be easily calculated and gives a good estimate about the precision and completeness of measurements; it is strongly recommended to perform such simple checks as a standard comparable to routine calibrations of equipment.

Practically all biological liquids are 'dirty' from an analytical point of view. The matrix (of the analyte) is most often the dominating component, dictates certain analytical procedures and limits the accuracy of results to be obtained. This necessitates, in many cases, tedious and time consuming separation, and possibly also enrichment techniques prior to quantitation (e.g. Widmer 1986). Enzymes are preferred analytical recognition elements (e.g. in biosensors) because of their generally high specificity for substrates, i.e. the analytes to be determined. Quantitation, however, is normally performed with unspecific sensors and by measurement of disappearing cosubstrates (e.g. O_2) or formed products (e.g. H_2O_2, H^+). This last step is widely prone to interferences from side components of the analytical matrix (e.g. H_2S may be liberated from proteinaceous sources and interfers with an H_2O_2 measurement although the H_2O_2 may be exclusively formed by an enzymic oxidase reaction in stoichiometric quantities according to the analyte's concentration). Besides these major effects, nonspecificity of methods or sensing devices is of paramount interest. A commonly underestimated but typical example is the quantification of protein(s) with several (e.g. Lowry or Bradford) methods. Calibration with different purified enzymes reveals the different sensitivity (and can be easily checked; this is a recommended, instructive experiment for practical training courses).

When batch or other transient experiments must be followed, a large dynamic range of the analytical tools is required. This is different to the medical applications where one order of magnitude may well be sufficient (e.g. glucose in blood). Only one type of commercially available sensors fulfills this requirement: the pH electrode (linear over 10 pH units is no longer a problem). All other sensors and analytical methods require dilution (or enrichment, if the limit of detection is high). High accuracy of manual dilution is possible but the trade offs are exact calibration of glassware and the long time consumed. Flow systems like FIA (flow injection analysis) solve this problem excellently; their accuracy can be determined by calibration, their speed and reproducibility, i.e. precision, are extremely high, e.g. $< 1\%$ for a $1 : 1000$ dilution in less than $100 \, s$ (Garn et al. 1988).

The indirect identification of state variables, fluxes and yield coefficients by model simulations represents an important and interesting extension of the information gained from direct measurements. The availability of a verified model is, of course, a necessary requirement. Structured models pave the way to investigate intracellular fluxes, segregated models open the possibility to study and estimate intercellular fluxes and relationships. These variables are normally not accessible by direct chemical analyses of cultures, i.e. in vivo.

Conclusions

Biological experiments are definitely reproducible, provided a major influence of evolutionary mechanisms can be ruled out. In fact, the often cited character of irreproducibility or high variability has its roots either in non appropriate equipment, in badly developed methodology or simply in ignorance. The hardware, however, is being improved; this affords higher investment. Continuation of scientific studies with improper hardware is just 'saving money at the wrong place'. Current methodologies are oriented to the possibilities of available equipment. New, improved hardware necessitates further, continuous development of methodologies, for instance the full exploitation of transient experiments rather than steady state or balanced growth situations only. Ignorance is a matter of training and the intellectual horizon, sometimes also of lazyness; at least, the latter must be eliminated. A good starting point is to measure as many variables as possible and to actively make most of them parameters by closed loop control. This is not a matter of intention, it is the result of openmindedness, good experimental design and high quality equipment and instrumentation.

Outlook

Much descriptive information about different biosystems has been collected in the past and a good fraction has also been validated. Prediction is currently hampered by the less developed understanding of dynamics: what are the driving forces and mechanisms that keep things go in living cells and populations? Investigation of dynamics in unperturbed living systems promises to be an interesting challenge; the present developments of methodologies and techniques support the expectations that this task can be performed more and more effectively and successfully.

Acknowledgement

Data for Figs 3 to 5 were kindly provided by G. Locher and T. Münch (PhD theses to be published in 1992, Institute for Biotechnology, ETH Zürich).

References

Domach MM, Leung SK, Cahn RE, Cocks GG & Shuler MM (1984) Computer model for glucose-limited growth of a single cell of *Escherichia coli* B/r-A. Biotechnol. Bioeng. 26: 203–216

Garn MB, Gisin M, Gross H, King P, Schmidt W & Thommen C (1988) Extensive flow-injection dilution for in-line sample pretreatment. Analytica Chimica Acta 207: 225–231

Gschwend-Petrik M (1983) Dynamische Untersuchungen zur Stoffwechselregulation und Ethanolbildung bei *Saccharomyces uvarum*. PhD thesis 7441, ETH Zürich, CH

Harder A & Roels JA (1982) Application of simple structured models in bioengineering. Adv. Biochem. Eng. 21: 56–107

Holst O, Hakanson H, Miyabayashi A & Mattiasson B (1988) Monitoring of glucose in fermentation processes using a commercial glucose analyzer. Appl. Microbiol. Biotechnol. 28: 32–36

King R (1991) On-line supervision and control of bioreactors. In: Reuss M, Chmiel H, Gilles ED & Knackmuss HJ (Eds) Biochemical Engineering – Stuttgart (pp 69–77). G. Fischer, Stuttgart

Klingenberg M (1974) Nicotinamid-adenin-dinucleotide (NAD, NADP, NADH, NADPH). In: Bergmeyr HU (Ed) Methoden der enzymatischen Analyse, 3 Auflage (s. 2094–2127). VCH, Weinheim

Lee ID & Palsson BO (1990) Reducing complexity of the model for human erythrocyte metabolism. AIChE 1990 Annual Meeting (lecture and extended abstract 60 C)

Locher G, Sonnleitner B & Fiechter A (1990) Pattern recognition: a useful tool in technological processes. Bioproc. Eng. 5: 181–187

Palsson BO & Joshi A (1987) On the dynamic order of structured *Escherichia coli* growth models. Biotechnol. Bioeng. 29: 789–792

Pirt SJ (1965) The maintenance energy of bacteria in growing cultures. Proc. Roy. Soc. B. 163: 224–231

Postma E, Scheffers WA & van Dijken JP (1988) Adaptation of the kinetics of glucose transport to environmental conditions in the yeast *Candida utilis* CBS 621: a continuous-culture study. J. Gen. Microbiol. 134: 1109–1116

Schügerl K (Ed), (1991) Analytische Methoden in der Biotechnologie. Vieweg, Braunschweig, D

Sonnleitner B, Fiechter A (1988) High performance bioreactors: a new generation. Analytica Chimica Acta 213: 199–205

Stouthamer AH & Bettenhaussen CW (1975) Determination of the efficiency of oxidative phosphorylation in continuous cultures of *Aerobacter aerogenes*. Arch. Microbiol. 102: 187–192

Theobald U, Baltes M, Rizzi M & Reuss M (1991) Structured metabolic modelling applied to dynamic simulation of the

Crabtree- and Pasteur-effect in baker's yeast. In: Reuss M, Chmiel H, Gilles ED, Knackmuss HJ (Eds) Biochemical Engineering – Stuttgart (pp 361–364). G Fischer, Stuttgart

Wehnert G (1989) Entwicklung einer kombinierten Fluoreszenz-/Streulichtsonde zur on-line und in situ Biomasseab-schätzung bei Fermentationsprozessen. PhD thesis, University of Hannover, D

Widmer HM (1986) Analytik im Spiegel der Zeit. Chimia 40 (7–8): 250–251

Antonie van Leeuwenhoek **60**: 145–158, 1991.

Quantifying heterogeneity: flow cytometry of bacterial cultures

Douglas B. Kell[1], Hazel M. Ryder[1], Arseny S. Kaprelyants[2] & Hans V. Westerhoff[3,4]
[1] Department of Biological Sciences, University College of Wales, Aberystwyth, Dyfed SY23 3DA, UK
[2] Bakh Institute of Biochemistry, USSR Academy of Sciences, Leninskii Prospekt 33, 117071 Moscow, USSR
[3] Division of Molecular Biology, The Netherlands Cancer Institute, Plesmanlaan 121, 1066 CX Amsterdam,
The Netherlands; [4] E.C. Slater Institute for Biochemical Research, University of Amsterdam,
Plantage Muidergracht 12, 1018 TV Amsterdam, The Netherlands

'In contrast to standard microbiological, genetic or biochemical techniques, this method provides information on individual cells, and not just average values for the population. This ability to analyze individual cells is invaluable in studying the distribution of cell parameters in a polydisperse population, and gives access to information that cannot be obtained in any other way.'

Boye & Løbner-Olesen 1990

'Flow cytometry has revolutionized the study of the cell cycle of eukaryotes. It is also possible to apply the flow cytometry principles to bacteria. . . . The importance of the flow cytometry results should not be underestimated. They provide a crucial link in the analysis of the division cycle. . . . While other experiments have substantially supported the initial membrane-elution results, the flow cytometry results determine the pattern of DNA replication without any perturbations of the cell.'

Cooper 1991

Key words: analysis, bacteria, cytofluorometry, flow cytometry, heterogeneity

Abstract

Flow cytometry is a technique which permits the characterisation of *individual* cells in populations, in terms of distributions in their properties such as DNA content, protein content, viability, enzyme activities and so on. We review the technique, and some of its recent applications to microbiological problems. It is concluded that cellular heterogeneity, in both batch and continuous axenic cultures, is far greater than is normally assumed. This has important implications for the quantitative analysis of microbial processes.

Introduction and scope

Based on work by Maxwell and Boltzmann, Gibbs developed the concept of an *ensemble*, as a collection of particles possessed of the same energy (mean and time-averaged distribution).Since that time, the treatment of macroscopic systems as ensembles of microscopic particles that, averaged over time, are identical has underpinned most of even modern thermodynamics (see e.g. Welch & Kell 1986; Westerhoff & van Dam 1987). Implicitly, microbial physiologists have normally followed the same path: we describe our cultures as having a certain growth yield or respiratory rate or internal pH or rate of glucose catabolism or whatever, with the implicit supposition that this represents a full description of these variables. However, this would be true only if our cells were not only identical but *at equilibrium*, constituting what thermodynamicists call an ergodic system. Since we know that growing cells are certainly non-equilibrium in character, it is usual, even within the framework on non-equilibrium thermodynamics, to ascribe a 'local' equilibrium to the macroscopic parameters and variables (forces and fluxes) in which we are interested, thus permitting us to refer to them as possessing a 'sharp' value. This approach is generally thought acceptable (but cf. Welch & Kell 1986;

Kamp et al. 1988) since the numbers of molecules participating in say glycolysis, or the ATP 'pool', even in a single cell, means that spontaneous thermodynamic fluctuations in their 'instantaneous' value will normally be negligible in the steady state.

The greater problem, which is the focus of the present article, is that the distribution of properties of *cells* in a culture is much more heterogeneous than we normally credit or assume. Whilst of course one appreciates that for reasons connected with the cell cycle alone (see e.g. Mitchison 1971; Donachie et al. 1973; Lloyd et al. 1982, Cooper 1991) there will be a distribution of properties such as cell size and macromolecular content, our problem is more acute than this: a culture with a respiratory rate of 100 nmol.(min.mg dry weight)$^{-1}$ might be made up of an ensemble of cells which all possessed this property or of a mixture in which half of the cells respired at 200 nmol.(min.mg dry weight)$^{-1}$ and half were metabolically inert (or of course a myriad of other possibilities encompassed by these extremes (Kell 1988)). Similar statements may be made for all possible parameters and variables! Such differences, in cultures which appear macroscopically identical, probably underlie the so-called 'problem of scale-up' (Kell 1987). In general, then, a full(er) description of the quantitative behaviour of a microbial culture, the topic of this Special Issue, would require that we describe our cultures not only in terms of the mean or macroscopic values of its parameters and (especially) variables but also in terms of their *distribution between individual cells*. Whilst it is not yet possible to do this for all parameters and variables of interest, it *is* now possible to *begin* this task. This is primarily due to technical advances, especially (but not exclusively) in the area of flow cytometry and its application to bacterial cultures.

Thus the purpose of this article is to outline the principles of flow cytometry, to illustrate existing and potential applications in quantitative microbial physiology, and to point out some of the conceptual and practical difficulties accompanying the analysis of heterogeneity.

Principles of flow cytometry

In flow analysis generally, perhaps best known to microbiologists via its implementation in the Coulter counter (see Harris & Kell 1985), cells are constrained (usually hydrodynamically) in a path or flowing stream and pass, *one at a time* to a sensor which analyses the property of interest at the single-cell level. The most straightforward output of the instrument is then a plot of the number of cells possessing a certain property at a certain magnitude as a function of that magnitude. This certainly qualifies as quantitative microbiology. In the Coulter counter the property is the cell volume (or more strictly the volume surrounded by the cytoplasmic membrane), and this instrument is widely exploited by those studying the cell cycle (see e.g. Lloyd et al. 1982). However, we would stress again that although the concept of heterogeneity underpins studies of the cell cycle there has been but little attempt to integrate such measurements with those more conventionally employed by microbial physiologists (but cf. Neidhardt et al. 1990; Cooper 1991).

Though an important (and not at all recent) development, the microbiological variables which may be measured by the Coulter counter are really limited to the cell volume, and it is with *optically-based* instruments that the power of flow analysis is revealed. To begin with, the extent of low-angle light scattering by a cell depends largely (though not always linearly; Salzman 1982, Davey et al. 1990a) on the mass or volume of the cell. In a generalised flow cytometer (Fig. 1) (Melamed et al. 1979; Shapiro 1988), individual particles pass through an illumination zone, typically at a rate of some 1000 cells.s^{-1}, and appropriate detectors, gated electronically, measure the magnitude of a pulse representing the extent of light scattered. The magnitudes of these pulses are sorted electronically into 'bins' or 'channels', permitting the display of histograms of number of cells *vs* channel number. The angular-dependence of scattered light provides further information on the nature of the scattering particles, and in favourable cases may be selective towards different organisms (Steen 1990). In addition, and more importantly,

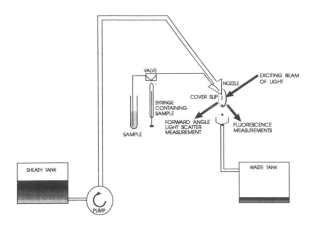

Fig. 1. The principle of flow cytometry. A pump passes fluid through a narrow tube, into which a slower-moving sample is injected as appropriate. Hydrodynamic focussing causes the sample to be constrained to the middle of the sheath fluid. In the type of system illustrated, based on the design of Steen et al. (1989) and optimised for work with microorganisms, the sample impinges on a microscope slide, and illumination is provided by a mercury arc lamp. Forward and right-angle light-scattering events are detected using photomultipliers, as (*via* suitable filters) is fluorescence, and are stored on a computer. In more traditional (and expensive) instruments, the source of illumination is a laser, and there is no cover slip; the particles pass through the zone of illumination in a jet, and may subsequently be sorted electrostatically. In either case, the computer may be used to gate measurements so that only particles scattering light or fluorescing above and/or below a critical amount are counted.

appropriate fluorophores may be added to the cell suspension. These may be stains which bind to (or react with) particular molecules such as DNA, RNA or protein, fluorogenic substrates which reveal distributions in enzymatic activity, indicators which change their property as a function of pH_{in} or which are taken up in response to membrane energisation, or, increasingly, antibodies (or oligonucleotides) tagged with a fluorescent probe. Clearly the possibilities are limited only by the ingenuity of the experimenter, and to avoid simply cataloguing these in the body of the text, we tabulate some of the better-known ones in Table 1. Whilst details should be sought in the references cited, it is worth mentioning that some, but not all, of these reagents require that the cells be fixed or permeabilised with ethanol (70 %), formaldehyde or glutaraldehyde. Several very useful overviews of technical aspects

of flow cytometry (Melamed et al. 1979, Muirhead et al. 1985; van Dilla et al. 1985; Shapiro 1988; Ormerod 1990; Darzynkiewicz & Crissman 1990; Melamed et al. 1990), and its application to microorganisms (Olson et al. 1986; Scheper et al. 1987; Frelat et al. 1989; Robertson & Button 1989; Steen et al. 1989; Boye & Løbner-Olesen 1990; Patchett et al. 1990; Pinder et al. 1990; Shapiro 1990; Steen 1990), are also available.

Is a culture ever in a steady state?

From a thermodynamic point of view, it is usually desirable to study (biological) systems in the steady state (Caplan & Essig 1983; Westerhoff & van Dam 1987). Under these conditions, all transients have died down, and all variables such as metabolite concentrations or the fluxes through pathways of interest are unchanging in time. These are the types of conditions normally treated by metabolic control analysis (see Kell et al. 1989: Westerhoff et al. this issue), and those usually assumed to hold, for instance, in the mid-exponential phase of batch cultures and in any chemostat culture whose dilution rate has not altered for some 5 or more volume changes. There is, of course, some arbitrariness about the definition of a steady state, since a true or global steady state implies that, after normalisation, *no* variable is changing (by a measurable amount) in time, and it is evident that the more variables one measures *on individual cells* the less chance will there be of ever persuading a culture to come to an *observable* (let alone true) steady state.

With a growing appreciation of the complexity of the dynamics of nonlinear systems (e.g. Glansdorff & Prigogine 1971; Gleick 1987; Moon 1987; Glass & Mackey 1988; Wolpert & Miall 1990) naturally comes the realization that quite small changes in a parameter that might normally be considered insignificant can have far-reaching consequences for the (time-)evolution of the system. In what follows, we wish to illustrate some of the unexpectedly complex dynamics of microbial cultures that have been observed, and to indicate the possibility, and means, of exploiting flow cytometry for their analysis.

Growth in batch cultures may be exponential but not balanced

When a batch culture is inoculated with a small inoculum that has been grown on the same medium, it is generally observed that after a short lag period the culture enters an exponential phase of growth (at its μ_{max} for the medium and other parameters such as temperature) that will continue until medium constituents are exhausted or toxic end-products accumulate to inhibit cell growth and division. The usual criterion for assessing the exponentiality of growth is to take measurements of the optical density or of the cell numbers in the culture; if a straight line occurs when these are plotted semi-logarithmically against time the culture is deemed to be exponential, and the growth taken to be balanced (in that μ_{max} represents a boundary value, and it would be a remarkable coincidence if major changes in the metabolic make-up of the cell were unaccompanied by changes in μ_{max}). At least two recent studies show that this is a highly dangerous practice.

Skarstad et al. (1983) studied the growth of *Escherichia coli* in batch culture, and measured both light scattering and DNA content of individual cells (the latter using a combination of ethidium bromide and mithramycin) via flow cytometry. Even when the culture growth was strictly exponential as judged by cell numbers, however, the

Table 1. Some determinands amenable to analysis by flow cytometry.

Determinand	Stain or reagent	Selected reference(s)
DNA	Hoechst 33258, Hoechst 33342, Ethidium bromide, propidium iodide, DAPI, Acridine orange., Chromomycin A_3, Mithramycin, Olivomycin	Darzynkiewicz 1979; Steen et al. 1982; Muirhead et al. 1985; Seo et al. 1985; Seo & Bailey 1987; Frelat et al. 1989; Sanders et al. 1990; Steen et al. 1990
RNA	Ethidium Bromide, Propidium iodide, Acridine Orange, Pyronin Y, Thioflavin T.	Darzynkiewicz 1979; Tanke 1990; Waggoner 1990
Protein	Fluorescein isothiocyanate, Rhodamine 101 isothiocyanate (Texas Red)	Hutter & Eipel 1978; Steen et al. 1982; Miller & Quarles 1990
Chlorophyll Phycoerythrin Carotenoids	Autofluorescent	Olson et al. 1986; Robertson & Button 1989; Cunnigham 1990 An et al. 1991
Enzyme activities	Substrates linked with: naphthoyl-, fluorescein-, umbelliferyl-, coumaryl- and rhodamine groups β-galactosidase	Dolbeare & Smith 1979; Kruth 1982 Srienc et al. 1986; Wittrup & Bailey 1988
Antigens	Fluorescently-labelled antibodies	Ingram et al. 1982; Steen et al. 1982; Frelat et al. 1989; Srour et al. 1991
Nucleotide sequences	Fluorescently-labelled oligonucleotides	Amman et al. 1990; Bertin et al. 1990
Internal pH	Numerous	Rabinovitch & June 1990b; Waggoner et al. 1990
Membrane fluidity	Anthroyloxy-labelled fatty acids	Collins & Grogan 1991
Inclusion bodies poly-β-hydroxy-butyrate	Changes in light-scattering behaviour	Wittrup et al. 1988 Srienc et al. 1984
Cellular Morphology		Betz et al. 1984; Allman et al. 1990 Hunter & Asenjo 1990
pCa	Aequorin, Indo-1, Fluo-3	Rabinovitch & June 1990a
Membrane energisation	Oxonols, cyanine, rhodamine 123	Ronot et al. 1986; Rabinovitch & June 1990a,b; Shapiro 1990; Kaprelyants & Kell 1991

distributions of both DNA and light scattering were highly inconstant. They concluded (correctly) that this was the likely cause of the variability in the reported cell cycle parameters of slowly-growing batch cultures of this organism (see also Jepras 1991). Steen (1990) extended this study to include optical density measurements, and found that whilst both the OD and the cell number increased in a strictly exponential manner, *they did so with different doubling times*, that based on cell counts being the shorter (18 vs. 23 min). It is difficult to implicate changes in the nutritional status of the medium, since the highest optical density considered was only 1 % of that attained in stationary phase. The sins of the parents, one might say, extend even unto the third and fourth generations.

This issue of heterogeneity within cultures of what is notionally a single clone comes on top of that which one might expect to find between different strains (and one might comment that flow cytometry could be of taxonomic utility). Even within a supposedly tightly defined taxon, however, Allman et al. (1991) found, using similar methodology to that described in the previous paragraph, that strains of *Escherichia coli* K-12 as a group are rather dissimilar to each other with respect to the pattern of their DNA replication, a finding consistent with the rapid changes in populations which are evidenced by the polymorphisms observable in this organism (Krawiec & Riley 1990).

Due in particular to their possession of a number of cofactors, cells are autofluorescent when excited with light of appropriate wavelengths. Fig. 2 shows the distribution in both light-scattering and autofluorescence of cells of *Micrococcus luteus* grown in batch culture. At the exciting and emission wavelengths used, the autofluorescence is ascribable predominantly to reduced flavin and pyridine nucleotides. Whilst one might have expected that larger cells would have a greater autofluorescence, the data observed (Fig. 2) show that there are functionally two populations of cells, with 'low' and 'high' autofluorescences that are not correlated with cell size, in what is by normal microbiological criteria an axenic culture multiplying exponentially under conditions of balanced growth. The protocol of the experiment (Fig. 2) suggests that the heterogeneity observed is not due to *genotypic* differences between the cells, whilst the fact that the accumulation of the flow cytometric data takes time does not permit one to exclude the existence of *oscillations* in the pyridine nucleotide concentrations in individual cells.

The steady state in chemostat cultures

Based on the kinetics of exponential washout, it is usually assumed that after a change in dilution rate, a culture attains the steady state characteristic of the new dilution rate after some 5 volume changes. Certainly this is a longer period than that considered in the usual batch culture, but one may doubt that even this is sufficient if *distributions* of properties are considered. To date, we are not aware of any studies that have looked carefully at this question, and flow cytometry obviously opens up many possibilities in this area. A strong pointer is given by the work of Rutgers et al. (1987), who found that the steady-state glucose concentration in glucose-limited chemostat cultures continued to decrease for as much as 50 generations after a change in dilution rate, long after the steady-state biomass level (and hence Y_{glu}) had been reached. Cells taken from the culture during this period showed a continuous increase in μ_{max} and cell size (measured with a Coulter counter), and a decrease in K_s. Based on the relevant kinetics, it was argued that the changes were likely to be genotypic in nature, and one may certainly state that cells which have high rates of 'spontaneous' mutation will eventually outcompete those which do not, since the former will eventually acquire beneficial mutations (Chao & McBroom 1985). Broadly similar data were obtained by Höfle (1983).

The above analysis is but one example of the long-term dynamics of continuous cultures; those in which the growing organism harbours a plasmid are of course notoriously complex (e.g. Caulcott et al. 1987; Weber & San 1990), and the selection pressures easy to construe (Westerhoff et al. 1983). However, conventional selection pressures can hardly explain the *oscillatory* behaviour of continuous cultures (Heinzle et al. 1982; Koizumi & Aiba

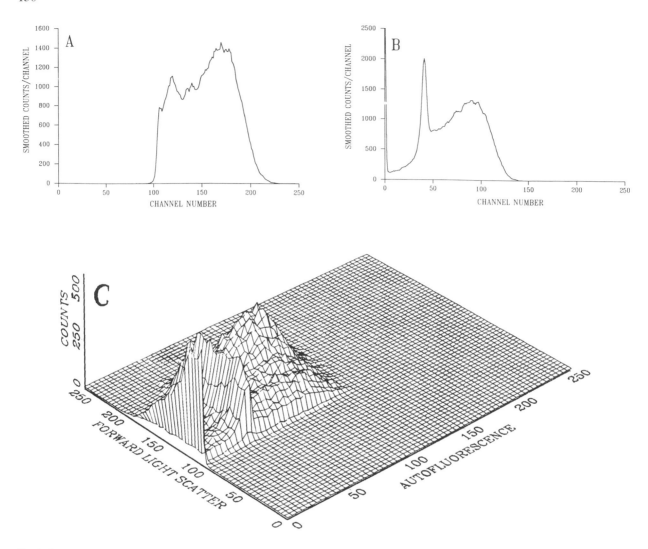

Fig. 2. Flow cytometric behaviour of the light-scattering *(A)* and autofluorescence *(B)* of a batch culture of *Micrococcus luteus*, and of their 2-dimensional distribution *(C)*. A single colony was selected following repeated streaking onto agar and growth in broth. Cells were grown in batch culture on 1.3 % Nutrient E broth, to stationary phase, and inoculated to give an optical density (680 nm) of 0.11. The data illustrated are from a sample taken when the OD_{680} of the culture was 1.93 (i.e. after more than 4 doublings had taken place). Flow cytometry was performed as described (Davey et al. 1990a,b, Kaprelyants & Kell 1991, Markx et al. 1991a,b), save that fluorescence was measured using a filter block with the following characteristics: excitation 395–440 nm, bandstop 460 nm, emission 470 nm and above. In *(A)* and *(B)* data were passed through a 3-point smoothing filter prior to plotting (Davey et al. 1990a). The photomultipler voltages for the light-scattering and fluorescence channels were respectively 450V and 950V, and the scales are logarithmic. Very similar flow cytometric data were obtained in samples taken at all stages of the growth of the culture. In addition, samples were taken periodically throughout the culture and streaked out to ensure that the culture remained axenic.

1989), and it is evident that we have a long way to go before we can claim a good understanding of these highly nonlinear processes. Experimentally tiny fluctuations in what are ostensibly parameters, such as oxygen tension and the concentrations of inhibitory molecules, may effect substantial chang-

es in steady-state variables such as dry weights, and whilst the importance of 'perfect' mixing has been known for many years (e.g. Sinclair & Brown 1970), recent studies show that the relevant micros-cale (50–300 mm) below which turbulence is not manifest, even in highly-agitated fermentors, is far greater than that of the dimensions of typical microorgansims (Fowler & Dunlop 1989; Dunlop & Ye 1990), providing a substantial contribution to heterogeneity in CSTRs. Indeed, although the mathematical analysis of bacterial size distributions is rather highly developed (e.g. Harvey 1983), the size of a microorganism is actually something that is only rarely considered a contributor to fitness or selection. Indeed, the analysis of selection in chemostats (see Kubitschek 1974; Dykhuizen & Hartl 1983), and in more complex ecosystems (e.g. Robertson & Button 1989), constitutes a fundamental topic, which is undoubtedly of biotechnological importance and which is conveniently carried out using flow cytometric procedures.

Life, death and other states

It may be taken that the most fundamental question which a microbial physiologist might ask about a cell is whether it is alive or not. This turns out to be far from simple to answer. It is well known, especially in Nature, that the number of cells observable by direct counts greatly exceeds the number capable of forming colonies (and thereby considered 'viable') (see Postgate 1976; Poindexter 1981; Morita 1982; Mason et al. 1986; Kjellberg et al. 1987; Roszak & Colwell 1987; Morita 1988; Matin et al. 1989; Gottschal 1990). However, not all the non-'viable' cells are 'dead', since many of these 'non-culturable' cells may be resuscitated by preincubation in a suitable nutrient broth prior to plating out. We may refer to such cells as 'vital'. The question then arises, for instance in slowly-growing cultures, as to what causes a cell to pass from the status of viable through vital to dead, and whether cells of each type in such cultures, though nominally homogeneous, coexist.

Several workers have studied the decline in ATP, adenylate energy charge, and/or the ability to accumulate lipophilic cations in starving cells or in cells grown at low dilution rates (Horan et al. 1981; Jones & Rhodes-Roberts 1981; Zychlinski & Matin 1983; Otto et al. 1985; Poolman et al. 1987), generally finding that none of these bioenergetic parameters could be correlated with the loss of viability (as judged by plate counts). However, these types of experiments possess the following, insurmountable problem: they represent bulk or ensemble measurements and it is therefore not possible (given our ignorance about the 'critical' values of these, if any, for the individual cell) to distinguish whether a decrease in ATP levels or in the uptake of the tetraphenylphosphonium cation, say, is due to the irreversible death of a proportion of the cells or an identical decrease of these parameters, unaccompanied by death, in all cells (or of course any combination of these extreme possibilities (Kell 1988)). In recent work, we have shown by flow cytometry (Kaprelyants & Kell 1991) that *Micrococcus luteus* cells grown in a chemostat at a low dilution rate (and even those grown in batch culture at μ_{max} (Fig. 3)) are *extremely* heterogeneous with respect to their ability to accumulate the lipophilic cationic dye Rhodamine 123. In particular, and in contrast to earlier suggestions based on bulk measurements, we found (Kaprelyants & Kell 1991) that cell viability, and resuscitation, *could* be quite well correlated with the ability of individual cells to accumulate the dye. Indeed, it was possible in part to relate the degree to which individual cells accumulated rhodamine 123 and the distinguishable physiological states ('viable', 'non-viable' and 'non-viable but resuscitable') exhibited by cells in the culture. Thus flow cytometry of cells stained with Rh 123 (or other appropriate dyes) allows one rapidly to distinguish not only 'viable' and 'non-viable' cells but the *degree of viability of individual cells* reflecting the heterogeneity of a culture observable following sub-lethal starvation, stress or injury.

Implications of heterogeneity for the analysis of microbial behaviour

From the experiments just described, it is obvious that the flow cytometric approach gave an answer that was exactly opposite to that which had previously been opined by others on the basis of cognate macroscopic experiments on *cultures*. Since these other workers had expected a correlation to exist between (say) adenylate energy charge or the up-

152

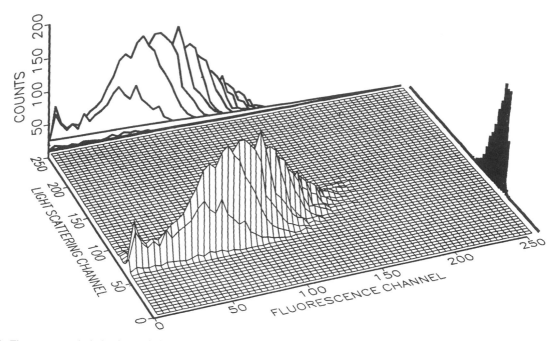

Fig. 3. Flow cytometric behaviour of the light-scattering and fluorescence of a batch culture of *Micrococcus luteus* stained with rhodamine 123. Cells were grown in batch culture on lactate minimal medium as described (Kaprelyants & Kell 1991), to an optical density (680 nm) of 1.8. Cells were stained with rhodamine 123 (final concentration 0.26 μM), incubated at room temperature 10 min, and flow cytometry was performed as described (Kaprelyants & Kell 1991). The optical characteristic of the relevant filter are: Excitation 470–495 nm, band-stop 510 nm, emission 520–550 nm. The photomultipler voltages for the light-scattering and fluorescence channels were respectively 500V and 650V, and the scales are logarithmic.

take of the tetraphenylphosphonium cation and viability on plates, it is important to understand the relevant points.

An example may serve to illustrate this. Suppose one wishes to understand how a physiological flux or process in a microbe depends on the intracellular concentration of its substrate. We assume that there is a single reaction determining the rate of the process (i.e. a reaction with a control coefficient of 1 on the overall process), that this reaction is insensitive to the concentration of its product, and that it follows simple Michaelis-Menten kinetics. In order to examine whether regulation of the process occurs only through the concentration of the substrate of the reaction, or if there is additional allosteric regulation, the relationship between the intracellular substrate concentration and the process rate is determined experimentally in a population of cells for two series of physiological transitions in which the concentration of S is modulated by two separate means, and data concerning the flux and

the (average) substrate concentration obtained. The idea is that, if the relationship between the rate of the process rate and concentration is not the same for the two series of experiments, the hypothesis that regulation is only through the substrate is falsified. We shall now show why, if the culture is heterogeneous, one would in fact obtain an *artefactual* falsification of the hypothesis.

Let us compare (i) a homogeneous population of cells in which the substrate concentration happens to be equal to the K_m of the flux-controlling enzyme (and the rate of the process therefore at $V_{max}/2$) to (ii) a heterogeneous population with the same average substrate concentration. An extreme case of heterogeneity would be constituted by a population in which half the cells contain the substrate at zero concentration, whereas the other half contain it at a concentration equal to twice the K_m. This gives rise to an average rate of the process for the heterogeneous population of $(0 + 2/(2+1))/2$ times $V_{max} = V_{max}/3$. Now, if in one of the series of experiments, only 50% of the cells are affected by the physiological transition chosen, whereas in the other physiological transition, all cells are equally affected, but to the same average substrate concentration, the two sets of transitions will reveal different overall fluxes at the same average substrate concentration. If the possibility of heterogeneity were not considered, this could easily lead to the conclusion, that there is (normally) additional regulation, which differs between the two physiological transitions. Clearly, quantitative approaches (such as flow cytometry), which analyse the *distribution* of cell properties between members of a population, are essential for the analysis of microbial physiology whenever the possibility of sigificant heterogeneity exists.

The essence of the problem is that one is trying, typically, to correlate a rate of change (v) of a certain variable with respect to the value of a certain property (p), and that a correlation may be expected between the mean values v and p *only* if v is kinetically of first order with respect to p. Whilst the previous example used the relationship between an internal substrate concentration (a variable) and a certain flux, the problem also exists when p is a parameter. Indeed, when one studies the extent to which the activity of an enzyme determines growth rate of a cell, for instance, one has to consider the possibility that the activity of that enzyme may be distributed heterogeneously between individual cells. Similarly, the heterogeneity (in viability) of chemostat populations at low dilution rates can lead to substantial errors in the estimation of Monod coefficients (Sinclair & Topiwala 1970).

Other cases in which unsuspected heterogeneity may be expected to have signifcant effects upon the kinetic analyses of microbial processes include fluctuating systems (Westerhoff et al. 1986), membrane bioenergetics (see the experiments with rhodamine 123, above) and stochastic systems in which small numbers of repressor molecules (say) may control gene expression.

Given the general nonlinearity of biochemical reactions, and the existence of thresholds, one may anticipate that the discovery of important heterogeneities will be the rule, and not the exception, when cultures are investigated by flow analysis.

Future prospects: cataloguing complexity

In the above, we have concentrated mainly on illustrating heterogeneities in the distributions of but a few parameters in cultures, not least because they are easy to display graphically in a 2-D plot. However, the ability to discriminate (sub)populations of cells will increase as the number of the measured parameters increases. Even a modest extension of this philosophy (of looking at many independent parameters on each cell) is likely to end by showing us that our cultures consist, when viewed in multidimensional space, of many more populations than we normally consider, a fact which follows simply from the properties of normal distributions (Williams 1956)). This does not of itself seem to be an intellectually useful goal, and it is not possible, in a simple way, to visualise the distribution of populations in multi-dimensional space. We therefore wish to end by discussing qualitatively the types of analytical advance that *could* benefit the practising microbial physiologist.

Technically, it is now fairly straightforward to acquire several measurements on an individual cell during its passage through a flow cytometer, and Kachel et al. (1990), for instance, describe an 8-parameter system based on a simple personal computer. Robinson et al. (1991) go even further, and acquire a set of multicolour immunofluorescence data from a given sample incubated with 11 tubes containing multiple phenotypic markers. One may imagine that the exploitation of diode array detectors in flow cytometry will produce even more data which one might wish to exploit for the analysis of heterogeneity! Since preparing a table listing the magnitude of each of the many determinands for each of the cells studied does not convey the observations in a digestible form, how are we to extract the salient features of an n-parameter data set, manipulate them mathematically, and encapsulate them in 2 (or at most 3) dimensions? Clearly the major problem is that of reducing the dimensionality of the representation of the data (Sychra et al.

1978), a problem similar in essence to some of those that are being faced and solved by bacterial taxonomists (Goodfellow et al. 1985; Austin & Priest 1986) and analytical chemists (Massart et al. 1988; Levy et al. 1991).

Conventional approaches to reducing the dimensionality of multivariate data, such as principal components analysis (PCA) (see Chatfield & Collins 1980; Causton 1987; Flury & Riedwyl 1988) and the related canonical variate analysis, work by transforming correlated variables exhibited by the cell, organism or molecule of interest into uncorrelated ones, and projecting these transformed variables onto a two-dimensional plane. Different populations are thus separated to a greater or lesser extent, depending on the method used and the structure of the data. In PCA, the two largest principal components are usually plotted against each other, and can show the clustering or otherwise of individual cells in the population according to these principal components. Statistical analysis permits statements to be made concerning the extent (%) to which (say) the first two principal components account for the variance in the sample.

In PCA, which is in fact but a subset of the Universe of possible pattern recognition algorithms, we are trying to construct a relation between several 'input' properties (the measured determinands on each cell) and some output property (for instance the cell's taxonomic status, or even *if* it is a cell and not a piece of debris). The conceptual disadvantages with the above PCA approach, however, are that (i) it assumes that all variables of interest (inputs) are linearly related to each other, which is *a priori* unlikely, and (ii) as a linear mapping it is inevitably associated with a loss of information. Thus, nonlinear mappings may in general be expected, and are found (Aoyama & Ichikawa 1991; Rose et al. 1991), to give much better discrimination or classification.

Neural networks (in this context, more strictly, *artificial* neural networks) are collections of very simple 'computational units' which can take a numerical input and transform it into an output (see McClelland & Rumelhart 1988; Kohonen 1989; Pao 1989; Simpson, 1990). The inputs and outputs may be to and from the 'external world' or to other units within the network. The way in which each unit transforms its input depends on the so-called 'connection weight' (or 'connection strength') and 'bias' of the unit, which are modifiable. The output of each unit to another unit or to the external world then depends on both its strength and bias and on the weighted sum of all its inputs, which are transformed by a (normally) nonlinear weighting function referred to as its activation function. The great power of neural networks stems from the fact that it is possible to present ('train') them with known inputs (and outputs) and to provide them with some form of learning rule which may be used, iteratively, to modify the strengths and biases until the outputs of the network as a function of the inputs correspond to the desired ('known' or 'true') outputs. The trained network may then be exposed to 'unknown' inputs which it will then be able to relate to the appropriate outputs.

A neural network therefore consists of at least 3 layers, representing the inputs and outputs and one or more so-called 'hidden' layers. It is, in particular, the totality of weights and biases of the interactions between inputs and outputs and the hidden layer(s) which reflect the underlying structures of the system of interest, even if its actual (physical) structure is not known. By training up a neural network with known data, then, it is possible to obtain outputs that can accurately predict the behaviour of systems, such as the (continuing) evolution of a time series, even if it is (deterministically) chaotic (Wolpert & Miall 1990). Although the training may be lengthy, the great advantage is that, once trained, interrogating the network is practically instantaneous and no lengthy, iterative processes are required.

In the present context, it is clear that trained neural networks have the potential to reduce the dimensionality of a graphical display by arbitrary amounts, since one may have all the parameters that one measures on a cell as the input nodes to the network and two output nodes representing the X- and Y-coordinates of a 2D-plot. The only question then is how one trains the network. In fact, for this, one simply needs 'standards' which one may assign, arbitrarily, to specific classes (well-separated and appropriately-chosen (x,y) coordinates in one's reduced display), which one can then use to train the network using the dataset of inputs and arbitrarily-assigned outputs. (Alternatively, simply to discriminate *n* types of cells or subpopulations one would provide *n* outputs, which take the value of 1 if the cell is of the stated type, otherwise zero.) After the network has been trained, which may be a lengthy process, one may test it on samples used it in the training set (to check its performance against what was a *known* input) and then, of course, on unknown populations of interest. Thus, by combining flow cytometry with the abilities of trained neural networks, one may expect to be able to provide accurate classifications of cell populations that are easy both to visualise and to comprehend.

Concluding remarks

We have seen that the technique of flow cytometry, though a relatively recent development in microbiology, has allowed penetrating insights into hitherto unrecognised aspects of the physiology of microorganisms. The observations have shown that not all problems of microbial physiology are amenable to correct analysis by performing conventional macroscopic measurements on representative samples of whole cultures. Indeed, one might hazard that a reevaluation of some fundamental tenets may be forthcoming when cellular heterogeneity is taken properly into account. The quantification of heterogeneity therefore constitutes a crucial component of the quantitative analysis of microbial metabolism.

Acknowledgements

DBK and HMR thank the Science and Engineering Research Council for financial support of this work. The work of ASK in Aberystwyth was supported by the Royal Society, under the terms of the Royal Society/USSR Academy of Sciences exchange agreement. HVW is supported by the Netherlands Organization for Scientific Research (NWO). DBK is most grateful to Dr David Causton for useful discussions and comments on the manuscript.

References

Allman R, Hann AC, Phillips AP, Martin KL & Lloyd D (1990) Growth of *Azotobacter vinelandii* with correlation of Coulter cell size, flow cytometric parameters, and ultrastructure. Cytometry 11: 822–831

Allman R, Schjerven T & Boye E (1991) Cell-cycle parameters of *E. coli* K-12 strains determined using flow cytometry. J. Bacteriol. (in press)

Amman RI, Binder BJ, Olsen RJ, Chisholm SW, Devereux R & Stahl DA (1990) Combination of 16S rRNA-targetted probes with flow cytometry for analysing mixed microbial populations. Appl. Env. Microbiol. 56: 1919–1925

An G-H, Bielich J, Auerbach R & Johnson EA (1991) Isolation and characterization of carotenoid hyperproducing mutants of yeast by flow cytometry and cell sorting Bio/Technology 9: 70–73

Aoyama T & Ichikawa H (1991) Basic operating characteristics of neural networks when applied to structure-activity studies. Chem. Pharm. Bull. 39: 358–366

Austin B & Priest F (1986) Modern Bacterial Taxonomy. Van Nostrand Reinhold, Wokingham

Bertin B, Broux O & van Hoegaerden M (1990) Flow cytometric detection of yeast by *in situ* hybridization with a fluorescent ribosomal RNA probe. J. Microbiol. Methods 12: 1–12

Betz JW, Aretz W & Härtel, W (1984) Use of flow cytometry in industrial microbiology for strain improvement programs. Cytometry 5: 145–150

Boye E & Løbner-Olesen A (1990) Flow cytometry: illuminating microbiology. The New Biologist 2: 119–125

Caplan SR & Essig A (1983) Bioenergetics and Linear Nonequilibrium Thermodynamics. The Steady State. Cambridge/Massachusetts, Harvard University Press

Caulcott CA, Dunn A, Robertson HA, Cooper NS, Brown ME & Rhodes PM (1987) Investigation of the effect of growth enviroment on the stability of low-copy-number plasmids in *Escherichia coli*. J. Gen. Microbiol. 133: 1881–9

Causton DR (1987) A Biologist's Advanced Mathematics (pp 48–72). Allen & Unwin, London

Chao L & McBroom SM (1985) Evolution of transposable elements: an IS10 insertion increases fitness in *Escherichia coli*. Mol. Biol. Evol. 2: 359–369

Chatfield C & Collins AJ (1980) Introduction to Multivariate Analysis (pp 57–81). Chapman & Hall, London

Collins JM & Grogan WM (1991) Fluorescence quenching of a series of membrane probes measured in living cells by flow cytometry. Cytometry 12: 247–251

Cooper S (1991) Bacterial Growth and Division. Biochemistry and Regulation of Prokaryotic and Eukaryotic Division Cycles. Academic Press, San Diego

Cunningham A (1990) Fluorescence pulse shape as a morphological indicatoor in the analysis of colonial microalgae by flow cytometry. J. Microbiol. Meth. 11: 27–36

Darzynkiewicz Z (1979) Acridine Organge as a molecular probe in studies of nucleic acids *in situ*. In: Melamed MR, Mullaney

PF & Mendelsohn ML (Eds) Flow Cytometry and Sorting (pp 285–316). Wiley, New York

Darzynkiewicz Z & Crissman HA (Eds) (1990) Flow Cytometry. Academic Press, New York

Davey CL, Dixon NM & Kell DB (1990a) FLOWTOVP: a spreadsheet method for linearising flow cytometric light-scattering data used in cell sizing. Binary 2: 119–125

Davey CL, Kell DB & Dixon NM (1990b) SKATFIT: A program for determining the mode of growth of individual microbial cells from flow cytometric data. Binary 2: 127–132

Dolbeare F & Smith R (1979) Flow cytoenzymology: rapid enzyme analysis of single cells. In: Melamed MR, Mullaney PF & Mendelsohn ML (Eds) Flow Cytometry and Sorting (pp 317–333). Wiley, New York

Donachie WD, Jones NC & Teather N (1973) The bacterial cell cycle. Symp. Soc. Gen. Microbiol. 23: 9–45

Dunlop EH & Ye SJ (1990) Micromixing in fermentors: metabolic changes in Saccharomyces cerevisiae and their relationship to fluid turbulence. Biotechnol. Bioeng. 36: 854–864

Dykhuizen DE & Hartl DL (1983) Selection in chemostats. Microbiol. Rev. 47: 150–168

Flury B & Riedwyl H (1988) Multivariate Statistics: A Practical Approach (pp 181–233). Chapman & Hall, London

Fowler JD & Dunlop EH (1989) Effects of reactant heterogeneity and mixing on catabolite repression in cultures of Saccharomyces cerevisiae. Biotechnol. Bioeng. 33: 1039–1046

Frelat G, Laplace-Builhe C & Grunwald D (1989) Microbial analysis by flow cytometry: present and future. In: Yen A (Ed) Flow Cytometry: Advanced Research and Clinical Applications, Vol 1 (pp 64–80). CRC Press, Boca Raton

Goodfellow M, Jones D & Priest FG (Eds) (1985) Computer-assisted Bacterial Systematics. Academic Press, London

Gibbs, JW (1902) Elementary Principles in Statistical Mechanics. New York, Scribner

Glansdorff P & Prigogine I (1971) Thermodynamic Theory of Structure, Stability and Fluctuations. Wiley-Interscience, London

Glass L & Mackey MC (1988) From Clocks to Chaos. The Rhythms of Life. Princeton University Press, Princeton

Gleick J (1987) Chaos. Making a New Science. Viking Press, New York

Gottschal JC (1990) Phenotypic response to environmental change. FEMS Microbiol. Ecol. 74: 93–102

Harris CM & Kell DB (1985) The estimation of microbial biomass. Biosensors J. 1: 17–84

Harvey JD (1983) Mathematics of microbial age and size distributions. In: Bazin MJ (Ed) Mathematics in Microbiology (pp 1–35). Academic Press, London

Heinzle E, Dunn IJ, Furukawa K & Tanner RD (1982) Modelling of sustained oscillations in continuous culture of Saccharomyces cerevisiae. In: Halme A (Ed) Modelling and Control of Biotechnical Processes (pp 57–65). Pergamon Press, Oxford

Höfle M (1983) Long-term changes in chemostat cultures of Cytophaga johnsonae. Appl. Env. Microbiol. 46: 1045–1053

Horan NJ, Midgley M & Dawes EA (1981) Effect of starvation on transport, membrane potential and survival of Staphylococcus epidermidis under anaerobic conditions. J. Gen. Microbiol. 127: 223–230

Hunter JB & Asenjo JA (1990) A population balance model of enzymatic lysis of microbial cells. Biotechnol. Bioeng. 35: 31–42

Hutter K-J & Eipel HE (1978) Flow cytometric determination of cellular substances in algae, bacteria, moulds and yeasts. A. van Leeuwenhoek 44: 269–282

Ingram M, Cleary TJ, Price BJ, Price RL & Castro A (1982) Rapid detection of Legionella pneumophila by flow cytometry. Cytometry 3: 134–137

Jepras RI (1991) Applications of photon correlation spectroscopy and flow cytometry to microbiology. PhD thesis, Centre for Applied Microbiology and Research

Jones KL & Rhodes-Roberts ME (1981) The survival of marine bacteria under starvation conditions. J. Appl. Bacteriol. 50: 247–258

Kachel V, Messerschmidt R & Hummel P (1990) Eight-parameter PC-AT based flow cytometric data system. Cytometry 11: 805–812

Kamp F, Welch GR & Westerhoff HV (1988) Energy coupling and Hill cycles in enzymatic processes. Cell Biophys. 12: 201–236

Kaprelyants AS & Kell DB (1991) Rapid assessment of bacterial viability and vitality using rhodamine 123 and flow cytometry. J. Appl. Bacteriol. (in press)

Kell DB (1987) The principles and potential of electrical admittance spectroscopy: an introduction. In: Turner APF, Karube I & Wilson GS (Eds) Biosensors; Fundamentals and Applications (pp 427–468). Oxford University Press

Kell DB (1988) Protonmotive energy-transducing systems: some physical principles and experimental approaches. In: Anthony CJ (Ed) Bacterial Energy Transduction (pp 429–490). London: Academic Press

Kell DB, van Dam K & Westerhoff HV (1989) Control analysis of microbial growth and productivity. Symp. Soc. Gen. Microbiol. 44: 61–93

Kjellberg S, Hermansson M, Marden P & Jones GW (1987) The transient phase between growth and non-growth of heterotrophic bacteria, with emphasis on the marine environment. Annu. Rev. Microbiol. 41: 25–49

Kohonen T (1989) Self-Organization and Associative Memory, 3rd edition. Springer, Heidelberg

Koizumi J & Aiba S (1989) Oscillatory behaviour of population density in continuous culture of genetically-engineered Bacillus stearothermophilus. Biotechnol. Bioeng. 34: 750–754

Krawiec S & Riley M (1990) Organization of the bacterial chromosome. Microbiol. Rev. 54: 502–539

Kruth HS (1982) Flow cytometry: rapid analysis of single cells. Anal. Biochem. 125: 225–242

Kubitschek HE (1974) Operation of selection pressure on microbial populations. Symp. Soc. Gen. Microbiol. 24: 105–130

Levy GC, Wang S, Kumar P & Borer P (1991) Multidimensional nuclear magnetic resonance spectroscopy and modeling of complex molecular structures: a challenge to today's computer methods. Spectroscopy International 6: 22–34

Lloyd D, Poole RK & Edwards SW (1982) The Cell Division Cycle. London, Academic Press

Markx GH, Davey CL & Kell DB (1991a) The permittistat: a novel type of turbidostat. J. Gen. Microbiol. 137: 735–743

Markx GH, Davey CL & Kell DB (1991b) To what extent is the value of the Cole-Cole α of the β-dielectric dispersion of cell suspensions accountable in terms of the cell size distribution? Bioelectrochem. Bioenerg. 25: 195–211

Mason CA, Hamer G & Bryers JD (1986) The death and lysis of microorganisms in environmental processes. FEMS Microbiol. Rev. 39: 373–401

Massart DL, Vandeginste BGM, Deming SN, Michotte Y & Kaufman L (1988) Chemometrics. Elsevier, Amsterdam

Matin A, Auger EA, Blum PH & Schultz JE (1989) The genetic basis of starvation survival in non-differentiating bacteria. Annu. Rev. Microbiol. 43: 293–316

McClelland JL & Rumelhart DE (1988) Explorations in Parallel Distributed Processing; A Handbook of Models, Programs and Exercises. MIT Press, Cambridge, Massachusetts

Melamed MR, Mullaney PF & Mendelsohn ML (Eds) (1979) Flow Cytometry and Sorting. New York, John Wiley

Melamed MR, Lindmo T & Mendelsohn ML (Eds) (1990) Flow Cytometry and Sorting, 2nd edition. New York, Wiley-Liss

Miller JS & Quarles JM (1990) Flow cytometric identification of microorganisms by dual staining with FITC and PI. Cytometry 11: 667–675

Mitchison JM (1971) The Biology of the Cell Cycle. Cambridge University Press, Cambridge

Moon FC (1987) Chaotic Vibrations. Wiley, New York

Morita, RY (1982) Starvation-survival of heterotrophs in the marine environment. Adv. Micr. Ecol. 6: 171–198

Morita, RY (1988) Bioavailability of energy and its relationship to growth and starvation survival in nature. Can. J. Microbiol. 34: 346–441

Mitchison, JM (1971) The Biology of the Cell Cycle. Cambridge University Press, Cambridge

Muirhead KA, Horan PK & Poste G (1985) Flow cytometry: present and future. Bio/Technology 3: 337–356

Neidhardt FC, Ingraham JL & Schaechter M (1989) Physiology of the Bacterial Cell. A Molecular Approach. Sinauer Associates, Sunderland, Massachusetts

Olson RJ, Vaulot D & Chisholm SW (1986) Marine phytoplankton distributions measured using shipboard flow cytometry. Deep-Sea Res. 32: 1273–1280

Ormerod MG (Ed) (1990) Flow Cytometry: A Practical Approach. IRL Press, Oxford

Otto R, Vije J, Ten Brink B, Klont B & Konings WN (1985) Energy metabolism in Streptococcus cremoris during lactose starvation. Arch. Microbiol. 141: 348–352

Pao Y-H (1989) Adaptive Pattern Recognition and Neural Networks. Addison-Wesley, Reading, Massachusetts

Patchett RA, Back JP & Kroll RG (1990) Enumeration of bacteria by use of a commercial flow cytometer. J. Appl. Bacteriol. 69: (6), xxiii

Pinder AC, Purdy PW, Poulter SAG & Clark DC (1990) Validation of flow cytometry for rapid enumeration of bacterial concentrations in pure cultures. J. Appl. Bacteriol. 69: 92–100

Poindexter, JS (1981) Oligotrophy: fast and famine existence. Adv. Microbial Ecol. 5: 63–89

Poolman B, Smid EJ, Veldkamp H & Konings WN (1987) Bioenergetic consequences of lactose starvation for continuously cultured Streptococcus cremoris. J. Bacteriol. 149: 1460–1468

Postgate JR (1976) Death in microbes and macrobes. In: Gray TRG & Postgate JR (Eds) The Survival Of Vegetative Microbes (pp 1–19). Cambridge, Cambridge University Press

Rabinovitch PS & June CH (1990a) Measurement of intracellular ionized calcium and membrane potential. In: Melamed MR, Lindmo T & Mendelsohn ML (Eds) Flow Cytometry and Sorting, 2nd edition, Ch 32 (pp 651–668). New York, Wiley-Liss

Rabinovitch PS & June CH (1990b) In: Ormerod JM (Ed) Flow Cytometry: A Practical Approach (pp 161–185). IRL Press, Oxford

Robertson BR & Button DK (1989) Characterizing aquatic bacteria according to population, cell size, and apparent DNA content by flow cytometry. Cytometry 10: 70–76

Robinson JP, Durack G & Kelley S (1991) An innovation in flow cytometry data collection and analysis producing a correlated multiple sample analysis in a single file. Cytometry 12: 82–90

Ronot X, Benel L, Adolphe M & Mounolou J-C (1986) Mitochondrial analysis in living cells: the use of rhodamine 123 and flow cytometry. Biology of the Cell 57: 1–8

Rose VS, Croall, IF & MacFie HJH (1991) An application of unsupervised neural network methodology (Kohonen topology-preserving mapping) to QSAR analysis. Qant. Struct.-Act. Relat. 10: 6–15

Roszak DB & Colwell RR (1987) Survival strategies of bacteria in the natural environment. Microbiol. Rev. 51: 365–379

Rutgers M, Teixeira de Mattos MJ, Postma PW & van Dam K (1987) Establishment of the steady state in glucose-limited chemostat cultures of Klebsiella pneumoniae. J. Gen. Microbiol. 133: 445–453

Salzman, GC (1982) Light scattering analysis of single cells. In: Catsimpoolas N. (Ed) Cell Analysis, Vol 1 (pp 111–143). Plenum Press, New York

Sanders CA, Yajko DM, Hyun W, Langlois RG, Nassos PS, Fulwyer M & Hadley WK (1990) Determination of guanine-plus-cytosine content of bacterial DNA by dual laser flow cytometry. J. Gen. Microbiol. 136: 359–365

Scheper T, Hitzmann B, Rinas U & Schugerl K (1987) Flow cytometry for Escherichia coli for process monitoring. J. Biotechnol. 5: 139–148

Seo J-H & Bailey JE (1987) Cell cycle analysis of plasmid-containing Escherichia coli HB101 populations with flow cytometry. Biotechnol. Bioeng. 30: 297–305

Seo J-H, Srienc F & Bailey JE (1985) Flow cytometry analysis of plasmid amplification in Escherichia coli. Biotechnol. Progr. 1: 181–188

Shapiro HM (1988) Practical Flow Cytometry, 2nd edition. Alan R. Liss, New York

Shapiro HM (1990) Flow cytometry in laboratory microbiology: new directions. ASM News: 584–588

Simpson PK (1990) Artificial Neural Systems: Foundations, Paradigms, Applications and Implementations Pergamon Press, New York

Sinclair CG & Brown DE (1970) Effect of incomplete mixing on the analysis of the static behaviour of continuous cultures. Biotechnol. Bioeng. 12: 1001–1017

Sinclair CG & Topiwala HH (1970) Model for continuous culture which considers the viablity concept. Biotechnol. Bioeng. 12: 1069–1079

Skarstad K, Steen HB & Boye E (1983) Cell cycle parameters of slowly growing *Escherichia coli* B/r studied by flow cytometry. J. Bacteriol. 154: 656–662

Srienc F, Arnold B & Bailey JE (1984) Characterization of intracellular accumulation of poly-β-hydroxybutyrate (PHB) in individual cells of *Alcaligenes eutrophus* H16 by flow cytometry. Biotechnol. Bioeng. 26: 982–987

Srienc F, Campbell JL & Bailey JE (1986) Flow cytometry analysis of recombinant *Saccharomyces cerevisiae* populations. Cytometry 7: 132–141

Srour EF, Leemhuis T, Brandt JE, van Besien K & Hofmann R (1991) Simultaneous use of rhodamine 123, phycoerthyrin, texas red and allophycocyanin for the isolation of human haematopoietic progenitor cells. Cytometry 12: 179–183

Steen HB (1990) Flow cytometric studies of microorganisms. In: Melamed MR, Lindmo T & Mendelsohn ML (Eds) Flow Cytometry and Sorting, 2nd edition, Ch 29 (pp 605–622). Wiley-Liss, New York

Steen HB, Boye E, Skarstad K, Bloom B, Godal T & Mustafa S (1982) Applications of flow cytometry on bacteria: cell cycle kinetics, drug effects, and quantitation of antibody binding. Cytometry 2: 249–257

Steen, HB, Lindmo, T & Stokke, T (1989) Differential light-scattering detection in an arc lamp-based flow cytometer. In: Yen A (Ed) Flow Cytometry: Advanced Research and Clinical Applications, Vol 1 (pp 64–80). CRC Press, Boca Raton

Steen HB, Skarstad K & Boye E (1990) DNA measurements of bacteria. Meth. Cell. Biol. 33: 519–526

Sychra JJ, Bartels PH, Bibbo M & Wied GL (1978) Dimensionality reducing dislays in cell image analysis. Acta Cytol. 21: 747–752

Tanke HJ (1990) In: Ormerod JM (Ed) Flow Cytometry: A Practical Approach (pp 187–207)

van Dilla MA, Langlois RG, Pinkel D & Hadley WK (1983) Bacterial characterization by flow cytometry. Science 220: 620–622

Waggoner AS (1990) Fluorescent probes for cytometry. In: Melamed MR, Lindmo T & Mendelsohn ML (Eds) Flow Cytometry and Sorting, 2nd edition, Ch 12 (pp 209–225). Wiley-Liss, New York

Weber AE & San K-U (1990) Population dynamics of a recombinant culture in a chemostat under prolonged cultivation. Biotechnol. Bioeng. 36: 727–736

Welch GR & Kell DB (1986) Not just catalysts: the bioenergetics of molecular machines. In: Welch GR (Ed) The Fluctuating Enzyme (pp 451–492). Wiley, New York

Westerhoff HV, Hellingwerf KJ & van Dam K (1983) Efficiency of microbial growth is low, but optimal for maximum growth rate. Proc. Natl. Acad. Sci. 80: 305–9

Westerhoff HV, Tsong TS, Chock PB, Chen Y & Astumian RD (1986) How enzymes can capture and transmit free energy from an oscillating electric field. Proc. Natl. Acad. Sci. 83: 4734–4738

Westerhoff HV & van Dam K (1987) Thermodynamics and Control of Biological Free Energy Transduction. Elsevier, Amsterdam

Williams RJ (1956) Biochemical Individuality. Wiley, New York

Wittrup KD & Bailey JE (1988) A single-cell assay of β-galactosidase activity in *Saccharomyces cerevisiae*. Cytometry 9: 394–404

Wittrup KD, Mann MB, Fenton DM, Tsai LB & Bailey JE (1988) Single-cell light scatter as a probe of refractile body formation in recombinant *Escherichia coli*. Bio/Technology 6: 423–426

Wolpert DM & Miall RC (1990) Detecting chaos with neural networks. Proc. R. Soc B. 242: 82–86

Zychlinski E & Matin A (1983) Effect of starvation on cytoplasmic pH, protonmotive force and viability of an acidophilic bacterium *Thiobacillus acidophilus*. J. Bacteriol. 153: 371–374

Antonie van Leeuwenhoek **60**: 159–174, 1991.

Microbial growth dynamics on the basis of individual budgets

S.A.L.M. Kooijman, E.B. Muller & A.H. Stouthamer
Biological Laboratory, Free University, De Boelelaan 1087, 1081 HV Amsterdam, The Netherlands

Key words: Surface area coupled uptake, energy reserves, division triggers, rods, filaments

Abstract

The popular theories for microbial dynamics by Monod, Pirt and Droop are shown to be special cases of a model for individual budgets, in which growth and maintenance are on the expense of reserve materials. The dynamics of reserve materials is a first order process with a relaxation time proportional to cell length; maintenance is proportional to cell volume, and uptake, which depends hyperbolically on substrate density, is proportional to cell volume as well. Because of the latter, population dynamics depends on the behaviour of the individuals in a simple way, such that the cell volume distribution has no quantitative effect.

When uptake is proportional to the surface area of the cell, which is realistic from a physical point of view, the relation between the individual level and the population one becomes more complicated and the cell size and shape distribution affects population dynamics. It is shown how the changing shape of rods modifies uptake and, consequently, growth.

The concept of energy conductance, defined as the ratio of the maximum surface area specific uptake and the volume specific energy reserve has been introduced in the analysis of microbial dynamics. The first tentative results indicate that the value for *E. coli* is close to the mean value for a wide variety of animals.

Properties of the model for cell suspension at constant substrate densities are analyzed and tested against a variety of experimental data from the literature on both the individual and the population level.

Introduction

By studying population responses to experimental manipulations, microbiologists hope to understand subcellular physiological mechanisms. Successful theories by Monod, Droop and Pirt describe population growth at relatively high substrate uptake levels very well. At low substrate levels these theories are less successful.

Several models for cell growth have been proposed which relate it to the DNA duplication cycle. Growth rate of volume or surface is assumed to be proportional to the number of copies of a hypothetical growth controlling gene present in the cell. Other models let the growth rate duplicate at a fixed time prior to division. See e.g. Zaritsky et al. (1982) for a review. One problem with these types of model is that it is hard to see how the cell can change its physiology under conditions of energy limited growth such that the uptake from the environment can keep pace with growth. Kubitschek (1990a,b) indicated

that membrane-associated transport systems also regulate growth. Another problem is that such models can not explain in a simple way why species like cocci clearly show a decrease of the growth rate during the cell cycle, rather than an increase (Knaysi 1941; Mitchison 1961).

An implicit but unrealistic assumption in the Droop and the Pirt models is that uptake is proportional to cell volume. From a physical point of view, it seems more appropriate to link uptake with surface area. See Koch (1985) for a discussion. In the giant cells of some Antarctic foraminiferans, DeLaca et al. (1981) could measure this directly. When uptake is proportional to surface area, cell size and shape distributions become of importance for population dynamics. At high substrate levels, the population dynamics closely resembles that predicted by volume related uptake, but at low substrate levels a much richer behaviour is possible.

The aim of the paper is to show consequences of surface bound uptake in combination with division at a fixed time after exceeding a threshold volume by the cell. We will restrict the present discussion to suspensions of heterotrophic prokaryotes and protists that divide into two, without sexual stages and whose metabolic rate is only limited by the supplied energy. For the moment we will assume that the microbes are isomorphic during their development, but then we will show how the theory can be extended to include organisms like rods which change in shape during growth. After the formulation of a dynamic energy budget (DEB) model for isomorphs, we will focus on situations of constant substrate density, where mortality is of minor importance.

Individual microbes

Growth and maintenance

See Table 1 for a list of frequently used symbols. The uptake of substrate is usually taken to depend hyperbolically on its concentration X in the environment. When the uptake is also proportional to the surface area of a cell of volume W, we arrive at

$$I = \{I_m\} f W^{2/3} \text{ with } f = X/(K + X) \qquad (1)$$

where K is the saturation constant and $\{I_m\}$ a proportionality constant with dimension: amount of substrate per time per squared length. It should be mentioned here that different definitions of the term saturation constant are in use. Owens and Legan (1987) define it as the substrate density at which the population growth rate is half the maximum value, rather than the uptake rate. We will show that both definitions only coincide in the simple Monod model.

We let substrate convert to an abstract quantity called energy, with fixed efficiency $\{A_m\}/\{I_m\}$. The uptake of some (polymeric) substrates involves the excretion of extracellular enzymes. Besides this energetic cost, the uptake itself can be an energy consuming process. We here assume that these costs are proportional to the uptake rate, so that they only affect the value of $\{A_m\}/\{I_m\}$.

The energy taken up adds to an intra-cellular pool of reserve energy S, which may be small. We refer to Dawes (1976, 1986) for a discussion on the central role of energy reserves in microbial physiology and their chemical composition. Related ideas on the dynamical role of intracellular pools can be found in Tsai and Lee (1990) and in the two and three compartment models of Harder and Roels (1982). Suppose further that the use of this energy reserve follows a simple first order dynamics when expressed as energy density, i.e. energy reserve per cell volume $[S] = S/W$. The relaxation time of this dynamics is

taken proportional with the cell length, so with $W^{1/3}$. One reason for this is that when the reserve materials are not dissolved in the cytoplasm, but located on a surface or on a fixed number of nuclei, the reserves become less accessible in a larger individual, just proportional to a length measure. A second reason is that it results in homeostasis, i.e. the chemical composition of the cell growing at constant substrate density does not change during its development. However, the reserve levels still depend on the substrate density. It is convenient to express the energy density as fraction of the maximum energy density, $e = [S]/[S_m]$, which is reached at long exposure to high substrate densities. Likewise, it will prove to be convenient to introduce the compound parameter $v = \{A_m\}/[S_m]$, with dimension length per time and refer to it as the energy conductance: the (maximum) energy flux across a surface relative to the (maximum) energy storage capacity. The dynamics of the energy density then becomes

$$\frac{de}{dt} = vW^{-1/3}(f - e) \qquad (2)$$

From the second term in (2), we see that the energy utilization rate, C, defined as the outflow of energy from the reserves, is given by

$$C \equiv -\frac{dS}{dt}\bigg|_{f=0} = -\frac{d[S]}{dt}\bigg|_{f=0} W - [S]\frac{dW}{dt}$$

$$= [S_m]\left(-\frac{de}{dt}\bigg|_{f=0} W - e\frac{dW}{dt}\right)$$

$$= e[S_m]\left(vW^{2/3} - \frac{dW}{dt}\right) \qquad (3)$$

Following the implicit assumption of e.g. Pirt (1965), maintenance is taken proportional to cell volume, with constant ζ. This seems realistic when e.g. protein turnover plays an important role while the amount of protein is proportional to cell volume. We assume that the volume-specific cost for growth is constant as well, with proportionality constant η. Since the utilized energy flow equals the drain to growth plus maintenance, so $C = \zeta W + \eta\frac{dW}{dt}$, substitution of (3) gives

$$\frac{dW}{dt} = \frac{W^{2/3}ev - W\frac{\zeta}{[S_m]}}{e + \frac{\eta}{[S_m]}} = \frac{W^{2/3}ev - Wam}{e + a} \qquad (4)$$

where the dimensionless energy investment ratio $a = \eta/[S_m]$ is the cost for growth as a fraction of the maximum energy density and the maintenance rate coefficient $m = \zeta/\eta$ is the cost for maintenance relative to that for growth. It has dimension per time.

Table 1. Parameters and main variables of the DEB model.

Parameters	Variables	Dimension	Interpretation
	t	time	time
	X	weight.length^{-3}	substrate density
	I	weight.time^{-1}	uptake rate of substrate
	f	-	$\frac{X}{X+K}$: scaled functional response
	W	length^3	cell volume
	L	length	$W^{1/3}$: cell volume$^{1/3}$
	l	-	L/L_m: cell volume$^{1/3}$/max. cell volume$^{1/3}$
	$[S]$	energy.length^{-3}	stored energy density
	e	-	$[S]/[S_m]$: energy reserves/max.energy reserves
K		weight.length^{-3}	saturation constant
δ		-	aspect ratio: cell length/ cell diameter
$\{I_m\}$		weight.length^{-2}.time^{-1}	max. surface area-specific uptake rate
$\{A_m\}$		energy.length^{-2}.time^{-1}	max. surface area-specific assimilation energy
$[S_m]$		energy.length^{-3}	maximum stored energy density
v		length.time^{-1}	$\{A_m\}/[S_m]$: energy conductance
ζ		energy.length^{-3}time^{-1}	volume-specific maintenance costs
η		energy.length^{-3}	volume-specific costs for growth
m		time^{-1}	ζ/η: maintenance rate coefficient
a		-	$\eta/[S_m]$: energy investment ratio
W_∞		length^3	ultimate volume if $W_\infty > 0$
W_m		length^3	W_∞ at $f = 1$
W_1		length^3	W_d for $f = 1$
W_d		length^3	volume at division
W_a		length^3	volume at initiation DNA duplication
$\mathcal{E}W$		length^3	mean cell volume
L_m		length	$W_m^{1/3}$
l_d		-	l at division
l_1		-	l_d for $f = 1$
l_a		-	l at initiation DNA duplication
$\mathcal{E}l$		-	mean l
t_a		time	DNA duplication time
μ		time^{-1}	population growth rate

It is convenient to introduce the scaled cell length l as the ratio of the cubic root of the actual cell volume, W, to that of the cell volume ultimately reached by cells that do not divide when living at high substrate densities. The latter volume, W_m, can be obtained from (4) through $\frac{dW}{dt} = 0$ for $e = 1$, giving $W_m^{1/3} = \frac{v}{am}$. So we define $l = W^{1/3}am/v$, which is like e a dimensionless quantity between 0 and 1. In terms of l, (2) and (4) become

$$\frac{de}{dt} = am\frac{f - e}{l} \qquad (5)$$

$$\frac{dl}{dt} = \frac{am}{3}\frac{e - l}{e + a} \qquad (6)$$

When the substrate density is constant for a sufficiently long period, so $e = f$, and when the cell divides at scaled length l_d, (6) can be readily integrated, yielding

$$l(t) = f - (f - l_d2^{-1/3})\exp\{-t\gamma_1\} \qquad (7)$$

where $\gamma_1 = \frac{am}{3(f+a)}$ and t is the time since division. Note that the ultimate scaled length when the cell would not divide is $l_\infty = f$, so for population growth, we must have that $l_\infty > l_d$. In the section on division triggers, we will discuss how l_d depends on substrate density. There we will also need the inverse of $l(t)$, giving

$$t(l) = 3\frac{f + a}{am}\ln\frac{f - l_d2^{-1/3}}{f - l} \qquad (8)$$

An important difference between structural biomass and reserves is in the way they are temporary. Structural biomass is subjected to maintenance while reserves are subjected to continuous supply and utilization. If the costs of the transformation of 'old' protein to 'new' protein is about equal to that of energy reserve to 'new' protein and if the main part of the maintenance costs is related to protein turnover, the mean lifetime of structural biomass is m^{-1}. From (2) we observe that the mean lifetime of a compound from the reserves is $W^{1/3}/v$.

The assumption that the size-specific cost of growth is constant heavily relies on the concept of homeostasis for the structural biomass. The first order dynamics for the energy reserves practically implies homeostasis for the energy reserves. Since the reserve density fluctuates with environmental conditions, the distinction between structural biomass and energy reserves allows for a change of cell composition with environmental conditions of a special kind. RNA, mainly consisting of rRNA, is an example of a compound which is known to be more abundant in the cell when it grows at a high rate (Koch 1970). Within the DEB model, we can only account for this relation when (part of the) RNA is included into the energy reserves. This does not seem unrealistic, because when the cell experiences a decline in substrate density, so a decline in energy reserves, it is likely to gain energy through the degradation of ribosomes (Dawes 1976). Their dynamics is most easy to follow, when they would constitute a fixed fraction of the energy reserves, see the section on tests.

The mean translation rate of a ribosome is proportional to the ratio of the rate of protein synthesis and the energy reserves $e[S_m]W$. The rate of protein synthesis is proportional to the growth rate plus part of the maintenance rate which is higher the lower the growth rate in bacteria (Stouthamer et al. 1990). When the population growth rate is not close to zero, the mean peptide elongation rate is dominated by cell growth. Using the present theory for rods, it should be proportional to μ/f for a culture growing at rate μ. See the section on tests.

The oxygen consumption rate (in aerobic environments excluding oxygen use related to substrate uptake) is probably proportional to the total metabolic activity and thus with the utilization rate of the reserves. It depends on the type of reserves. To find the utilization rate, we substitute (4) into (3) and get

$$C = \frac{ea[S_m]}{e + a}\left(vW^{2/3} + mW\right) \qquad (9)$$

At constant substrate density, it is a weighted sum between a surface area and a volume, which explains why the weight-specific oxygen consumption rate is frequently found to be about proportional to weight to the power -0.25.

Division triggers

We here follow the widely accepted hypothesis by Donachie (1968) in the definition of the division trigger: A cell divides at a fixed time, say t_a, after exceeding a threshold volume, say W_a, as long as the interval between subsequent divisions is long enough. The mechanism proposed by Moreno et al. (1990) for (eukaryotic) somatic cells is via the growth bounded accumulation of the cdc25 mitotic inducers, which level results in activation of p34$^{\text{cdc2}}$ protein kinase on reaching a critical level. This mechanism makes it plausible that for shorter inter division times, the cell starts a new DNA duplication cycle when the cell size exceeds size $2W_a$, $4W_a$, $8W_a$ etc. In a dynamical environment, where (2) and (4) are supposed to apply, the implementation of this trigger is not simple. At constant substrate densities, the scaled cell length at division, l_d, and the division interval, $t(l_d)$, can be obtained directly.

When i is an integer such that $2^{i-1} < W_d/W_a < 2^i$, we can solve W_d from

$$t_a = it(W_d) - t(2^{i-1}W_a) \qquad (10)$$

For this purpose, we substitute (8), and obtain for $l_a = W_a^{1/3}am/v$:

$$(f - l_d)^i - (f - l_d 2^{-1/3})^{i-1}(f - 2^{(i-1)/3}l_a) \times$$
$$\exp\frac{-t_a am}{3(f + a)} = 0 \qquad (11)$$

For $l_d^3 < 3l_a^3$, so $i = 1$, the solution is simply

$$l_d = f - (f - l_a)\exp\frac{-t_a am}{3(f + a)} \qquad (12)$$

Obviously, we must have that $f > l_a$, else the cell will not divide at all. When $t_d \gg t_a$, we have $l_d = l_a$.

Changing shapes

It has been assumed that the shape of the organism does not change during development, which means that surface area is proportional to $W^{2/3}$ and length to $W^{1/3}$. This applies not only when the whole surface area is used for uptake, but also when the uptake is over a fixed fraction of it. Think e.g. of *Paramecium* where the uptake is through the cell mouth. The proportionality constant of the surface area with $W^{2/3}$ is included into the energy conductance, which thus has the more exact interpretation of the maximum energy flux across a unit of surface area per maximum energy density times the ratio of the relevant surface area of an individual of a certain volume and that volume to the power 2/3. In this way, we can postpone to think about the size of relevant surface areas till we want to make suborganismal comparisons.

Cooper (1989) argues that *Escherichia* grows at constant substrate density in length only, while the diameter-length ratio at division remains constant when we compare growth at different substrate densities. We can allow for changes in shapes by multiplying the surface related substrate and energy uptake by the ratio of real surface area and the isomorphic one. To this end, we select the shape and size at division at a high substrate density as a reference, at which the cell length is L_1, the diameter δL_1, the surface area A_1 and the volume W_1, say. The shape is idealized by a cylinder with hemispheres at both ends and we assume that all of the surface is equally used for uptake. The length is $L = \frac{\delta}{3}\left(\frac{4W_d}{(1-\delta/3)\delta^2\pi}\right)^{1/3} + \frac{4W}{\pi\delta^2}\left(\frac{(1-\delta/3)\delta^2\pi}{4W_d}\right)^{2/3}$, so $L_d = \left(\frac{4W_d}{(1-\delta/3)\delta^2\pi}\right)^{1/3}$. The surface area is $A = L_d^2\frac{\pi}{3}\delta^2 + \frac{4W}{\delta L_d}$.

The volume and surface area at division at high substrate densities are respectively $W_1 = L_1^3\frac{\pi}{4}\delta^2(1 - \delta/3)$ and $A_1 = L_1^2\pi\delta$. Expressed in volumes, rather than lengths, the sought dimensionless correction function becomes

$$M(W) = \frac{A}{A_1}\left(\frac{W_1}{W}\right)^{2/3}$$
$$= \frac{\delta}{3}\left(\frac{W_d}{W}\right)^{2/3} + \left(1 - \frac{\delta}{3}\right)\left(\frac{W}{W_d}\right)^{1/3} \qquad (13)$$

and has to be applied to the uptake in parameter $\{I_m\}$ in (1) and to the assimilation parameter $\{A_m\}$ and thus the energy conductance v in (2) and (4). Since the size at division depends on substrate density, the shape changes at changing substrate densities in a complex way, (Cooper 1989). We here confine ourselves to constant substrate densities, where (4) becomes for $e = f$ and $W_\infty = \frac{W_d\delta/3}{\frac{am}{vf}W_d^{1/3}-1+\delta/3}$:

$$\frac{dW}{dt} = \frac{vf}{f + a}W_d^{2/3}\delta/3 +$$
$$- \left(\frac{am}{f + a} - \frac{vf}{f + a}W_d^{-1/3}(1 - \delta/3)\right)W$$
$$= \frac{fvW_d^{2/3}\delta/3}{(f + a)W_\infty}(W_\infty - W) \qquad (14)$$

For $W(0) = W_d/2$, the volume-time curve thus becomes

$$W(t) = W_\infty - (W_\infty - W_d/2)\exp\{-t\gamma_2\} \qquad (15)$$

where $\gamma_2 = \frac{fvW_d^{2/3}\delta/3}{(f+a)W_\infty}$. The interpretation of W_∞ is the ultimate volume when the cell would not divide. That is to say, as long as $W_\infty > 0$. For those values, $W(t)$ is convex and of the same type as we found in (7) for the lengths of isomorphs. Note that the volume, i.e. the cubed length, of isomorphs grows skewly S-shaped. When W_∞ is positive, the cell will only be able to divide when $W_\infty > W_d$, thus when $fv > amW_d^{1/3}$. Writing $l_d = W_d^{1/3}am/v$ as before, we arrive at the same condition, $f > l_d$, as we obtained for isomorphs with $f = l_a$ as lower bound. For $fv(1 - \delta/3) = amW_d^{1/3}$ the volume of rods is growing linearly at rate $\frac{vf}{f+a}W_d^{2/3}\delta/3$. For $W_\infty < 0$, $W(t)$ is concave, tending to an exponential growth curve. It no longer has an ultimate size, when the individual would not divide. W_∞ then no longer has the interpretation as an ultimate size, but this does not invalidate the equations.

In order to obtain the size at division, W_d, from (10), we need the inverse of (15):

$$t(W) = \frac{(f + a)W_\infty}{fvW_d^{2/3}\delta/3}\ln\frac{W_\infty - W_d/2}{W_\infty - W} \qquad (16)$$

Substitution into (10) learns that W_d has to be found numerically.

In more detail, affairs are a bit more complicated due to the synthesis of extra cell wall material at the end of the cell cycle. Since the cells are growing in length only, the growth of surface material is directly tied to that of cytoplasm material. Straightforward geometry learns that $\frac{d}{dt}A = (16\pi\frac{1-\delta/3}{\delta W_d})^{1/3}\frac{d}{dt}W$. So the energetic costs for growth can be partitioned like $\eta = \eta_W + \eta_A(16\pi\frac{1-\delta/3}{\delta W_d})^{1/3}$, where η_A denotes the energetic costs for the material in a unit surface area of cell wall and η_W that for the material in a unit volume of cytoplasm. At the end of the cell cycle, when cell volume is twice the initial one, the surface material has to increase from $A(W_d)$ to $2A(W_d/2) = (1 + \delta/3)A(W_d)$. This takes time, of course. If all incoming energy is used, apart from maintenance, for the synthesis of this wall material, the change in surface area is given by $\frac{d}{dt}A = \frac{1}{\eta_A}(f\{A_m\}A - \zeta W_d)$. So, $A(t) = (A(0) - \frac{\zeta W_d}{f\{A_m\}})\exp\{tf\{A_m\}/\eta_A\} + \frac{\zeta W_d}{f\{A_m\}}$. The time it takes for the surface area to reach $(1 + \delta/3)W_d^{2/3}$, starting from $A(0) = W_d^{2/3}$, equals

$$t_A = \frac{a_A}{fv}\ln 2\frac{W_\infty - W_d/2}{W_\infty - W_d} \qquad (17)$$

where $a_A = \eta_A/[S_m]$. For the time interval between subsequent divisions, we have to add $t(W_d)$, giving

$$t_d = \frac{a_A}{fv}\ln 2 + (\frac{a_A}{fv} + \frac{(f+a)W_\infty}{fvW_d^{2/3}\delta/3})\ln\frac{W_\infty - W_d/2}{W_\infty - W_d} \qquad (18)$$

In this paper we assume that a_A is sufficiently small to be neglected.

When $\delta = 0.6$, the shape just after division is a sphere like in cocci, so this is the upper limit for δ. This value is obtained by equating the volume of a cylinder to that of a sphere with the same diameter. When $\delta \to 0$, the shape tends to a filament, which grows exponentially because the length, the surface area as well as the volume all grow at the same rate in these organisms. When we idealize the filament as a cylinder with fixed diameter and neglect the contribution of the caps this time, the shape correction function becomes $M(W) = (W/W_1)^{1/3}$. Note that W_1 is used as a reference volume, rather than W_d, because the diameter is now assumed to be independent from substrate density. This transforms (4) for $e = f$ into

$$\frac{dW}{dt} = \frac{fvW_1^{-1/3} - am}{f + a}W \qquad (19)$$

which gives for $W(0) = W_d/2$

$$W(t) = \frac{1}{2}W_d\exp\{t\gamma_3\} \qquad (20)$$

where $\gamma_3 = \frac{fvW_1^{-1/3} - am}{f+a}$. The inverse is thus

$$t(W) = \frac{f + a}{fvW_1^{-1/3} - am}\ln\frac{2W}{W_d} \qquad (21)$$

When the proposed division trigger still applies, we find from (10):

$$W_d = W_a\exp\{t_a\frac{fvW_1^{-1/3} - am}{f + a}\} \qquad (22)$$

Contrary to isomorphs and rods, the size at division for such filaments no longer plays a role in the population growth rate, as we will see.

Population characteristics

Population growth rate

When the substrate density is constant for a sufficiently long period, the population will grow exponentially with rate $\mu = t_d^{-1}\ln 2$, where the division interval t_d is given in (8), (16) or (21). Substitution gives the expressions listed in Table 2 for $L = W^{1/3}$ and $l = Lam/v$.

The scaled length at division, l_d, is a function of f. It has to be solved from (10). For $f = 1$, this scaled length has been denoted by l_1 and the unscaled one by L_1. When the maintenance rate is negligibly small we get for filaments the model by Droop (1965); when, on the contrary, the maximum capacity of the reserves is negligibly small we arrive for filaments at the model by Pirt (1965). The model by Monod is a special case of that by Pirt for a negligibly small maintenance rate.

Mean cell size

For the purpose of testing this theory and estimating the parameters, it is in practice easier to measure the mean length, rather than the length at division. In the steady state of exponential population growth, the stable age distribution is for $t \in (0, t_d)$ given by $g_t dt = 2\mu\exp\{-\mu t\}dt = \frac{2\ln 2}{t_d}2^{-t/t_d}dt$. From (6), we get the mean length for isomorphs

$$\mathcal{E}l = \frac{2\ln 2}{t_d}\int_0^{t_d}2^{-t/t_d}l(t)dt = f - \frac{f - (2^{2/3} - 1)l_d}{1 + \frac{1}{\ln 2}\ln\frac{f - l_d/\sqrt[3]{2}}{f - l_d}} \qquad (23)$$

Table 2. The population growth rates for the various special cases of the DEB model.

	isomorphs		rods		filaments
μ	$\frac{am}{f+a}$	$\frac{\frac{1}{3}\ln 2}{\ln\frac{f-l_d2^{-1/3}}{f-l_d}}$	$\frac{(1-\delta/3)f/l_d-1}{(f+a)/am}$	$\frac{\ln 2}{\ln\frac{2(1-l_d/f)}{1-l_d/f+\delta/3}}$	$\frac{f/l_1-1}{(f+a)/am}$
$\frac{\mu}{\mu_{\max}}$	$\frac{1+a}{f+a}$	$\frac{\ln\frac{1-l_12^{-1/3}}{1-l_1}}{\ln\frac{f-l_d2^{-1/3}}{f-l_d}}$	$\frac{1+a}{f+a}\frac{(1-\delta/3)f/l_d-1}{(1-\delta/3)/l_1-1}$	$\frac{\ln\frac{2(1-l_1)}{1-l_1+\delta/3}}{\ln\frac{2(1-l_d/f)}{1-l_d/f+\delta/3}}$	$\frac{1+a}{f+a}\frac{f/l_1-1}{1/l_1-1}$
$\frac{\mu}{\mu_{\max}}\big|[S_m]\to 0$	$\frac{\ln\frac{1-l_12^{-1/3}}{1-l_1}}{\ln\frac{f-l_d2^{-1/3}}{f-l_d}}$		$\frac{(1-\delta/3)f/l_d-1}{(1-\delta/3)/l_1-1}$	$\frac{\ln\frac{2(1-l_1)}{1-l_1+\delta/3}}{\ln\frac{2(1-l_d/f)}{1-l_d/f+\delta/3}}$	$\frac{f/l_1-1}{1/l_1-1}$
$\frac{\mu}{\mu_{\max}}\big|\zeta\to 0$	$f\frac{1+a}{f+a}\frac{L_1}{L_d}$		$f\frac{1+a}{f+a}\frac{L_1}{L_d}$		$f\frac{1+a}{f+a}$
$\frac{\mu}{\mu_{\max}}\big|[S_m],\zeta\to 0$	$f\frac{L_1}{L_d}$		$f\frac{L_1}{L_d}$		f

Analogous to (23), we find for the mean volume of rods at steady state

$$\mathcal{E}W = W_\infty\left(1 - \left(1 + \ln\frac{W_\infty - W_d/2}{W_\infty - W_d}/\ln 2\right)^{-1}\right) \quad (24)$$

Note that the mean length increases less steep with increasing substrate density or μ than length at division, because the mean age reduces.

Stable size distribution

The stable size distribution has an intimate relationship with the growth of individuals as has been recognized (Collins & Richmond 1962; Harvey et al. 1967; Voorn & Koch 1986; Marr et al. 1969). It can be most easily expressed in terms of the survivor function. The survivor function of the stable age distribution is given by $G_t(t) \equiv \int_t^{t_d} g_t(s)\,ds = 2\exp\{-\mu t\}-1$. That of the stable size distribution is $G_W(W) = G_t(t(W)) = 2\exp\{-\mu t(W)\}-1$. The probability density is thus

$$g_W(W) = 2\mu\exp\{-\mu t(W)\}\frac{dt}{dW} \quad (25)$$

For isomorphic organisms, $t(W)$ is given in (8). Substitution in the survivor function for the stable size distribution gives

$$G_l(l) = 2^{1+\ln\frac{f-l}{f-l_d2^{-1/3}}/\ln\frac{f-l_d2^{-1/3}}{f-l_d}} - 1 \quad (26)$$

The same can be done for rods and filaments, which leads to

$$G_l(l) = 2^{\ln\frac{1-l_d/f}{(1-l_d/f-\delta/3)(1/l_d)^3+\delta/3}/\ln\frac{2(1-l_d/f)}{1-l_d/f+\delta/3}} - 1 \quad (27)$$

and

$$G_l(l) = (l_1/l)^3 - 1 \quad (28)$$

respectively. These relations are of importance for testing assumptions about the growth process by means of the stable size distribution. Actual stable volume distributions reveal that the volume at division, W_d, is not identical for all individuals, but has some scatter, which is close to a normal distribution (Koch & Schaechter 1962). When $\Psi(W)dW$ denotes the probability density of the number of baby cells of volume W, i.e. cells of an age less than an arbitrarily small period Δt, and $\Phi(W)dW$ the probability density of the number of mother cells of volume W, i.e. cells which will divide within the period Δt, Painter and Marr (1968) derived that

$$g_W(W) = \frac{dt}{dW}\mu\exp\{-\mu t(W)\}\times$$
$$\int_{W_{\min}}^W \exp\{\mu t(X)\}(2\Psi(X)-\Phi(X))\,dX \quad (29)$$

where $W_{\min}$ is the smallest possible cell volume and μ satisfies $\int_{W_{\min}}^{W_{\max}}\exp\{\mu t(X)\}(2\Psi(X)-\Phi(X))\,dX = 0$ (Voorn & Koch 1986). When mother cells divide in two equally sized daughter cells, we have $\Psi(W) = 2\Phi(2W)$. So, when division occurs at W_d for sure, $\Psi(W)dW = (W = W_d/2)$ and $\Phi(W)dW = (W = W_d)$, where the terms within brackets denote booleans which take the value 1 when true or 0 when false. Substitution into (29) gives (25). When Φ is log-normal, we do not have any problems with tails when $W_\infty < 0$. Substitution into (29), together with (16) and $\mu t(W_d) = \ln 2$ gives for rods:

$$g_W(W) = \frac{2^{-H(W)/H(W_d \exp\{\sigma^2/2\})} \ln 2}{(W_\infty - W)H(W_d)\sqrt{2\pi\sigma^2}} \times$$

$$\int_0^W 2^{H(W)/H(W_d e^{\sigma^2/2})} \times$$

$$\left(2e^{-\frac{(\ln 2X/W_d)^2}{2\sigma^2}} - e^{-\frac{(\ln X/W_d)^2}{2\sigma^2}}\right) d\ln X$$

$$(30)$$

with $H(W) = \ln \frac{W_\infty - W_d/2}{W_\infty - W}$. Although the expression looks massive, it has only three parameters: W_d, W_∞ and σ^2. Additional knowledge about δ can be used to uncover $\frac{am}{vf}$ from W_∞ and knowledge about μ can be used to uncover $\frac{f+a}{fv}$.

Yield & specific uptake rate

The stable size distribution is also of importance in connection with the concept of yield, defined as the ratio of the biomass formed and the total substrate consumed. Ignoring the scatter in W_d among individuals, we have for isomorphs $Y = \frac{\mu \mathcal{E} W}{\{I_m\} f \mathcal{E} W^{2/3}} = \frac{\mu L_m \mathcal{E} l^3}{\{I_m\} f \mathcal{E} l^2}$, where $\mathcal{E} W^i$ denotes the expected volume^i of a randomly drawn individual, i.e. $\mathcal{E} W^i = \int_{W_d/2}^{W_d} W^i g_W(W) \, dW$. These integrals can be written out explicitly for isomorphs, but the result is line filling. See Kooi & Kooijman (1991) for its application in microbial food chains.

For rods, where we have to multiply $\{I_m\}$ with the shape correction function and obtain $Y = \frac{\mu W_d^{1/3} \mathcal{E} W}{\{I_m\} f} \left(W_d \delta/3 + \mathcal{E} W(1 - \delta/3)\right)^{-1}$. The mean volume $\mathcal{E} W$ is given in (24). For filaments we find $Y = \frac{\mu W_1^{1/3}}{\{I_m\} f} = \frac{v - \mu W_1^{1/3}}{\{I_m\} a} \frac{\mu}{\mu + m}$ which reduces to $Y = \frac{\{A_m\}}{\eta \{I_m\}} \frac{\mu}{\mu + m}$ for $[S_m] \to 0$. So $\frac{1}{Y} = \frac{\{I_m\}}{\{A_m\}} \eta + \frac{\{I_m\}}{\{A_m\}} \frac{\zeta}{\mu}$ is linear in $\frac{1}{\mu}$, which is well known (Pirt, 1965). The slope still depends on the substrate-energy conversion $\{A_m\}/\{I_m\}$. Pirt (1965) found a wide range of 0.083 to 0.55 h^{-1} for two bacterial species growing on two substrates, aerobically and anaerobically. The quotient of slope and intercept amounts to $\frac{\zeta}{\eta} = m$, the maintenance rate coefficient, which does not depend on the substrate-energy conversion. The mentioned range for these quotients was 0.0393 to 0.0418 h^{-1}. Being very narrow indeed, this strongly supports the interpretation that the main difference was in the relation between cell and environment, but not in the intracellular energetics.

Although the expressions for isomorphs, rods and filaments look very different, their numerical behaviour is

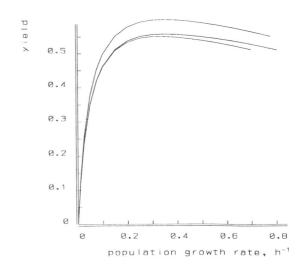

Fig. 1. The yield of isomorphs (middle curve), cocci ($\delta = 0.6$, upper curve) and filaments ($\delta = 0.03$, lower curve) as a function of the population growth rate. The parameters were $l_d = 0.05$, $\{I_m\}/L_m = 0.07$h^{-1}, $a = 3$ and $m = 0.05$h^{-1}.

most similar, as is illustrated in Fig. 1. When we choose the parameters freely, the match can be made even closer. This suggests that the changing shape is only relevant in connection with specialized problems.

The specific uptake rate for the growth limiting substrate, $q = \mu/Y$, is closely related to the yield. We will use it in the next section. The specific uptake rate for oxygen for rods is on the basis of (9) proportional to $\frac{a[S_m]f}{(f+a)W_d^{1/3}\mathcal{E} W} \left(W_d v \frac{\delta}{3} + \mathcal{E} W((1 - \frac{\delta}{3}) + mW_d^{1/3})\right)$.

Tests against experimental data

The cell length, volume and wet weight (which corresponds with volume assuming a specific density of 1 cm^3/g) of widely different isomorphs (a ciliate, a heliozoan and a rhizopod) and a filament (a fungus) are given in Fig. 2 as a function of time. The least squares estimates (with s.d.) of the fitted equations (7) and (20), respectively were:

Source	S. & Synga. (1925)	Synga. (1935)	Prescott (1957)	Trinci et al. (1990)
Temp. (°C)	17		23	25
org.	*Param.*	*Actino.*	*Amoeba*	*Fusarium*
$W_d^{1/3}$	17.53 a.u. (0.38)	3.46 μm (0.050)	2.14 ng$^{1/3}$ (0.012)	3.7 μm (0.55)
$W_\infty^{1/3}$	29.69 a.u. (0.62)	4.30 μm (0.022)	2.79 ng$^{1/3}$ (0.016)	
γ_1, γ_2 (h)	0.187 (0.024)	0.33 (0.042)	0.098 (0.0065)	0.30 (0.044)

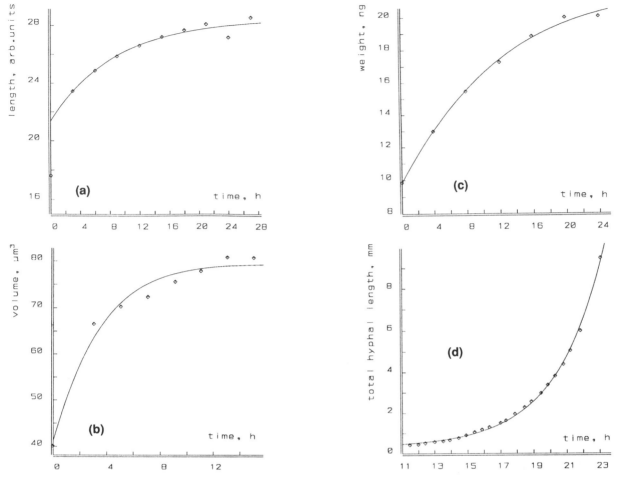

Fig. 2. Cell length, volume and weight as a function of time in the isomorphs *Paramecium caudatum* (a), *Actinophrys sp.* (b) and *Amoeba proteus* (c) and the filament *Fusarium graminearum* (d). Data from resp. Schmalhausen and Syngajewskaya (1925), Syngajewskaya (1935), Prescott (1957) and Trinci et al. (1990). The parameters are given in the text.

The unit for the cell length of *Paramecium* was divisions of ocular micrometer. Although it might seem odd to classify *Amoeba* as an isomorph, it does not change its surface area to volume ratio during growth in a way other than isomorphs. In constant environments, cell growth for isomorphs is strikingly different from exponential growth for filaments, which has been found for mycelia of e.g. *Penicillium* (Pirt & Callow 1960). Trinci et al. (1990) also showed that the number of hyphal tips is growing exponentially as well, with the same rate. When uptake only occurs in these tips, it means that the uptake area is proportional to volume, which explains the exponential growth. The distinction between the individual and the population level for this type of filaments becomes problematic. The DEB model implies that there is no need to make the distinction for these organisms.

The cell volume of rods and cocci are given in Fig. 3 as a function of time. The parameter values used and the least squares estimates (with s.d.) of (15) were

Source	Collins & R. (1962)	Kubitschek (1990a)	Mitchison (1961)	Mitchison (1961)
Org.	*B. cereus*	*E. coli*	*S. faecalis*	*S. faecalis*
Temp. (°C)	35	37	17	40
δ	0.196	0.28	0.6	0.6
W_d (μm^3)	3.481	0.856	1.3	1.45
	(0.086)	(0.010)		
$\frac{vf}{f+a}$ (($\frac{\mu}{min}$))	0.070	0.039	0.069	0.115
	(0.060)	(0.008)	(0.144)	(0.026)
$\frac{am}{vf}$ ($\frac{1}{\mu\,min}$)	0.334	0.936	0.906	0.856
	(0.288)	(0.032)	(0.010)	(0.011)

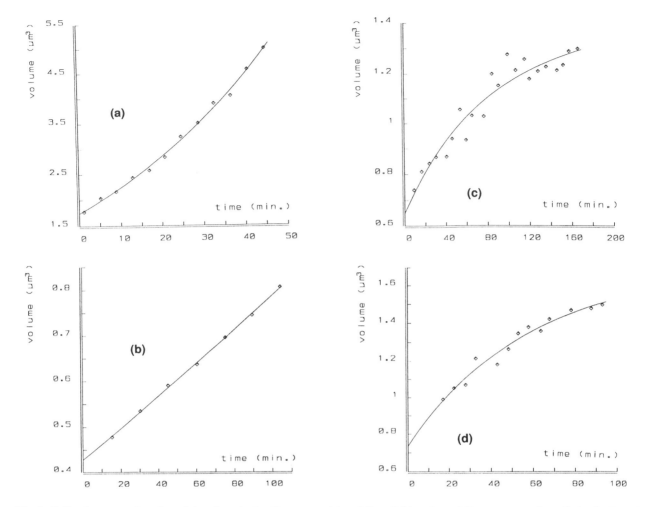

Fig. 3. Cell volume as a function of time in rods *Bacillus cereus* (a) and *E. coli* (b) and cocci *Streptococcus faecalis* (c, d). Data from resp. Collins and Richmond (1962), Kubitschek (1990a) and Mitchison (1961). The parameters are given in the text.

Although exponential cell growth during one division interval has been frequently reported for *E. coli*, the data from Kubitschek (1990a) show that his cultures grew almost linear in volume in the media he used. The DEB model for rods allows linear growth as well as growth which is close to exponential. The theory suggests that it depends on the medium chosen. Our theory also predicts that bacteria hold an intermediate position between isomorphs and filaments, with cocci deviating most among the rods from exponential growth. This is confirmed by the presented data from Mitchison (1961).

Figure 6 shows that the mean cell volume of *E. coli* grows almost exponentially as a function of the population growth rate. (24) has been fitted, where W_d was solved from (10) using (16). As can be expected, the energetic

parameters are not well fixed by these data. The start of new forks is predicted at $\mu = 1$, 1.9 and 2.55 h^{-1} on the basis of the present parameter values.

RNA as a fraction of dry weight is given in Fig. 4. The model description is based on the assumption that (a small) part of RNA is a fixed fraction of the structural biomass, while the main part is a fixed fraction of the stored energy reserves, so proportional to f. Both the structural biomass and the energy reserves contribute to the dry weight, so the RNA concentration is given by $\frac{\gamma + \alpha\beta f}{1 + \beta f}$. The parameters of Fig. 6 were: used to relate μ to f.

The peptide elongation rate is plotted in Fig. 5 for *E. coli* at 37°C. If we neglect the contribution of maintenance to protein synthesis, the elongation rate at constant substrate density is proportional to the ratio of the growth rate $\frac{dW}{dt}$

and the stored energy $[S_m]fW$. For a typical rod, i.e. a rod of size $\mathcal{E}W$ given in (24), substitution of (14) shows that the elongation rate should be proportional to μ/f at population growth rate μ. It allows the estimation of the parameter l_b, which is hard to obtain in another way.

The stable length distribution of *E. coli* is plotted in Fig. 7. A log-normal distribution of the number of mother cells has been assumed according to (30). Although there are only three free parameters, with one related to the growth process, W_∞, it is not well fixed by these data. The range of values less than 0 produce all very similar distributions, when W_d is chosen appropriately.

The population growth rate is plotted against substrate concentration in Fig. 8; three rods growing on glucose and one isomorph growing on bacteria. Only the maintenance rate constant m and the saturation constant K have been least squares estimated from these data, conditionally under chosen values for $a = 3$, $l_d = 0.03$ and $\delta = 0.3$ for the three rods and $l_d = 0.2$ for the isomorph. This is because these data are not very informative with respect to the other parameters. The least squares fitted parameters (with s.d.) were

Source	Schulze & L. (1964)	Senn (1989)	Rutgers (1987)	Taylor (1978)
Org.	*E.coli*	*E.coli*	*K.pneumo.*	*C.campylum*
Temp. (°C)	30	37	35	20
m h$(^{-1})$	0.04 (6.7E-4)	0.033 (7.6E-4)	0.056 (2.4E-3)	0.064 (6E-3)
K (mg/l)	103 (4)	0.090 (0.013)	1.28 (0.16)	19.69 10^6/ml (3.47)
Gluc. (mg/l)	0	0.012 (5.7E-3)	0.080 (0.026)	

The first observation is that the data from Schulze and Lipe (1964) fit better than those from Senn (1989) and Rutgers et al. (1987), which points to the danger of looking hard to a single data set. With the data of Schulze and Lipe, Westerhoff and van Dam (1987) obtained a bad fit of the hyperbola, which is close to ours. They proposed a linear relation between the population growth rate and the log of the substrate concentration on the assumption of a constant product concentration. Senn found this relation also the best fitting one. The reason for the bad fit of the hyperbola is that it can not rise steeply combined with a gradual approach to its asymptote. However, it is very difficult to measure very low levels of glucose when the flux as well as the biomass concentration is very high. When we assume that there could be a small unmeasured background concentration of glucose the growth data indicate

a tiny fraction of a mg/l (see the last line in the table of parameters) – the fit can be improved substantially.

The second observation is that there exists a factor of 1000 between the glucose concentrations of equal growth. Senn suggested that difference in pretreatment could be the cause. The cultures of Senn (1989) had a 3 month period

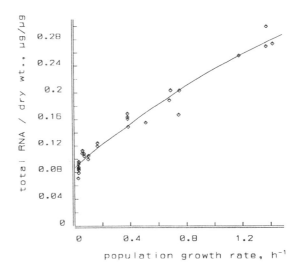

Fig. 4. The concentration of RNA as a function of the population growth rate in *E. coli*. Data from Koch (1970). The least squares estimates of the parameters (with s.d.) were $\alpha = 0.44(0.05)$, $\beta = 20.7(5.4)$ and $\gamma = 0.077(0.005)$.

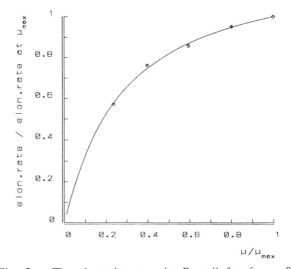

Fig. 5. The elongation rate in *E. coli* for $\delta = 0.3$, $l_d = 0.24(0.019)$, $a = 32.4(91.9)$. Data from Bremer and Dennis (1987). Both the elongation rate and the population growth rate are expressed as fractions of their maximum value of $\mu = 1.73$ h^{-1} with an elongation rate of 21 aa s^{-1} rib^{-1}.

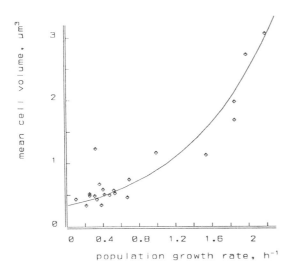

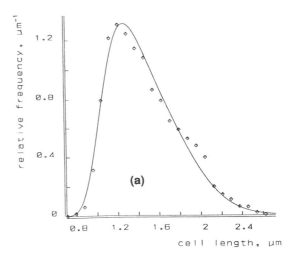

Fig. 6. The mean volume of *E. coli* as a function of the population growth rate at 37°C. Data from Trueba (1981). For a chosen aspect ratio $\delta = 0.28$, a maintenance rate coefficient $m = 0.05$ h^{-1} and $a = 1$, the least squares estimates (with s.d.) of the size at start of DNA replication was $W_a = 0.454(0.069)$ μm^3, the time required for division was $t_a = 1.03(0.081)$ h and the energy conductance $v = 31.3(32)$ $\mu m/h$.

of glucose limited growth prior to the experiment, while Schulze and Lipe (1964) used a short adaptation period. Rutgers at al. (1987) reported a similar adaptation period over 50–100 generations for glucose limited *Klebsiella pneumoniae* cultures.

The third observation is that the shape of the curve for isomorph (the ciliate *Colpidium campylum* growing on *Enterobacter aerogenes*) and rods are very similar, despite the difference in mathematical expression.

The specific uptake rate of glucose by *K. pneumoniae* is given in Fig. 9. Both the structural biomass and the energy reserves are assumed to contribute to the measured standing crop, here given in C-moles. The relation fitted was

$$q = \alpha f W_d^{-1/3}(W_d\frac{\delta}{3}/\mathcal{E}W + 1 - \frac{\delta}{3})(1 + \beta f)^{-1} \quad (31)$$

where $\alpha = 8.9(0.13)$ $\frac{\mu m.mole}{h.C-mole}$ and $\beta = 0.04(0.16)$ have been least squares estimated. Using (10) and (24), six other parameters are needed to fix the relation and were taken as follows: $\delta = 0.22$, $t_a = 0.8$ h, $W_a = 0.45$ μm^3, $a = 3$, $m = 0.05$ h^{-1} and $v = 20$ μ/h.

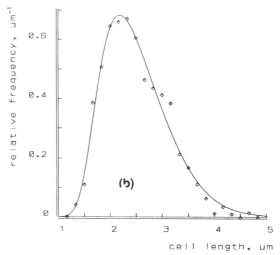

Fig. 7. The probability density of the length of *E. coli* B/r A (a) and K (b) at a population growth rate of 0.38 and 0.42 h^{-1}, respectively at 37°C. Data from Koppes et al. (1978). For an aspect ratio of $\delta = 0.3$, the parameters were $W_d = 0.506$ μm^3, $W_\infty = -0.001$ μm^3 and $\sigma^2 = 0.026$ and $W_d = 2.324$ μm^3, $W_\infty = -1$ μm^3 and $\sigma^2 = 0.044$.

Discussion

We believe that the DEB model is not just one addition to the existing voluminous collection of models for bacterial growth, because of its relevance to animal nutrition, development and growth (Kooijman 1986 a,b,c 1988; Evers & Kooijman 1989; Zonneveld & Kooijman 1989, 1991; van Haren & Kooijman 1991, submitted). This reduces its ad hoc character and opens the possibility for comparison with data on non-bacteria. The comparison of the population dynamics for different species is only possible when

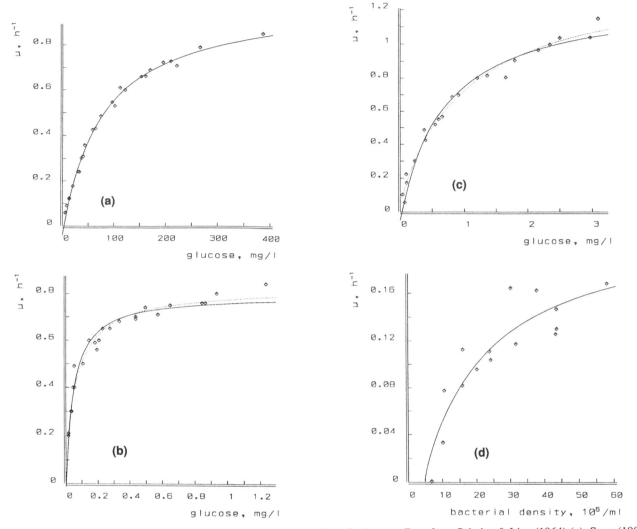

Fig. 8. The population growth rate as a function of the concentration of substrate. Data from Schulze & Lipe (1964) (a), Senn (1989) (b), Rutgers et al. (1987) (c) and Taylor (1978) (d). For parameter values, see text. The dotted curves allowed for a small unmeasured background glucose concentration to show that the goodness of fit is very sensitive for such an error in measurement.

the parameters determining this dynamics are in the same dimension. For animals, the mean energy conductance for some 250 species has been found (Kooijman 1988) to be 0.433 mm/d at 20°C. Data on embryo development (Zonneveld & Kooijman 1991) show only little variation around this value among several different species. The first tentative estimate for *E. coli* is 31 μm/h= 0.72 mm/d (see Fig. 6) at 37°C. This is in the same range. More research is needed on this point.

The key of our approach is in modelling individuals and evaluating consequences. Since decisions about uptake and division are taken by the individual, this level of organization seems to be the most natural one to model from a system theoretical point of view. So, the individuals in a population are no longer identical. Such populations are therefore called structured populations (Metz & Diekmann 1986). Models for such populations are called corpuscular (Harder & Roels 1982) or segregated models (Fredrickson et al. 1967; Ramkrishna 1979). Esener et al. (1982) introduced the term structured models for models that describe the activity of organisms by more than one variable. Since they do not deal with individuals, we would like to call such models multivariate models for non-structured populations.

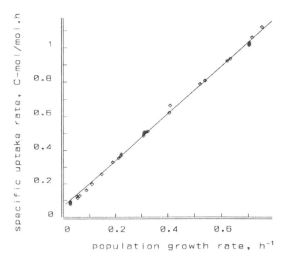

Fig. 9. The specific uptake rate of glucose by *K. pneumoniae* at 35°C. Data from Esener et al. (1983).

The structured population approach taken here has some characteristics of the microscopic one, which deals with the subcellular machinery in detail and, the macroscopic one, which deals with element and energy balances at population level. See e.g. Esener et al. (1982, 1983) for a discussion. Apart from the changes in shape in rods during transients, the DEB model presented here also applies in dynamic environments. We agree with Zaritsky et al. (1982) that phenomena during transients are most suitable to test competing model formulations. Having formulated the growth dynamics of individuals, that of the population follows and can be evaluated using formulations in terms of partial differential equations (Metz & Diekmann 1986; Kooi & Kooijman 1991). We disagree with the approach taken by Zaritsky et al. (1982) and Grover et al. (1980), where the additional assumption at population level has been made that the shift-up in total cell mass is instantaneous. Additional assumptions of this type lead to inconsistencies in model formulations and invalidates subsequent model tests. Except in the case that individuals are growing exponentially, the dynamics of the population depend in a rather complex way on the properties of individuals. It is standard to assume exponential growth of individuals implicitly, by directly relating subcellular phenomena to population characteristics, see e.g. Marr (1991). Sometimes this assumption is made explicitly, see e.g. Cooper, 1991. Such models would fail the tests presented in this paper on the growth of individuals and require complex assumptions for the mechanism of nutrient uptake at the molecular level.

There are four good reasons to tie uptake to surface area rather than to volume. The fit with experimental data is excellent (see above), it links up with a simple physical mechanism (Aksnes & Egge 1991; Button 1991), the dimensions of the parameters allow a comparison with animal and plant physiology and population dynamics (Kooijman 1986b), and it explains why oligotrophic conditions select for small cell sizes (Morita 1982; Button 1991). With a volume bound uptake, small individuals are not better compared with large ones. With a surface area bound uptake, individuals with a larger cell length than $f\{A_m\}/\zeta$ can not survive. The assumption that the relaxation time of the storage dynamics is proportional to length can be made consistent with a class of models for the dynamics of cellular components, where surface-bound enzymes catalyze reactions whose rates depend on substrate and product concentrations in the cytoplasm.

The concept of energy storage allows the description of phenomena like delays in response to instantaneous changes in the environment, as has been frequently reported (Kubitschek 1990b; Marr 1991) and a rapid uptake of substrate which is only later followed by an increase in biomass. It is closely related to the idea of intracellular precursor pools for synthesis, that seem to link up naturally with molecular mechanisms (Button 1991; Marr 1991). However, this improvement in performance comes with a price to be paid. Because the chemical composition of such a storage will deviate from the rest of the biomass, the whole cell no longer has a constant chemical composition, which complicates macroscopic mass balances. Although the role of storage is primarily of the 'hidden variable' type as introduced by Harder and Roels (1982), we do not believe that it is feasible to identify a limited number of compounds that are not species-specific, which are readily measurable and supposed to be in complete control of cellular dynamics.

From a biochemical point of view, the DEB model is very simplistic. From an applications point of view, the number of parameters is large enough to require different data sets on a single species to allow estimation. We showed that some population performances are quite robust for details in individual behaviour, suggesting that in some applications sensible simplifications apply. For other applications the DEB model has to be extended to include product formation. One particular extension seems very attractive: a fixed fraction of utilized energy is spent on product formation. In this case all growth descriptions given here still apply, but some parameter values and interpretations change. This implementation exactly corresponds to the energy allocation to reproduction in animals as proposed in Kooijman (1986a,b,c).

Acknowledgements

We like to thank Wyanda Yap for her assistance in digitizing and fitting of the data and Henk van Verseveld, Karel van Dam, Hans de Hollander and Luc Mur for stimulating discussions.

References

Aksnes DL & Egge JK (1991) A theoretical model for nutrient uptake in phytoplankton. Mar. Ecol. Prog. Series 70: 65-72

Bremer H & Dennis PP (1987) Modulation of chemical composition and other parameters of the cell by growth rate. In: Neidhardt FC (Ed) Escherichia coli and Salmonella typhimurium (pp 1527-1542). Am. Soc. Microbiol., Washington

Button DK (1991) Biochemical basis for whole-cell uptake kinetics: specific affinity, oligotrophic capacity, and the meaning of Michaelis constant. Appl. Envir. Microbiol. 57: 2033-2038

Collins JF & Richmond MH (1962) Rate of growth of Bacillus cereus between divisions. J. Gen. Microbiol. 28: 15-33

Cooper S (1989) The constrained hoop: an explanation of the overshoot in cell length during a shift-up of Escherichia coli. J. Bact. 171: 5239-5243

Cooper S (1991) Bacterial Growth and Division. Academic Press, London

Dawes EA (1976) Endogenous metabolism and the survival of starved prokaryotes. In: Gray TR & Postgate JR (Eds) The Survival of Vegetative Microbes (pp 19-53). Cambridge Univ. Press, Cambridge

Dawes EA (1986) Microbial Energetics. Blackie, Glasgow

DeLaca TE, Karl DM & Lipps JH (1981) Direct use of dissolved organic carbon by agglutinated benthic foraminifera. Nature 289: 287-289

Donachie WD (1968) Relationship between cell size and time of initiation of DNA replication. Nature 219: 1077-1079

Donachie WD, Begg KG & Vicente M. (1976) Cell length, cell growth and cell division. Nature 264: 328-333

Droop MR (1983) 25 years of algal growth kinetics. Botanica Marina 26: 99-112

Esener AA, Veerman T, Roels JA & Kossen NWF (1982) Modeling of bacterial growth: formulation and evaluation of a structured model Biotechnol. Bioeng. 24: 1749-1764

Esener AA, Roels JA & Kossen NWF (1983) Theory and applications of unstructured growth models: kinetic and energetic aspects. Biotechnol. Bioeng. 25: 2803-2841

Evers EG & Kooijman SALM (1989) Feeding and oxygen consumption in daphnia magna: a study in energy budgets. Neth. J. Zool. 39: 56-78

Fredrickson AG, Ramkrishna D & Tsuchiya MM (1967) Statistics and dynamics of procaryotic cell populations. Math. Biosci. 1: 327-374

Grover NB, Zaritsky A, Woldringh CL & Rosenberger RF (1980) Dimensional rearrangement of rod-shaped bacteria following nutritional shift-up. I. Theory. J. Theor. Biol. 86: 421-439

Harder A & Roels JA (1982) Application of simple structured models in bioengineering. Adv. Biochem. Eng. 21: 55-107

Harvey RJ, Marr AG & Painter PR (1967) Kinetics of growth of individual cells of Escherichia coli and Azotobacter agilis. J. Bacteriol. 93: 605-617

Knaysi G (1941) A morphological study of Streptococcus faecalis. J. Bacteriol. 42: 575-586

Koch AL (1970) Overall controls on the biosynthesis of ribosomes in growing bacteria. J. Theor. Biol.28: 203-231

Koch AL (1985) The macroeconomics of bacterial growth. In: Fletcher M & Floodgate GD (Eds) Bacteria in Their Natural Environment. Special Publ. Soc. Gen. Microbiol. 16: 1-42

Koch AL & Schaechter M (1962) A model for the statistics of the division process. J. Gen. Microbiol. 29: 435-454

Kooi BW & Kooijman SALM (1991) Existence and stability of microbial prey-predator systems. (submitted)

Kooijman SALM (1986a) Population dynamics on the basis of budgets. In: Metz JAJ & Diekmann O (Eds) The Dynamics of Physiologically Structured Populations (pp 266-297). Springer Lecture Notes in Biomathematics. Springer-Verlag, Berlin

Koppes LJH, Woldringh CL & Nanninga N (1978) Size variations and correlation of different cell cycle events in slow-growing Escherichia coli. J. Bacteriol. 134: 423-433

Kubitschek HE (1990a) Cell volume increase in Escherichia coli after shifts to richer media. J. Bact. 172: 94-101

Kubitschek HE (1990b) Cell growth and abrupt doubling of membrane proteins in Escherichia coli during the division cycle. J. Gen. Microbiol. 136: 599-606

Marr AG (1991) Growth rate of Escherichia coli. Bacterial Reviews 55: 316-333

Marr AG, Painter PR & Nilson EH (1969) Growth and division of individual bacteria. Symp. Soc. Gen. Microbiol. 19: 237-261

Metz JAJ & Diekmann O (Eds) (1986). The Dynamics of Physiologically Structured Populations. Springer Lecture Notes in Biomathematics. Springer-Verlag, Berlin

Mitchison JM (1961) The growth of single cells. III. Streptococcus faecalis. Exp. Cell. Res. 22: 208-225

Moreno S , Nurse P & Russell P (1990) Regulation of mitosis by cyclic accumulation of $p80^{cdc25}$ mitotic inducer in fission yeast. Nature 344: 549-552

Morita RJ (1982) Starvation-survival of heterotrophs in the marine environment. Adv. Microb. Ecol. 6: 171-198

Owens JD & Legan JD (1987) Determination of the monod substrate saturation constant for microbial growth. FEMS Microbiol. Rev.46: 419-432

Painter PR & Marr AG (1968) Mathematics of microbial populations. Annu. Rev. Microbiol. 22: 519-548

Pedersen S (1984) Escherichia coli ribosomes translate in vivo with variable rate. EMBO J. 3: 2895-2898

Pirt SJ (1965) The maintenance energy of bacteria in growing cultures. Proc. Roy. Soc. B 163: 224-231

Pirt SJ & Callow DS (1960) Studies of the growth of Penicillium chrysogenum in continuous flow culture with reference to penicillin production. J. Appl. Bact. 23: 87-98

Prescott DM (1957) Relations between cell growth and cell division. In: Rudnick D (Ed) Rhythmic and Synthetic Processes in Growth (pp 59–74). University Press, Princeton

Ramkrishna D (1979) Statistical models of cell populations. Adv. Biochem. Eng. 11: 1-45

174

Rutgers M, Teixeira de Mattos MJ, Postma PW & Dam K van (1987) Establishment of the steady state in glucose- limited chemostat cultures of *Kleibsiella pneumoniae*. J. Gen. Microbiol. 133: 445-453

Schmalhausen II & Syngajewskaja E (1925) Studien über Wachstum und Differenzierung. I. Die individuelle Wachtumskurve von *Paramaecium caudatum*. Roux Arch. 105: 711-717

Schulze KL & Lipe RS (1964) Relationship between substrate concentration, growth rate, and respiration rate of *Escherichia coli* in continuous culture. Arch. Mikrobiol. 48: 1-20

Senn HP (1989) Kinetik und Regulation des Zuckerabbaus von Escherichia coli ML 30 bei tiefen Zuckerkonzentrationen. Ph-D thesis, Techn. Hochschule Zurich

Stouthamer AH, Bulthuis BA & Verseveld HW van (1990) Energetics of growth at low growth rates and its relevance for the maintenance concept. In: Poole RK, Bazin MJ & Keevil CW (Eds) Microbial Growth Dynamics (pp 85-102). Irl Press, Oxford

Syngajewskaja E (1935) The individual growth of protozoa: *Blepharisma lateritia* and *Actinophrys sp*. Trav. de l'Inst. Zool. Biol. Acad. Sci. Ukr. 8: 151-157

Taylor WD (1978) Growth responses of ciliate protozoa to the abundance of their bacterial prey. Microb. Ecol. 4: 207-214

Trinci APJ, Robson GD, Wiebe MG, Cunliffe B & Naylor TW (1990) Growth and morphology of *Fusarium graminearum* and other fungi in batch and continuous culture. In: Poole RK, Bazin MJ & Keevil CW (Eds) Microbial Growth Dynamics (pp 17-38). Irl Press, Oxford

Trueba FJ (1981) A morphometric analysis of *Escherichia coli* and other rod-shaped bacteria. Ph-D Thesis, University of Amsterdam

Tsai SP & Lee YH (1990) A model for energy-sufficient culture growth. Biotechnol. Bioeng. 35: 138-145

Voorn WJ & Koch AL (1986) Characterization of the stable size distribution of cultured cells by moments. In: Metz JAJ & Diekmann O (Eds) The Dynamics of Physiologically Structured Populations (pp 430-473). Springer Lecture Notes in Biomathematics. Springer-Verlag, Berlin

Westerhoff HV, Dam K van (1987) Thermodynamics and control of biological free-energy transduction. Elsevier, Amsterdam

Zaritsky A, Woldringh CL, Grover NB, Naaman J & Rosenberger RF (1982) Growth and form in bacteria. Comments Mol. Cell. Biophys. 4: 237-260

Zonneveld C & Kooijman SALM (1989) The application of a dynamic energy budget model to *Lymnaea stagnalis*. Func. Ecol. 3: 269-278

Zonneveld C & Kooijman SALM (1991) The comparative kinetics of embryo development. (submitted)

Antonie van Leeuwenhoek **60**: 175–191, 1991.

Quantitative aspects of cellular turnover

Arthur L. Koch
Department of Biology, Indiana University, Bloomington, IN 47405, USA

Key words: growing systems, kinetics, murein wall, nucleic acid, protein, turnover

Abstract

Living organisms do not just grow by synthesizing cellular components. As part of the necessary steps for existence, some components are degraded after synthesis. Even for bacteria in balanced, exponential growth some substances, under some conditions, are turned over. In other phases of growth turnover can be much more extensive, but it is still selective. This review covers studies with animals as a way to put the studies on microorganisms in perspective. The history, the mathematics, and experimental design of turnover experiments are reviewed. The important conclusion is that most of the proteins during balanced growth are very stable in bacteria, although ribosomal proteins are degraded under starvation conditions. Another generalization is that the process of wall enlargement in general is associated with obligatory turnover of the peptidoglycan.

Introduction: why turnover?

In order for a cell to grow, it must convert substances available to it into substances that it can polymerize into the nucleic acids, proteins, carbohydrates, and lipids. This metabolism/anabolism (meta = change; ana = build up) is contrasted with catabolism (cata = breakdown), the process in which compounds are converted into less energetic substances and the resultant liberated energy trapped as an electrochemical gradient or forced to cause ATP formation, either of which are intermediate stages in the energy metabolism of the cell used to drive cell processes. The contrast of assimilatory and dissimilatory processes leaves out another important facet: turnover. Turnover implies the manufacture of cell substance and then its dissolution without net gain. Seemingly this is wasteful, fruitless, and futile (I use three words meaning the same thing for emphasis).

Several classes of reasons for turnover have been put forth. 'Wear-and-tear' is the mechanism that first comes to mind. If cell components are tools or structural entities and are damaged and degraded in their use. So the worn out ones are replaced with new ones while the defective units are degraded; then their component parts disposed of or reutilized.

A second potential mechanism is if an informational substance is continuously being degraded and resynthesized. This allows cell can quickly to redirect its metabolism by synthesizing a new version of the substance that has different properties. Then because of turnover mechanism already in place, the original version of the process will be quickly stopped. It was this kind of thinking that led to the concept of messenger RNA as a short lived molecule that programed the long lived ribosome, and thus allows a bacterium to adapt to altering conditions (Monod & Jacob 1961).

Yet a third potential mechanism is if some substances are only made as temporary scaffolds or jigs, i.e., objects useful only in that they serve a role in forming something else of more permanence. For example, some viruses have genes that code for proteins functioning only in the assembly of the virion (King 1980).

Finally, there is a mechanan only slightly different from the last mechanism, it is if some objects can only be made to function for a limited time and are then discarded. The general example is the protective covering of an organism that is shed or discarded when outgrown. Specific examples are the carapace of insects and crustaceans, the skin of snakes or even the skin of human infants. In micro-biological context, the prime example is the side wall of the Gram-positive rod, first studied in *Bacillus subtilis* (Koch et al. 1982; Koch & Doyle 1985; see Doyle & Koch 1987). The evidence is persuasive that wall precursors are secreted through the cell membrane continuously and diffusely (Koch 1988, 1990). By laying down new cross-linked lamellae enclosing the cell, it is insured of a strong, intact covering that can stretch as growth takes place. Still, over time the elastic limit for a layer is reached, then, the cell must have in place hydrolytic enzymes (i.e. autolysins) to cleave the now inadequate wall. The old wall is broken down and, often, the pieces are discarded, although sometimes the disassembled pieces are reused.

Multicellular organisms have additional types of turnover. Some cells in an organism may die, or be killed. Then they may autolyze (or be phagocytosed in a vertebrate host). Many kinds of cells on appropriate stimulation or stage of a developmental program commit suicide by the process that is now designated as 'aptotosis'. In this process the DNA is cleaved, starting initially at the interval between the nucleosomes. Thus the identifying mark of the process is generation of DNA fragments of about 140 nucleotides in length.

There are procaryotic equivalents of aptotosis. For example, the sporulating Gram-positive cell rearranges the genome of the mother cell, so that the cell would be unable to grow. Products needed for the development of the spore are produced, but the cell does so that that the mother cell genome deletes essential information. Of course this is immaterial because the mother cell, genome and all, disintegrates while releasing the spore (Stragier et al. 1989).

Many Gram-positive cells turnover their wall peptidoglycan. If a culture cannot sporulate, for example, because of a *spo* mutation, then autolysis when stationary growth phase is reached, destroys the cells and make such cultures difficult to maintain. Thus, in many ways, the culture of a microorganism should be thought of as the entity and not the individual cell. Instances where individual cells, carry out strategies that aid the culture but function to the detriment of the individual are known. This is exactly the same phenomenon that characterizes somatic cells of the higher animals and many plants that cannot reproduce but aid, to their detriment, the genomic cells that do. It is also related to the phenomena of sociobiology: a bee stings an intruder though that action is sure to kill the individual bee performing the altruistic act.

From these remarks it can be seen that there are commonalities as well as differences between turnover processes in various types of organisms and different cellular macromolecules. There are also interrelationships in the development of studies on different organisms. At first only the adult mammal, particularly the human, was considered. Later methods for the study of growing bacteria were developed. Next, the biochemistry of higher animals developed. Now, many aspects of the field have been approached. This essay focusses on crystallizing the ideas, approaches, and findings to give perspective for the benefit of the microbial physiologist on these concepts. Turnover is a field that is, in large part, old and has been largely forgotten.

Turnover in non-growing, steady-state systems

Early work

Ideas frequently arise long before methods develop for their adequate study; so it was for the field of turnover. In the twenties, the disparity between interesting concepts and available experimental techniques was extreme, but the imagination of the scientist was up to the task (Borsook & Keighley 1935). The logic at the time was as follows. An adult man under good nutrition remains at constant weight: so any excess production of biomass must be balanced by internal destruction. It could be asked:

> 'What is the partition in the steady-state case of biochemical reactions, into the generation of utilizable energy, into replacement for wear-and-tear, and into planned obsolescence?'

In this pre-isotope era, a tracer experiment was designed and carried out simply with the measurement of input in the diet and output in the excreta of nitrogen and sulfur under conditions of adequate caloric nutrition and chemically determined the Kjeldahl nitrogen. The experiment was to shift individuals from one diet to another diet which had different levels of nitrogen and then back to the original diet. On switching from a high to a low nitrogen diet sulfur balance was maintained although the qualitative nature of the supplied nitrogen changed. The nitrogen excretion continued and only slowly reached a lower plateau. Similarly, during the second part of the experiment there was a corresponding gradual return. Thus, turnover was first suggested.

Schoenheimer's work

Schoenheimer developed the concept of 'the dynamic state of body constituents' half a century ago (Schoenheimer 1942). According to his findings, although an adult animal had a constant composition from day to day, proteins were being continuously synthesized and broken down. He was able to demonstrate this because of the then recent developments that lead to the availability of heavy isotopes and mass spectrometers. He prepared diets for rats containing ^{15}N. Fed them for periods of time and then changed the diet to one containing only nitrogen of normal atomic abundance. He found that the liver protein nitrogen turned over with a half-life of several days.

The mathematics of steady-state turnover systems is important to our goal of studying growing systems. We will start with the simplest system: Schoenheimer's adult rats experiments. This will allow us to develop the mathematics to treat the more complex cases. Let us start with a convention

that capital roman letters (except the letter 'V') designate the atom percent excess, APXC, (i.e., the percent of the atoms that are of the heavy form in the biological specimen minus the same quantity for an appropriate unenriched sample).

For Schoenheimer's case, we could take A_o to be the APXC of the protein in the diet and B the APXC of the liver protein at any time. If the metabolic scheme were:

$$A \rightarrow B \rightarrow degradation$$

then B will change in proportion to the difference between A and B; this can be expressed mathematically by:

$$dB/dt = k_b(A_o - B) \qquad (1)$$

where k_b is the rate constant for the conversion of A to B. This differential equation is the same one that is met as the first example of kinetics taught in chemistry courses. It has the solutions:

$$B = A_o (1 - e^{-k_b t}) \qquad (2)$$

and

$$B = B_o e^{-k_b t}, \qquad (3)$$

where t is elapsed time. The first equation refers to the case where the APXC of the nitrogen fed in the diet was raised from zero to A at time zero and the second equation refers to the case where the APXC in the diet was reduced to zero. The second equation refers to the case where the APXC in the diet was reduced to zero when the APXC in the liver protein was B_o. These relationships are both shown in Fig. 1.

After Schoenheimer

After these first critical experiments, there was an explosion in the application of tracer methods to a variety biological systems for many different purposes (see Zak et al. 1979; Samarel et al. 1991). This largely resulted from the increase in availabil-

178

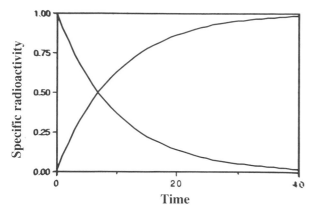

Fig. 1. Specific radioactivity during a pulse and chase of a steady-state system. The rising curve represents the increase in specific radioactivity of a single compartment model of a constant size, from the time that a radioactive precursor whose specific activity has been taken as unity is administered. The falling curve is the theoretical prediction if the compartment is fully labeled and at time zero the specific radioactivity of the precursor pool is suddenly reduced to zero. The half time for either build up or decay is ln 2 /k; for this example, k was set at 0.1 and therefore the half time is 6.9315 of the arbitrary time units.

ity of radioactive isotopes and the synthesis of biologically relevant labeled compounds, but it was also due do to improved instrumentation. The results on these studies led to mathematical techniques to analyze systems that were more complex than the one described above and led to a field called 'compartmental analysis'. An organism was considered to be composed of many physiological compartments. They could be chemical components or pools or tissues or parts of tissues, tracer movements could be followed by formulating a system of linear differential equations with one equation for every compartment. In the system described above only one equation (Equation 1) needed to be formulated. Often it was possible for the system of equations to be solved for the time course of the activity in each compartment. From a fitting process to the mathematical solutions to the differential equations, the values for the k's of the various transformations could be deduced. These studies are very important, but from the microbiological point of view were lacking in that they usually did not consider systems that are growing.

Turnover in non-exchanging growing systems

Heavy isotopes are used less often now and in the rest of this paper, only radioactive tracers will be considered. Now, the capital roman letters (except V) will be used as the symbol for the specific radioactivity of components at any time, and are the logical equivalent of APXC values. There is a difference, the estimation of specific activities requires two measurements instead of one. Specific radioactivity refers to the radioactivity per unit amount of material. The radioactivity may be variously expressed in units of Be (Becquerel), μCi (microCurie), cpm (counts per minute), or dpm (disintegrations per min) per unit amount. The amount of material can be measured in grams, or moles, or unit volume of culture, or it can be the entire animal; we will use small letters to designate the amount present in the system.

Figure 2 shows several metabolic interrelationships. All the six cases have importance in particular situations. Case 1 corresponds to the irreversible transformation of the isotope in the medium into a cellular constituent. Case 2 corresponds to the sequential precursor/product relationship. Case 3 corresponds formation and then degradation. Case 4 corresponds to the passage through a small intermediate pool. Case 5 corresponds to the situation in which the pool does not communicate to the outside. Finally, Case 6 corresponds to the metabolic situation where label enters two compartments, one reversibly and one terminally. Eventually the label leaves the first and accumulates in the second.

Instead of assuming that each process is characterized by a value of k, the formulation will be more fundamental, and developed as follows: Let V's designate the velocity of biosynthesis or degradation for metabolic processes; thus V's are the same type of quantities here as considered in enzyme kinetics. For the examples shown in Fig. 2, the velocity, V_b, might have the units of μmoles of component B made (from A or X) per min per individual growing animal or per unit volume of steady-state exponentially growing bacterial culture.

With these definitions we can treat growing systems easily. We equate two equations for the dif-

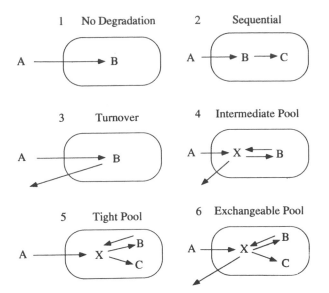

Fig. 2. Possible turnover situations. Six schemes for cellular metabolism are depicted. *(1.)* No Degradation. An isotopically labeled compartment of specific activity A present in the medium is irreversibly incorporated into cell compartment B. *(2.)* Sequential. The cellular compartment B is converted irreversibly to C. This is also known as the precursor/product scheme. *(3.)* Turnover. B is synthesized and excreted or secreted into the medium. It is assumed that the concentration and specific activity of the A compartment does not change as this occurs. *(4.)* Intermediate pool. This is a very common case; a species enters the cell into a small, but rapidly turning over pool that then serves as a source of cellular species. Such small pools are designated in this paper with the symbol X. X then is an obligatory intermediary for introduction of label into the usually more abundant and larger sized cell compartments. *(5.)* Tight pool. In this case, X serves as a precursor of both B and C, and allows label originally in B to enter C. The scheme is call 'tight' to indicate that there is no path whereby isotope can return to the medium. *(6.)* Exchangeable pool. This case is similar to case 5, but with the additional process in which the pool communicates with the medium. If exchange is sufficiently fast the specific activity of X approaches that of A and serves as a way for turnover to transfer some of the radioactivity of B through X into the medium.

ferential change in total amount of the tracer in the compartment of interest for the scheme of interest; one equation for Compartment B for Case 2 or Fig. 2 is:

$$d(Bb)/dt = AV_b - BV_c \quad (4)$$

(Note that each side of the equation has the units of 'change in amount of radioactivity in the compart-

ment of the system per unit time'.) Whether growing or not, the change in the amount of Component B, designated for a biological system is db, and is the velocity of formation minus the degradation of B; i.e., $db = V_b - V_c$. Then remembering the rule for the differentiation of a product, we obtain:

$$d(Bb)/dt = bdB/dt + Bdb/dt = bdB/dt + B(V_b - V_c) \quad (5)$$

for the second equation. Equating these equations and cancelling terms, the equation reduces to:

$$bdB/dt = AV_b - BV_b \quad (6)$$

Now note that b and V_b vary as growth proceeds since they refer to a unit volume of culture. If the system is undergoing balanced expansion, all components increase exponentially with the same growth rate constant as does the population's total biomass or the numbers of cells. For example, b must satisfy the differential equation:

$$db/dt = \mu b \quad (7)$$

where μ is the specific growth rate, just as the number of cells must satisfy $dN/dt = \mu N$. Therefore after integration and assignment of the constant of integration:

$$b = b_o e^{\mu t} \quad (8)$$

Similarly:

$$V_b = V_{bo} e^{\mu t} \quad (9)$$

Consequently, both will increase in proportion to each other. Thus, V_b/b is constant throughout balanced growth; let us call this ratio, k_b. If we substitute these relationships into the above equation and we also use the symbol A_o to designate the constant specific activity of the precursor that the experimenter supplied, then:

$$dB/dt = k_b(A_o - B) \quad (1)$$

In some sense we have gotten back to where

180

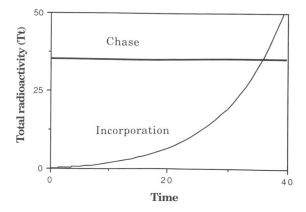

Fig. 3. Pulse and chase in a growing system that has no mechanism to transfer radioactivity back to the medium. The total radioactivity is shown for a hypothetical case in which the radioactivity is introduced at time zero for the curve marked 'incorporation' and the source of radioactivity is removed for the curve marked 'chase' after some incorporation has occurred. The incorporation curve will get steeper and steeper as the bacteria continue to grow exponentially. On a logarithmic plot the curve would become a straight line, but only after growth of many fold has occurred. For the chase situation the total radioactivity of all the components of the cell remains constant, because dilution in specific activity by growth is just compensated for by the increase in biomass.

Schoenheimer started, and again the problem is reduced to a mathematically soluble problem, because it relates the variable B to the variable t. Now we recognize that the k's are the ratio of the velocities of processes converting material in one compartment into another relative to the amounts in one of these compartments. In the case where B goes no further; i.e., there is no further interconversion, turnover, nor excretion then k will equal μ. Again we have the two interesting cases, step labeling and pulse-chase, shown in Fig. 1. In the step radioactive A is supplied at constant specific activity, A_o; in the chase part of the pulse-chase, the specific activity of A is reduced from A_o to zero by removing the supply of labeled compound and/or diluting the radioactivity with a high level of unlabeled compound at a time we will call t_o. Let us designate B_o as the radioactivity at the time of the chase. Because A = 0, the equation reduces to:

$$dB/dt = -k_b B, \text{ or } dB/B = -k_b t, \text{ or}$$
$$\ln B = -k_b t + \text{constant} \tag{10}$$

The latter form is usually rewritten:

$$B = B_o e^{-k_b t}, \tag{3}$$

where B_o turns out to be the constant of integration.

The solution for the step case when A quickly rises from 0 to A_o and B is zero at the time designated as zero; is:

$$B = A_o(1 - e^{-k_b t}) \tag{2}$$

This can be shown by defining $X = A_o - B$, substituting, then integrating, resubstituting, and evaluating the constant of integration.

These expressions are useful only when we have measured specific activities. In most microbiological applications, only the total activities are usually measured, which is much easier to do. The total activities (Bb, Cc, etc.) that we do measure can be derived from the relationships presented above and for incorporation are given by:

$$Bb = A_o b_o(e^{\mu t} - e^{(\mu - k)b)t}) \tag{11}$$

and for the chase, by:

$$Bb = B_o b_o \tag{12}$$

These relationships for total activities are shown in Fig. 3. This is the usual representation of tracer data, but it is one that obscures the underlying biology. A more useful plot is the isometric plot made famous by Jacques Monod in the late forties. As applied to isotope data, the axes of the plot are chosen as the total radioactivity and the biomass concentration. Such a graph of the same hypothetical data shown in Fig. 3 is shown in Fig. 4 for the pre-labeling period, the pulse, and the chase; the idealized data assumes that the biomass concentration was unity at the time the isotopically labeled compound was introduced into the balanced growing culture. It shows that this type of plot yields a straight line whose slope is the specific activity of the newly formed B. In our example the slope is 1, because A_o had been chosen as one.

Figure 2 also illustrates five other common metabolic situations in addition to the simple irreversible incorporation into one compartment. Case 2 deals with a sequential precursor/product relationship. If only the total activity incorporated into the cells were measured, then the graphs of the total activity versus either time or biomass would be the same as for the first case, as shown in Figs 1, 3, and 4. In fact, no matter how complex the metabolic scheme (which could have a large number of compartments, either in series, parallel, or in a pattern with very complicated interconnections) a Monod type plot (as in Fig. 4) still would be observed to have a straight line. The slope would be equal to the specific activity of the original precursor per unit biomass, assuming no competing *de novo* pathway. In order for any other result to be observed there must be some process(es) that allow the radioactivity in some form to escape from the cell as indicated for schemes 3, 4, and 6. Before dealing with these, further comment will be made on two aspects of the cases in which isotopic tracer enters the growing cell and never leaves.

For all the non-exchange cases the total activity of the sum of all the labeled species can be represented as

$$Tt = A_o b_o (e^{\mu t} - 1) = A_o (b - b_o) \quad (13)$$

Tt approaches pure exponentiality ($Tt = e^{\mu t}$) as a function of time, only after enough time has elapsed so that $e^{\mu t}$ becomes much greater than 1. So, even though bacteria may be growing exponentially and there is no degradation or secretion of labeled compounds, the total activity does not increase in a simple exponential relationship; i.e., in proportion to the number of organisms or biomass present. Contrary to some errors in the literature, that fact does not imply turnover.

After a perfect, complete chase the relationship for the total radioactivity is given by:

$$Tt = (Tt)_o \quad (14)$$

and thus *the total radioactivity in the culture remains constant after the tracer compound has been consumed or removed from the environment, if there is no excretion*, because the fall in specific activity is

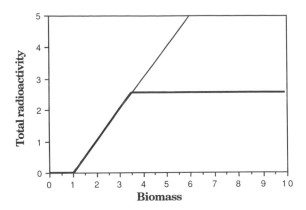

Fig. 4. Theoretical pulse and chase as an isometric plot. Plotted against the fold increase in cellular biomass and expressed as total radioactivity, the labeling kinetics are shown for a culture during the pre-labeling, pulse period, and chase period. A graph such as shown here would only have been obtained if radioactivity entered one or more cell compartments irreversibly. This situation would occur if there were no exchangeable pools before the major compartment(s) and if there were no degradation of the major compartments. In such a case there would be no lag when the incorporation of radioactivity started and no lag when the incorporation stopped; i.e., the graph would be composed of discontinuous line segments.

exactly compensated by the increase in amount per unit volume of culture. Many papers falsely assume that constancy of total activity implies no turnover. Evidently radioactivity may be moving from one compartment to another and will not be detected and appreciated unless a set of suitable experiments are carried out with different labeled precursors and/or the chemical and physiologically distinct compartments are separated from each other and their radioactivities independently determined.

(Methodological details are beyond the scope of this article. Still it must be mentioned that as a chase is carried with growing cultures of bacteria, the specific radioactivity becomes less and less and, consequently, the total radioactivity become more difficult to measure, and because of these difficulties it can fallaciously appear that turnover is occurring. This error happens in cases where the efficiency of counting is decreased because there has been lack of, or improper, correction for counting efficiency.)

When $Tt = A_o(b - b_o)$ or $Tt = (Tt)_o$ (Equations

13 and 14) are satisfied during a pulse or a chase, respectively, we might suspect that degradation (turnover) is negligible or does not occur, but in fact, the valid assertion is that excretion of labeled compounds was undetectable. If deviations from these expressions were found, we would suspect that there is turnover, but would need to double check our methodology. Then, such a conclusion would only be valid if our measurements showed conversion of one compartment to another. If not, turnover could not be excluded.

An important use of tracers is in cases where the label can be chased from the cells, however there are several cases (see Koch 1962) that commonly occur and confuse interpretation. These cases and their correct interpretation will be presented. The points to be made are not trivial because frequently experimental results have been misinterpretated largely due to failure to chose cases and use condition where the chase is effective; the onus to assess the efficacy of the chase is on the experimenter, particularly if he is concluding that there is no turnover.

Simulation of tracer kinetics during growth

With modern minicomputers it is easier to carry out simulations than to solve difficult systems of differential equations to show principles and to fit data. The present paper extends earlier work (Koch 1962, 1968, 1971a). Appendix I gives a simple program in BASIC used for the preparation of Figs 5–8. It can easily be extended for more complicated situations. The top few lines contain all the parameters that apply to a system where an external compound A enters compartment B of the cells. Depending on which velocities the user sets to zero, material may go to compartment C, may return to B, and the medium may receive material from B. The program computes the initial total activities of all cellular components from the initial amounts and specific activities. Starting from this initial state, the program simulates the passage of time; to do this for each time unit of duration 'del', it computes the change in the total activity of each component and the change in the amount of each. Then

the total activities and amounts are updated, and the specific radioactivities computed by dividing the total radioactivities by the amounts. The total cellular constituents are summed and the velocities upgraded on the assumption that they are proportional to the original velocities and the current biomass. After N such time cycles the computer lists the status of the system. It repeats another N cycles and only terminates after the time specified by the user. Both pulses and chases can be simulated by choosing which velocities have non zero values.

There is one minor difficulty in using this program: the initial amounts and velocities in a system of bacteria undergoing balanced growth are quantitatively connected with each other. Any randomly chosen set of parameters would usually correspond to non-balanced growth conditions. Usually one would be interested in balanced growth conditions. The connection between the constants can be deduced by inspection, but when the relation is not obvious the program can be run with the desired velocities and any set of arbitrary amounts. If the program is allowed to run for a long enough time, the amounts will end up in the proper proportion. This change in proportions is characteristic of balanced growth; i.e., the amounts will self-adjust during growth to achieve the ratio that permits balanced growth with the given reaction velocities. Then maintaining the proportions the amounts can be scaled down and the program rerun with the radioactivity applied as a pulse or chase as desired.

Figure 5 shows four panels allowing the comparison of uptake in four cases. All depicted the total radioactivity as a function of biomass in the tradition of Monod's isometric plot. Figure 5A shows the case $A \rightarrow B \rightarrow C$. This example is the precursor/product scheme of Fig. 2–2. It shows that Bb is curvilinear downward and Cc has only a slight lag. Thus, although neither the total radioactivity of B nor C is linearly related to biomass, their sum, Tt, is a linear function of biomass for this case because is occuring no secretion or excretion of tracer occurs. Figure 5B shows the case where C is formed from B and is turned over to reform B. Turnover could not be deduced from the total radioisotope (Tt = Bb + Cc) in the unit of culture that still shows a linear

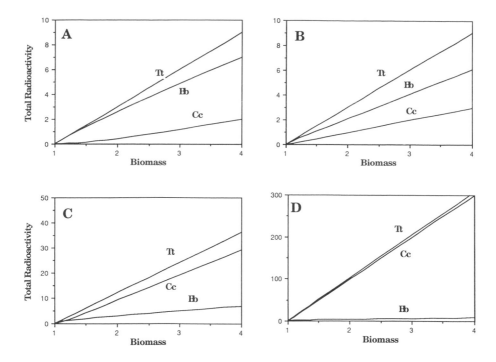

Fig. 5. Monod isometric plots for cases where there is no exchange with the medium. *(A)* corresponds to the precursor/product scheme shown as Case 2 in Fig. 2 when the ratio of the quantity of b to c is 2. There is only slight lag for incorporation into C. The sum of the two species is exactly a linear function of biomass or of $e^{\mu\tau}$. *(B)* corresponds to quite rapid turnover of C; the b/c ratio remains at 2, but the rate of breakdown rate is 10 times the rate of net accumulation of C. All that can be seen is that the graphs are even more closely linear for B and C. *(C)* is exactly the same as *(B)* except the size of Compartment C is 10 times larger; i.e., b/c =2/10. *(D)* is also exactly the same as *(B)* except the size of Compartment C is 100 times larger i.e., b/c = 2/100.

rise. One might have falsely concluded that both B and C were directly derived from the medium resource, except a slight concavity of the Bb curve. On the other hand, separate determination of the time course of the specific activity of B and C would have identified a turnover process; it would be very clearly shown if both a pulse and chase were carried out. Figure 5C is a similar example, but one in which the amounts of the two components are different: 10 times more c. Figure 5D is for the case in which the amount of c has been increased another 10 times. In either Fig. 5C or 5D, it would be impossible to exclude the pathway A → C, without involving B.

In cases where there is secretion, excretion, or exchange with the growth medium, turnover can usually be inferred from measurements of total activity much better in a chase than in a step type of experiment. First, consider step experiments. Fig. 6 shows the consequence of several metabolic

schemes. Figure 6A shows a case where isotope enters B, moves to C, but can retrace its steps back into the medium. It has been assumed that the specific activity of A in the medium remains constant and would not be appreciably diluted by any non-radioactive material otherwise made by the bacteria. Note the total of the total radioactivities is not linear but slightly concave. While Fig. 6A deals with a case where the amount of B relative to C is significant, b/c = 2/10, Fig. 6B deals with the frequently realistic case in which compound B is present in a small amount, b/c = 0.2/10. This is the case where the compartment in question is a small rapidly turning over pool. Virtually all the radioactivity resides in component B and to a very accurate approximation the total radioactivity is a linear function of biomass. As mentioned, it is for this reason that in Fig. 2 such small compartments were designated by X. This has been done in Fig. 6 as well. Now the effects of turnover are evident in

184

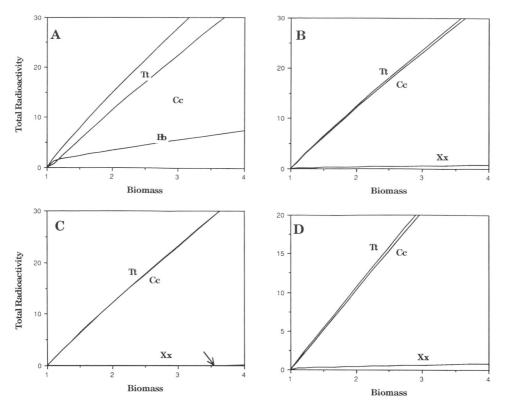

Fig. 6. Total radioactivity during growth when exchange occurs. The relative size of the compartments is: *(A)* b/c = 2/10; *(B)*, *(C)*, *(D)* 0.2/10. *(C)* corresponds to no turnover and *(D)* to a very slow rate of turnover. See text.

incorporation type experiments. Figure 6C shows that exchange of the pool is without significant effect if there is no turnover in the sense that C is formed and never degraded and that the intermediate pool, X, is small. Finally Fig. 6D is like 6B but with a much slower turnover process.

If there is either no turnover or no exchange, such experiments will fail to detect a difference in the total radioactivity. Fig. 7 shows three situations: both turnover and exchange, no turnover, and no exchange. There was almost no difference between the three when presented as a monod plot. However, Fig. 7 shows different behavior in the specific radioactivity. For the top curve (exchange only) the pool quickly achieves the same specific activity as the medium. For the curve lowest curve (turnover only), the absence of exchange forces a slower rise on the pool. Of course, curve marked (both), the time course was intermediate.

Chase experiments can in principle give clearer

evidence for turnover. Figure 8 shows the ratio of the true turnover rate to the observed rate for various degrees of turnover in curves A through D. Curve E shows the inhibiting effect of lowering the exchange rate 10-fold. Notice that even for Curve D, it would be difficult to estimate the half time for turnover. This emphasizes the need of very rapid exchange to accurately ascertain turnover.

De novo versus salvage pathways

In steady-state non-growing systems composed of many compartments with multiple interconnections, mathematical analysis has shown that the time course for any compartment is a complex sum of many exponential terms. Usually there are as many terms as there are compartments. Moreover the coefficients of time in the exponential terms are functions of many of the rate constants for the

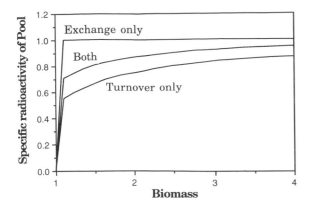

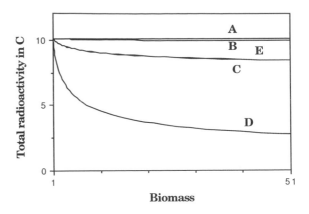

Fig. 7. Effect of turnover and exchange on the specific radio-activity of the pool. Three cases are shown with the same growth characteristics except presence or absence of exchange of X with the medium or turnover of C with the X, the small intermediary pool.

Fig. 8. Chase of radioactivity from cells. *(A)* no turnover; *(B)* turnover rate 1% of synthesis rate; *(C)* 10% of synthesis rate; *(D)*, *(E)* 100% of synthesis rate. *(E)* shows the effect of lowering the exchange rate 10-fold.

individual processes. This makes the analysis of kinetic data very difficult, arbitrary, and inaccurate.

If there is a rapidly turning over, small, metabolic pool schematized as Fig. 2, Scheme 2, labeled 'Intermediate Pool', the kinetic curves for uptake and release will look the same as for Scheme 1, but the interpretation is much different. Similarly, if an intermediate pool, X, is interposed in Scheme 3 to generate Scheme 4, where the amount of X, x, is very small and B is a metabolic *cul de sac*, this change cannot be detected from measurements of the specific or total activity of B. While the B species turns over, the intermediate pool turns over much more rapidly and the kinetics of the rise and fall of B will have the same mathematical form as before. Thus, in all these cases the specific radioactivity rises and falls according to equations with only one exponential term. In fact, a little algebra (Koch 1962) is sufficient to show that for the intermediate pool case, k_b instead of being equal to V_b/b or μ is instead $V_x V_b/(V_x + V_b)b$ or $\mu V_x/(V_x + V_b)$. Of course, if V_x is very large, then equilibration is fast and this expression reduces to V_b/b as intuitively expected. Although in this case the presence of the intermediate pool does not influence the isotope kinetics, if V_x is very small then there is much recycling and the increase or decrease in the total activity of B has no relationship to the rate of formation and degradation of B from the intermediate pool. Thinking that it does is a fatal mistake. V_x will be very small if X does not exchange with the medium.

There will be little exchange with the medium for the case where the cell only takes in the metabolite as needed. In Case 5-Tight Pool, the total activity present inside the cell, $(Xx + Bb + Cc)$, registers the increase in the amount of the sum of the species. So for example, if X is the nucleotide pool, B is messenger RNA and C is stable RNA (rRNA plus tRNA) then the measurement mainly registers increase in the stable RNA, though at short time the majority of the nucleic acid radioactivity would be in mRNA (Koch 1971a). Similarly if X were periplasmic precursors of murein and wall turnover occurred, wall turnover would not be observed if exchange with, or leakage to the outside the outer membrane was slow or non existent. In such a case any label from the degradation of murein would eventually be taken into the cytoplasm and reutilized, although not necessarily for murein synthesis.

For Case 6-Exchangeable Pools (Fig. 2), where there is of a rapidly turning over component like mRNA in the presence of more stable rRNA and tRNA, the uptake into total RNA may exhibit complex kinetics, but in short times be mainly a measure of mRNA synthesis.

In summary of this section, measurement of tracer kinetics is essential for the study of the physiology of growth, but there are precautions in the design and serious questions about interpretation. It can be seen from the kinetic dependency on EXP $[-\{V_xV_b/(V_x + V_b)b\}\,t]$ that arranging circumstances so that V_x and V_b become large is essential. So the goal of experimental design is to rearrange the isotope flow to convert the other cases into Case 6 with an effective chase, or to use an internal trap by manipulations of the cell's physiology to measure turnover in Case 5 by sensing C instead of A, or to choose a label like the carboxyl labeled glutamic acid that is treated like Case 3 instead of Case 5.

History of experiments on turnover in bacteria

A key element for the development of Central Dogma of Biology was the work carried out in Monod's laboratory at the Institut Pasteur in Paris in the late 1950s leading to a new concept: messenger RNA. Although ribosomes were involved in protein synthesis something else which was later to be called messenger RNA, carried the information and was rapidly turned over. The isometric plot (Monod 1958) provided the first inkling of this because a plot similar in shape to Fig. 4 was obtained in which the ordinate was amount of beta-galactosidase units per ml of culture and the abscissa was the optical density of the culture that is equivalent to dry biomass per ml. First an uninduced culture was followed; samples from it had very low levels of the enzyme. Then inducer was added and a rapid linear increase was observed. Finally, the inducer was removed and the total number of enzyme units per ml culture persisted as long as the experiment was followed, but no more were formed. This generated the same three phases as shown in Fig. 4. Because the three phases produced straight lines that were separated by abrupt discontinuities (i.e., the corners were not rounded), they concluded that the ribosomes could be reprogrammed very rapidly. This interpretation of the plot depended on the presumption that beta-galactosidase was metabolically stable.

To test the stability of beta-galactosidase, an experiment was conducted (Hogness et al. 1955) in which the cells were induced in the presence of a radioactive precursor and then chased in the absence of exogenous tracer, but either in the presence or in the absence of inducer. From the various samples, the enzyme was isolated, purified, and its specific activity determined. Then from the specific radioactivity and measurements of the extent of growth, the total radioactivity was computed. Similar experiments were also done in Spiegelman's laboratory in Urbana (Rotman & Spielgelman 1954). While different experimental methods were used, the conclusion was the same: Turnover of beta-galactosidase was very slight at best.

I, working at the Argonne National Laboratory near Chicago, felt that the experiments from both laboratories, though putting small limits on possible turnover, were not small enough to make valid any absolute conclusions about turnover in bacteria. Moreover, since only one enzyme was studied, these experiments were not general enough to test the question whether there was a systematic process whereby proteins were turned over in growing bacteria. My argument was that serum albumin in some mammals might have a half life of two week. But during two weeks a bacterium growing with doubling time of 60 min would have grown $10^{101.14}$ fold, which would be greater than the collective mass of the elephant. Consequently, in the short time available for the study of growing cultures, very much more sensitive methods would be needed to detect degradation rates of that order of magnitude. So I hit upon a strategy of 'the internal trap'.

Understanding how the internal trap work in this application requires a review of intermediary metabolism of purines. Bacteria in many cases make their own purines from 'one' carbon units and from glycine; the nitrogen comes from the glycine and from glutamine. Organisms, such as *E. coli*, have also a salvage pathway to utilize purines and any available purine nucleosides or nucleotides for their own nucleic acids. As we know today, the *de novo* pathway is controlled in several ways; most rapidly by feedback inhibition, but also by some degree of repression of the formation of enzymes of the pathway. So, it is possible to manipulate the bacterial purine metabolism by including or omitting purines in the growth medium.

The experiments (Koch & Levy 1955) to measure protein turnover were as follows: Actively growing bacteria were labeled with a ^{14}C precursor of both protein and purines, such as formate, uniformly labeled mixture of ^{14}C-glucose and ^{14}C-fructose, 3-^{14}C-labeled serine, 1-^{14}C-glycine, or 2-^{14}C-glycine. Radioactivity from these precursors would not only enter the purines, but also into proteins (for some of the tracer compounds they first would be made into amino acids). Glycine and serine would enter the amino acid pool directly, but in addition they and formate would feed the 'one' carbon pool that then led to purines and into amino acids like glycine, serine and histidine. Therefore, in the presence of the label but in the absence of adenine and guanine in the medium, the distribution of activity in amino acids isolated from proteins and the adenine and guanine isolated from the cells (which is largely derived from RNA) would measure the quantitative partitioning of the labeled precursor once it had entered the initial intermediary pool. These data were used to correct the increment of label into the purines in a chosen interval to give the rate of degradation of protein in the experiment proper.

The experiment, itself, was to grow bacteria in the presence of non-radioactive adenine and guanine, then at a certain point add the labeled precursor. Because of the inhibition of the *de novo* pathway by the purine mixture, the formation of labeled purines was very greatly decreased, while the conversion into protein may have increased to some minor degree. Then in a second phase of the experiment, the cells were centrifuged, washed, and resuspended in fresh medium with neither the labeled compound nor any exogenous purines. Although the cells continued to grow, they must turn off the feedback inhibition and restart the synthesis of their own purines via the *de novo* pathway. Consequently, if any proteins broke down to amino acids, these would enter the intermediary pool and be utilized for both protein and purines. It is expected that such regenerated amino acids would have the same probability of being utilized for purines versus protein as measured in the control experiments mentioned above. Therefore, a correction to estimate the rate of protein degradation could be made.

Although the experiment succeeded with several precursors, the most critical and clear results were obtained with 2-^{14}C-glycine. In the medium devoid of purines, it was found that the radioactivity had a 50% chance of entering the cellular purines. To make the experiment even more sensitive for measuring intracellular turnover, the cells were centrifuged and resuspended in fresh, warm medium at hourly intervals. This would to remove the contribution of: isotope present in pools that might exchange with the medium; any original labeled compound; isotope present in non-protein non-purine pools; and any possible dead or dying cells that might yield amino acids that might then be scavenged by living cells, but endure in the medium for some time.

It was found that indeed about 2% of the original protein label was turned over. This turnover occurred quickly, but probably this estimate is high because it is the combination of a small amount of contaminating glycine as well as the turnover of a few proteins that the cell actually does turnover quickly. Such proteins might be involved in the cell cycle or they could be proteins produced by aberrant genes that have abnormal chain termination codons. Another contribution to the rapid component is the degradation of the initial few amino acids in the nascent protein chains (Pine 1972). Although the first amino acid of all proteins added to the growing peptide chain is methionine, cells frequently process the nascent protein to remove the methionine and usually the next few amino acids down stream from the initiating AUG codon as well.

The conclusion from these experiments was that, aside from a very minor group of protein, the bulk of the proteins in a growing culture are very stable. Allowing for experimental variation, it was concluded that the first order rate constant for degradation had to be less that 10^{-3} per h. This corresponds to a half life of greater than 30 days. In such a time the biomass would have increased 10^{124}

whereas if the half life was 30 days the specific radioactivity would have been diluted by growth and turnover twice 10^{124} times.

Whereas the experiments just presented are very old, I believe that they are still the last word on the subject and proteins, generally, are very stable in growing bacteria. There has been an important study by Luktinhaus et al. (1979) using the 2-D gel systems of O'Farrell. It was found that very few proteins were degraded and lost during a chase. The studies of Luktinhaus et al. (1979) used cells synchronized by the best available procedure and showed that only a very few proteins were synthesized and degraded within the same cell cycle. While synthesis and degradation of these proteins may be very essential for the functioning of the cell cycle, their uniqueness and small amount points to the general stability of proteins formed by growing cells.

Turnover during starvation

When cells run out of nutrients they become desperate. There are many aspects of the cell physiology and genetics of starvation. Most will only be mentioned in passing, I will focus, however, on only the facet that was studied first. This is starvation-induced turnover, which was discovered and studied by Mandelstam in England in the late 1950s. He studied protein turnover by labeling *E. coli* with ^{14}C-leucine (Mandelstam 1958, 1960). Both the organism and tracer are good choices because leucine is not degraded nor converted into other cell constituents by *E. coli*. He followed the labeling period with a chase containing a very high level of non-radioactive leucine. On the unproven assumption that cold leucine would completely turnstile out the hot leucine from the cell's intermediary pool, he interpreted the release of leucine to the medium as showing protein turnover.

[Later we measured the effectiveness of the chase and found that it was fairly good under certain conditions (see Nath & Koch 1971).]

We also could quibble about the radioactivity that was released from growing cultures that was much higher than that expected from the experi-

mental studies presented above. What is to the point is his observation of much greater release when the bacteria were starved in a variety of ways, particularly of a compound serving the cells as both carbon and energy. Typically he found a release of about 5% per hour continuing for at least 4 h. Further analysis showed that the source of this labeled leucine was mainly from ribosomes. This makes good sense. Under starvation conditions the cell has no need for the number of ribosomes that work in a cell during rapid growth (10^4 per cell from cells growing in rich medium). Very few ribosome would suffice for the limited synthesis of protein that a starving cell must undertake when growing conditions are reinstated. So this internal resource would allow the cell to continue synthesis of those special proteins aiding the cell to survive during starvation.

Understanding how bacteria survive and evolve during nutritional challenges is a field under current study in a number of laboratories. In a nut shell, cells growing in a complete medium usually are able to store reserves as glycogen and in the large numbers of functioning ribosomes (Koch 1979). They deplete these stores in a few hours, still. The ability to synthesize new proteins in this interval in the absence of external resources allows them to form 'starvation proteins'. These constitute a set of 30–50 proteins that allow the cells to survive for an extended period. A simple experiment from Matin's laboratory at Stanford shows that protein synthesis in the next few hours after removal of a carbon source is crucial (Matin et al. 1989). A protein inhibitor only present within this period, such as chloramphenicol, greatly lowers the survival capability.

This leaves unanswered the very important question of how some mutants arise and grow after a long period of starvation. The genetic side of the story is becoming clear with the work of Hall (1988) who found an instance where the ejection of an insertion sequence activated an existing gene, allowing the bacteria to grow. Later he found (Hall 1990) that the bacteria have a mechanism so that at random certain cells greatly increase their mutation rate. Of course, this process usually fails and probably accelerates death, but rarely it succeeds and then can allow the cell to grow. From the perspective of this paper the point is that it is not understood what provides the means for proteins synthesis to take place under these severe conditions after extended starvation.

Besides engaging in new protein synthesis upon starvation, there are other cellular needs. Degradation of ribosomes provides not only amino acids, but purines, pyrimidines, ribose, ammonia, nitrogen, and phosphate. An anecdote concerning this is relevant. It was found in the late 1940s that when ^{32}P as phosphate was given to a growing culture it was incorporated into the nucleic acids of bacteria and found primarily in RNA. This is because cells that were grown in rich medium have high levels of RNA relative to DNA. On starvation, the ^{32}P was found in DNA. While this was taken as suggestive (but false) evidence that RNA was the precursor of DNA, what really occurred was the degradation of ribosomes and the reutilization of the component part for DNA synthesis. Again this is of survival value to the strain of organism because it allowed many more (but smaller) cells to be produced. These stationary cells have only very few ribosomes, but enough to grow when conditions improved.

We found that different strains of *E. coli* differ in their ability to turnover proteins. Thus both strains K12 and B degrade their ribosomes and use the nucleic acid part similarly, we could show that strain B, however, fails to reutilize an amino acid label as effectively as does K12 (Nath & Koch 1971).

The first conclusion from these studies of turnover during starvation is that the cell is selective in what is degraded, but degrades excess 'means of production' in hard times to form some 'product'; these utilizable resources are mostly tapped shortly after the starvation is imposed. The second conclusion is that the resources are used to form proteins to protect the cell from death as the results of the starvation and somehow allow even after a long period of starvation to make some new (mutant) protein that may allow a genetically changed cell to grow and survive.

Chronic semi-starvation

To study both cells in a steady state, where their properties are the same from time to time and culture to culture, and also study cells under severe challenge, the chemostat is the proper tool. For this reason we have used chemostat culture in many different ways. The major finding is that chronically semi-starved bacteria are a quite different kind of creatures than either rapidly growing or frankly starved bacteria.

It was found that cultures growing with a 10 h doubling time limited for glucose turn over their ribosomes (Coffman et al. 1971). Usually one thinks of ribosomal and transfer RNA as stable in contrast to messenger RNA, but under chronic starvation they are not. Similarly, protein synthesis is not continuously on-going. Instead, in slow growing cultures (doubling times of 13 h or greater) protein synthesis is episodic (Koch & Coffman 1973). This is as if during a period of synthesis resources are depleted and synthesis is stopped until some resources are regenerated or are sequestered from the medium. This may be a very inefficient mode and some proteins may be only partially made; then, these would be degraded back to amino acids.

After starvation only a few ribosomes remain in the cell. What rule would chemostat grown cells have about the synthesis and level of the machinery for protein synthesis? We were surprised at first at our finding that they made and had on hand functional ribosomes that would not be able to function under chemostat condition. Thus, chemostat cells with a nominal doubling time of 10 h were synthesizing protein continuously, but the rate that protein was being made could be raised several fold when the medium was enriched in less time than new ribosomes could be made (Koch & Deppe 1971; Alton & Koch 1974). On reflection, this shows again that the organism follows a strategy that is long term and are not restricted to short term tactics. It makes more that the number of ribosomes needed under the steady state conditions on the expectation that conditions would improve. Such an expectation is reasonable in terms to the past history of the organism living in a fluctuating environment.

The chemostat was also used to measure the level of resources for protein synthesis. Glucose-limited chemostats growing at different dilution rates were sampled and subjected to total starvation (for exogenous glucose) at different times they were subject to a 30 min pulse of inducer for beta galactosidase (Koch 1979). It was found that the ability to form beta galactosidase fell with a half life of about 30 min. Once this component was depleted the residual rate of synthesis was so small that it corresponded to the amount of synthesis that could be obtained from the resources produced by the hydrolysis of one cell when distributed to 100000 cells. A similar situation was found with phosphate starvation. But on the contrary, when cells had been growing in more rapid chemostats that would give the cells a greater doubling time than 10 hours the cells were can store reserves that allowed them to make beta-galactosidase for many hours at a constant rate.

Appendix

Simulation program in BASIC for turnover

```
00010 "Tracer simulation", Arthur L. Koch, 7/7/91
00020 VA=1:VB=.3:VNB=0:VC=.1:VNC=0
00030 A=1:B=2:C=1
00040 SA=1:SB=0:SC=0
00050 TIME=40:DEL=.01:N=100:MO=B+C:F=1:T=0:
      G=DEL:A$="###.####":DEF TAB=7
00060 AA=SA*A:BB=SB*B:CC=SC*C
00070 DEFSNG A-Z
00080 OPEN "O",1,"DATA"
00090 LPRINT:LPRINT:LPRINT
      A$;VA;USING A$;VB;USING A$;VB;USING
      A$;VNB;USING A$;VC;USING A$;VNC
00110 LPRINT USING A$;A;USING A$;B;USING A$;C
00120 LPRINT USING A$;SA;USING A$;SB;USING
      A$;SC
00130 LPRINT USING A$;TIME;USING A$;DEL;USING
      A$;N:LPRINT
00140 LPRINT USING A$;T;USING A$;B;USING
      A$;BB;USING A$;SB;
00150 LPRINT USING A$;F;USING A$;C;USING
      A$;CC;USING A$;SC
```

```
00160 PRINT #1,T CHR$(9) BB CHR$(9) SB CHR$(9) F
      CHR$(9) C CHR$ (9) CC CHR$ (9) CHR$ (9) SC
00170 FOR T = N*DEL TO TIME STEP N*DEL
00180   FOR I = 0 TO N
00190     DBB=(VB*SA−(VNB+VC)*SB+VNC*SC)*G
00200     DCC=(VC*SB−VNC*SC)*G
00210     DB=(VB+VNC−VC−VNB)*G
00220     DC=(VC−VNC)*G
00230     BB=BB+DBB:B=B+DB:SB=BB/B
00240     CC=CC+DCC:C=C+DC:SC=CC/C
00250     M=B+C:F=M/MO:G=F*DEL
00260   NEXT I
00270   TRONX:PRINT USING A$;SB;USING A$;SC
00280   LPRINT USING A$;T;USING A$;B;USING
        A$;BB;USING A$SB;
00290   LPRINT USING A$;F;USING A$;C;USING
        A$;CC;USING A$;SC
00300   PRINT #1,T CHR$(9) B CHR$(9) BB CHR$(9) SB
        CHR$(9) F CHR$(9) C CHR$(9) CC CHR$(9) CHR$
        (9) SC
00310 NEXT T
00320 CLOSE
00330 END
```

Parameters are defined in lines 20–40. V stands for velocity; S stands for specific activity; and single roman capital letters for the amounts. Formatting for creating a file, turning off the program if it gets into an infinite loop, and printing can be different than thoses given for the ZBASIC dialect used; relevant lines of code are 100–160 and 270–300. More complex scheme can readily be constructed by ading new parameters and augmenting line 190–250.

References

Alberts B, Bray D, Lewis J, Raff M, Roberts K & Watson JD (1989) Molecular Biology of the Cell. Garland Publishing, Inc, New York

Alton TH & Koch AL (1974) Unused protein synthetic capacity of *Escherichia coli* grown in phosphate-limited chemostats. J. Mol. Biol. 86: 1–9

Borsook H & Keighley GL (1935) The 'continuing' metabolism of nitrogen in animals. Proc. Roy. Soc. London Ser. B. 118: 488–521

Coffman RL, Norris TE & Koch AL (1971) Chain elongation rate of messenger and polypeptides in slowly growing *Escherichia coli*. J. Mol. Biol. 60: 1–19

Darnell J, Lodish H & Baltimore D (1990) Molecular Cell Biogy. Scientific American Books, New York

Doyle RJ & Koch AL (1987) The functions of autolysins in the growth and division of *Bacillus subtilis*. Crit. Rev. Microbiol. 15: 169–222

Hall BG (1988) Adaptive evolution that requires multiple spontaneous mutations. I. Mutations involving an insertion sequence. Genetics 120: 887–897

Hall BG (1990) Spontaneous point mutations that occur more often when advantageous than when neutral. Genetics 126: 5–16

Hogness DS, Cohn M & Monod J (1955) Studies on the induced synthesis of β-galactosidase in *Escherichia coli*: the kinetics and mechanism of sulfur incorporation. Biochim. Biophys. Acta 16: 99–116

King J (1980) In: Goldberg RF (Ed) Biological Regulation and Development, Vol 2 (pp 101–132). Plenum, New York

Koch AL (1962) The evaluation of the rates of biological processes from tracer kinetic data. I. The influence of labile metabolic pools. J. Theor. Biol. 3: 283–303

— (1968) The evaluation of the rates of biological processes from tracer kinetic data. II. RNA metabolism in growing bacteria. J. Theor. Biol. 18: 105–132

— (1971a) Evaluation of the rates of biological processes from tracer kinetic data. III. The *net* synthesis lemma and exchangeable pools. J. Theor. Biol. 32: 429–450

— (1971b) Evaluation of the biological processes from tracer kinetic data. IV. Digital simulation of nucleic acid metabolism in bacteria. J. Theor. Biol. 32: 451–469

— (1979) Microbial growth in low concentrations of nutrients. In: Shilo M (Ed) Strategies in Microbial Life in Extreme Environments (p 261–279). Dahlem Konferenzen-1978, Berlin

Koch AL, Higgins ML & Doyle RJ (1982) The role of surface stress in the morphology of microbes. J. Gen. Microbiol. 128: 927–945

Koch AL & Coffman R (1970) Diffusion, permeation, or enzyme limitation: A probe for the kinetics of enzyme induction. Biotech. and Bioeng. 12: 651–677

Koch AL & Deppe CS (1971) *In vivo* assay of protein synthesizing capacity of *Escherichia coli* from slowly growing chemostat cultures. J. Mol. Biol. 55: 549–562

Koch AL & Doyle RJ (1985) Inside-to-outside growth and the turnover of the Gram-positive rod. J. Theor. Biol. 117: 137–157

Koch AL & Levy HR (1955) Protein turnover in growing cultures of *Escherichia coli*. J. Biol. Chem. 217: 947–957

Lutkenhaus JF, Moore BA, Masters M & Donachie WD (1979) Individual proteins are synthesized continuously inthroughout the *Escherichia coli* cell cycle. J. Bacteriol. 138: 352–360

Mandelstam, J (1958) Turnover of protein in growing and nongrowing populations of *Escherichia coli*. Biochemical J. 69: 110–119

— (1960) The intracellular turnover of protein and nucleic acid in its role in bacterial differentiation. Bacteriol. Rev. 24: 289–308

Matin A, Auger EA, Blum PH & Schultz JE (1989) Genetic starvation survival in nondifferentiating bacteria Ann. Rev. Microbiol. 43: 293–316

Monod J (1958) An outline of enzyme induction. Recueil Travaux Chim. Pays-bas. 7: 569–585

Nath K & Koch AL (1971) Protein degradation in *Escherichia coli*. II. Strain differences in the degradation of protein and nucleic acid resulting from starvation. J. Biol. Chem. 246: 6956–6967

Pine MJ (1972) Turnover of intracellular proteins. Ann Rev. Microbiol. 26: 103–125

Rotman B & Spiegelman S (1954) On the origin of the carbon in induced synthesis of β-galactosidase in *Escherichia coli*. J. Bacteriol. 68: 419–429

Samarel AM (1991) In vivo measurement of protein turnover during muscle growth and atrophy. FASEB J. 5: 2020–2028

Schoenheimer R (1942) The dynamic state of body constituents. Cambridge

Stragier P, Kunkel B, Kroos L & Losick R. (1989). Chromosomal rearrangement generating a composite gene fro a developmental transcription factor. Science 243: 507–512

Zak R, Martin AF & Blough R (1979) Assessment of protein turnover by use of radioisotopic tracers. Physiol. Rev. 59: 407–447

Antonie van Leeuwenhoek **60**: 193–207, 1991.
© 1991 *Kluwer Academic Publishers. Printed in the Netherlands.*

Quantitative approaches to the analysis of the control and regulation of microbial metabolism

Hans V. Westerhoff[1,2], Wally van Heeswijk[1,2], Daniel Kahn[1] and Douglas B. Kell[3]
[1] *Division of Molecular Biology, The Netherlands Cancer Institute, Plesmanlaan 121, NL-1066 CX Amsterdam, The Netherlands;* [2] *E.C. Slater Institute for Biochemical Research, University of Amsterdam, Plantage Muidergracht 12, NL-1018 TV Amsterdam, The Netherlands*
[3] *Department of Biological Sciences, University College of Wales, Aberystwyth, Dyfed SY23 3DA, UK*

Key words: metabolic control analysis, physiology, rate limitation, modelling, modules

Abstract

Recently, a number of novel ways of considering the control, regulation and thermodynamics of microbial physiology have been developed and applied. We here present an overview of the new concepts involved, of their limitations and of the most recent attempts to deal with those limitations. We conclude that there no longer exist reasons of principle for vagueness in discussions of the control of microbial physiology and energetics. Further, the novel conceptual methods serve to remove part of the discordance between holistic and reductionistic views of microbial physiology.

1. Introduction

In a microbial cell of a few cubic microns, thousands of processes occur simultaneously at rather appreciable rates. For the fitness of the cell it is mandatory that the rates of these processes are well adjusted to each other. Whilst some of this adjustment occurs 'automatically' through mass-action effects, much involves more sophisticated control mechanisms. Biochemists and molecular biologists may study control at the level of single processes, whilst cell physiologists' may consider control and regulation at the integrated level of entire cells. In general, however, the sheer complexity of metabolism has precluded the establishment of strict relationships between the molecular and the cellular level. In recent years, however, concepts and methodologies have been developed that serve to relate the two levels. Because they have to deal with complexity, many of these methods implement mathematics, yet they have brought concepts that can also be grasped and employed more intuitively. Whilst giving some insight into the quantitative power of some of these methods, this overview will lean towards conveying some of the conceptual advances which they represent.

We shall begin by reviewing some of the *concepts* of a method rationally to analyze the control of metabolic fluxes in metabolic pathways, i.e., Metabolic Control Analysis (MCA; a review of applications of MCA to microbial physiology will be found elsewhere in this volume; Van Dam & Jansen 1991). Subsequently, we shall outline nonconventional aspects of MCA that pertain to chemostats. Next we shall point at two limitations inherent to conventional MCA, i.e., its emphasis on metabolism in cells of constant composition with immutable enzyme activities, and its limitation to small changes. Subsequently, we shall discuss some recent advances that lift these limitations. One is a recent extension of MCA that makes use of the fact that cellular physiology tends to be organized in terms of recognizable chunks ('modules') that each carry out their own task and regulate each other. We shall elaborate this method in some detail for the particular case of ammonia assimilation in *E. coli*, as it is regulated through a covalent modification cascade affecting the activity of glutamine synthetase. Finally, we shall discuss analysis methods that may be used whenever changes are not small, such as Mosaic Non Equilibrium Thermodynamics and Biochemical Systems Theory.

2. Metabolic control analysis and microbial physiology: the concepts

The metabolic control analysis (MCA) devised by Kacser, Burns, Heinrich, Rapoport, and Rapoport in the early 1970s, and more recently extended by others (see later), provides both a strategy and a formalism for the quantitative description of the control of metabolism under (mainly) steady-state conditions. It relates the 'local' kinetic properties of enzymes to their 'global' properties like their contribution to the control of variables such as fluxes and intermediary metabolite concentrations. From this point of view, it constitutes in principle an ideal formalism for attacking the problems on which this Special Issue is focussed, viz. Quantitative Aspects of Microbial Metabolism. To this end, we begin by providing a short review of the salient features of MCA.

MCA has been reviewed several times in recent years (e.g. Groen et al. 1982; Kell & Westerhoff 1986 a, b; Brand & Murphy 1987; Kacser & Porteous 1987; Westerhoff & van Dam 1987; Kell et al. 1989; Kell & Westerhoff 1990; Cornish-Bowden & Cárdenas 1990; Van Dam & Jansen 1991). Its chief distinction, from our point of view, is that it can relate the properties of *individual* components of a system (e.g. 'external' metabolites, or enzymes in a metabolic pathway) to their global behavior in contributing to the control of a metabolic flux.

Flux-control coefficients
We first define what we mean by a pathway. The operational definition used in the MCA is that a pathway is a system that consists of a flux from starting substrates at fixed concentrations (S in Fig. 1) to products that are also maintained at constant concentrations (P in Fig. 1). Alternatively, concentrations of substrates and products are such that changes in them do not affect enzyme activities in the pathway (because of saturation). The pathway is to include the production and utilization of all (generally allosteric) effector molecules ('internal effectors') acting on the pathway. This functional isolation of the pathway of interest from the rest of the cellular metabolism is based upon the fact that different parts of metabolism are or may be isolated kinetically from each other. In other words, in conventional MCA, we deal with metabolic steady states that are not affected by other responses of the system such as the induction of relevant gene products.

When studying a typical metabolic pathway, not least in terms of maximizing fluxes of biotechnological interest, it is traditional to ask questions such as 'which enzyme is rate-limiting'? The metabolic control analysis shows that the contribution of an individual enzyme to the control of flux through a pathway is both a *systemic* property and can be expressed in quantitative terms. Now we know, of course, that removing all of the enzyme in a pathway will reduce the flux to zero; however, this only tells us that the enzyme is *in* the pathway of interest. To obtain a meaningful analysis, therefore, we must determine the change in flux caused by a small (strictly an infinitesimal) change in enzyme activity (k_{cat} or V_{max}). To obtain a dimensionless number, we use the fractional change in enzyme activity and in flux. Thus, using the new, unified terminology (Burns et al. 1985), we define a flux-control coefficient $C_{e_i}^J$ as $((dJ/J)/(dv_i/v_i))_{ss} = (d \ln J/d \ln v_i)_{ss}$, where v_i is the activity of enzyme e_i, and the subscript ss (steady-state) implies that the comparison is made after the system has relaxed to its steady state(s). J represents the steady-state flux through the system. Here it is to be understood that the change in enzyme activity v_i is brought about by changing a parameter that affects that activity. The flux-control coefficient therefore equals the slope of a log-log plot of J *vs* v_i at the concentration (activity) of e_i prevailing.

Except for systems that exhibit 'channelling'

rather than pool behavior (see Welch & Keleti 1988; Kacser et al. 1990; Kell & Westerhoff 1990; Welch & Keleti 1990; Ovádi 1991), the sum of the flux-control coefficients of the enzymes in (or acting upon) a pathway equals 1. This relationship, known as the *flux-control summation theorem*, means that if we find that a particular enzyme has a flux-control coefficient of say 0.1, we know the rest of the flux control resides in other enzymes. It is also worth pointing out that if one considers branched pathways, the enzymes in the branch other than that containing the flux of interest ('reference flux') will tend to have negative flux-control coefficients (since increasing their activity will decrease the flux of interest); the sum of the flux-control coefficients of the enzymes in the branch of interest will therefore tend to exceed 1 (Kacser 1983).

Other control coefficients

We may also define control coefficients for the control of flux by the external (starting) substrate concentration (C_S^J) and by external modulators such as inhibitors (C_I^J), or for the control of intermediary metabolite concentrations ([X]) by enzyme activities (concentrations) ($C_{e_i}^X$). The latter are known as metabolite concentration-control coefficients, and have a summation equal to zero. Each of these coefficients are defined in a similar way to the flux-control coefficients, i.e. as (d ln superscript / d ln subscript)$_{ss}$.

Parameters and variables

The MCA lays great stress on the distinction between parameters and variables. Parameters are those factors which are set by the experimenter (typically temperature, pH and the starting or 'clamped' substrate concentrations) or by the system itself (typically in this case K_m, K_i and V_{max} values), *and which are unchanging during the course of an experiment*. As it stands, therefore (but see later), conventional MCA does not consider changes in enzyme concentrations caused for instance by the induction or repression of genes. Variables are those factors which attain a constant value only when the system attains a stable steady state. The most important variables are the flux J

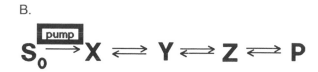

Fig. 1. (A) a sample metabolic pathway. S and P are supposed to be present at effectively fixed concentrations, whereas the concentrations of Y and Z, as well as the fluxes can vary.
(B) Microbial culture in chemostat written so as to be amenable to metabolic control analysis; the first reaction is that of the pump introducing substrate, present in the feed at concentration S_0, into the chemostat. The substrate concentration in the chemostat is now a variable and referred to by X.

and the concentrations of intermediary metabolites. We would also stress that variables cannot control fluxes, so that it is quite inappropriate to ascribe a control of a flux to a low concentration (relative to a K_m) of an intermediary metabolite, say; a particular metabolite concentration adopts a low value in a steady state because of the properties of the enzymes producing and consuming it.

Elasticity coefficients

Of course, enzyme activities do depend upon the concentrations of their substrates (and of other effector molecules), and the MCA describes these interactions in terms of so-called elasticity coefficients or elasticities. These are defined in a form that is mathematically very similar to that of the definition of the control coefficients; thus, the elasticity of enzyme e_i towards (the concentration of) substrate X is $\varepsilon_X^i = \partial \ln v_i / \partial \ln X$, i.e. the fractional change in enzyme turnover number caused by a fractional change in substrate concentration, subject to the important constraint that the change is carried out with all other parameters and variables held constant at their steady state values (Burns et al. 1985).

The connectivity theorems

Elasticities and control coefficients describe, in quantitative terms, respectively local and global properties of the metabolic system. The question then arises as to how these may be related to each other, since of course the behaviour of the metabolic system of interest does depend upon that of its constituent parts. The MCA formalizes this in terms of the so-called flux-control and concentration-control connectivity theorems. Perhaps the easiest way to think about the flux-control connectivity theorem is to imagine adding a non-competitive inhibitor to a steady-state pathway consisting of a linear series of metabolites (A ... --> ... E) whose metabolism is catalyzed by a series of enzymes (e_1 to e_4) obeying reversible Michaelis-Menten kinetics. If the inhibitor is a specific inhibitor of enzyme e_3, the first effect will tend to be a build-up of its substrate (C). This will either cause enzyme e_3 to speed up (if [C] was originally somewhere near the K_m of e_3) or will have no effect on the turnover of the enzyme (if e_3 was already saturated with respect to C). In the first case, from our definitions above, e_3 would have a high elasticity (large change in turnover for small change in substrate concentration) but a low flux-control coefficient (little change in pathway flux for a significant change in effective enzyme concentration or activity), whereas in the second case the converse would be true. More generally, the flux-control connectivity theorem shows that the sum of the products of the flux-control coefficients of the enzymes in a pathway and their elasticities towards a given metabolite is zero. Other theorems relate the metabolite concentration-control coefficients to the elasticities.

A number of approaches, some of which use matrix methods, have been devised which relate the control coefficients to the elasticities (e.g. Fell & Sauro 1985; Westerhoff & Kell 1987; Sauro et al. 1987; Reder 1988; Small et al. 1989; Cascante et al. 1989). For linear pathways, it is possible to express the control coefficients in terms of the elasticities alone, whereas branched pathways require, additionally, a knowledge of the flux-ratio at the branches (Westerhoff & Kell 1987; Sauro et al. 1987; Small & Fell 1989). For the pathway of Fig. 1A, the analysis is as follows (Westerhoff & Kell 1987; Sauro et al. 1987; Westerhoff & Van Dam 1987). First one writes the matrix **E**, which contains information both concerning the enzyme properties (elasticity coefficients, ε) and concerning the structure of the system (in the sense of where the flows flow):

$$E = \begin{bmatrix} 1 - \varepsilon_Y^1 & -\varepsilon_Z^1 \\ 1 - \varepsilon_Y^2 & -\varepsilon_Z^2 \\ 1 - \varepsilon_Y^3 & -\varepsilon_Z^3 \end{bmatrix}$$

Here the row of 1's reflects the fact that the example is that of a linear pathway. ε_Y^1 is the elasticity coefficient of the first reaction with respect to the concentration of metabolite Y, etc. Inversion of this matrix gives the matrix that gives all control coefficients:

$$C = \begin{bmatrix} C_1^J & C_2^J & C_3^J \\ C_1^Y & C_2^Y & C_3^Y \\ C_1^Z & C_2^Z & C_3^Z \end{bmatrix} = E^{-1}$$

C_1^J quantifies the control exerted by enzyme 1 on the steady state pathway flux. C_2^Z does this for the control exerted by enzyme 2 on the concentration of metabolite Z. This procedure can be extended to pathways of any complexity, where for the more complex cases, the method of Reder (1988) is the most systematic one (see also Holstein & Greenshaw 1991).

For a specific inhibitor, the flux-control coefficient of the inhibitor equals the flux-control coefficient of the target enzyme times the elasticity of the target enzyme towards the inhibitor. In other words, for a 'perfect' inhibitor, the flux-control coefficient equals the ratio of the initial slopes of normalized flux and normalized enzyme activity when plotted against the inhibitor concentration (Groen et al. 1982, Kell & Westerhoff 1986 a,b; Westerhoff & Kell 1988; Kell et al. 1989).

Measurement of flux-control coefficients

The measurement or estimation of flux-control coefficients follows in principle directly from their definition: one modulates the concentration or activity of an enzyme and measures the consequent change in flux, between steady-state conditions in which no other parameters have changed. Methods for doing this, with selected examples, include (see also van Dam & Jansen 1992): (a) titrations with specific metabolic inhibitors (Groen et al. 1982; Walter et al. 1987; Cornish et al. 1988), (b) variation of enzyme concentration by variation of their expression in diploid organisms (Flint et al. 1981; Middleton & Kacser 1983), (c) modulation of enzyme concentration by recombinant DNA methods in which the expression may be controlled by using a promoter of variable strength such as the *tac* promoter (Walsh & Koshland 1985), or by other molecular cloning methods (Heinisch 1986; Schaaf et al. 1989), and (d) variation of enzyme concentrations in systems reconstituted *in vitro* (Torres et al. 1986). Method (a) requires that the specificity of the inhibitors used is known, and preferably absolute, whilst the molecular genetic methods require that pleiotropic effects are absent, at least for the systems studied using conventional MCA.

Each of these methods suffers from the problem that as the flux-control coefficients become small, as they will indeed tend to do for long pathways, they become increasingly difficult to distinguish from zero, and in fact, with these approaches, values less than approximately 0.1 are probably not very reliable quantitatively. In say an inhibitor titration, the accuracy also depends upon how far one may inhibit the flux before the flux-control coefficient itself changes significantly (i.e. the curve bends round). In some pathways, such as that described in Fig. 5 of Kell et al. (1989), this may be a long way, whereas in other cases (e.g. Groen et al. 1982; Savageau & Voit 1982) the distribution of control depends strongly on the absolute flux. The biological significance of these very interesting differences is not yet understood. Statistical problems associated with the estimation of flux-control coefficients are discussed by Small (1988) and Small & Fell (1990).

Flux-control coefficients in supercomplexes

If one is trying to distinguish 'pool' from 'channelled' metabolism, a particularly interesting problem arises. To describe it we may imagine a 'perfect' (so-called 'static') channel (Keleti & Ovádi 1988), in which 'free' metabolites either do not exist or are not used (significantly) as substrates for 'their' enzymes due to unfavourable K_m values. In this case, the entire pathway and its intermediates

behave as a 'supercomplex' such that inhibiting one of the enzymes present (by say 1%) will inhibit the flux in direct proportion so that the enzyme would apparently have a flux-control coefficient of 1. If similar inhibitors were used for other enzymes in the complex, this would also be true for them, so that the flux-control summation theorem would appear to be violated when judged by these means (Kell & Westerhoff 1985, 1990; Westerhoff & Kell 1988); if the supercomplex contains n enzymes, the sum of the apparent flux-control coefficients would be n. In contrast, if one modulated the concentration of enzyme present in the system by *adding* enzyme (whether directly or by cloning), the enzyme added would not be able to participate in supercomplex formation, so that adding enzyme would not increase the flux and the flux-control coefficient would be zero! Heinisch (1986) increased the concentration of the phosphofructokinase (PFK) enzymes by cloning the 2 relevant structural genes, and acquired data which suggested that PFK has a rather low flux-control coefficient (although the data, in terms of the constancy of the flux from glucose to ethanol, are probably not good enough to exclude a value below approximately 0.2). If one were to carry out similar experiments for the rest of the glycolytic system (more than 13 enzymes), one would probably obtain similar data in each case. Schaaf et al. (1989) extended this study to include 8 glycolytic enzymes, to similar effect. One might therefore conclude *either* that the system exhibits pool behaviour, the distribution of control is rather homogeneous and the flux-control coefficients are too small to measure reliably, *or* that the system operates as a supercomplex. These possibilities may be distinguished by cloning both 'up' and 'down' in enzyme concentration.

Van Dam and Jansen (1991) give a detailed overview of the application of MCA to understanding the control of microbial metabolism in a number of systems.

3. Metabolic control analysis, chemostats and bioreactors

In conventional control analysis, a pathway is delimited by substrates and products that are kept at constant concentrations (see above). In standard MCA, asking to what extent microbial growth is determined by the concentration of the growth-limiting substrate, takes the following form: One grows the cells at one, fixed substrate concentration and determines the growth rate. One then increases the substrate concentration by p%, maintains it at the new level whilst measuring the new steady-state growth rate (p being taken as small as consistent with accurate experimentation). The percentage change in growth rate divided by p is the coefficient of growth control by that substrate.

For important growth substrates, the substrate concentrations at which this control coefficient significantly exceeds zero, are so low that it is experimentally unfeasible to maintain them constant throughout an experimental definition of the growth rate, especially because the latter requires the cells to reach a steady state. Moreover, a better way to grow cells under steady-state conditions is to grow them in a chemostat (Monod 1942). In a chemostat, however, it is the growth rate, rather than the substrate concentration, that is set by the experimenter; parameter and variable are reversed.

In chemostats then, the pertinent experiment is to manipulate the growth rate (by manipulating the dilution rate) and measure the relative change in concentration of the growth limiting substrate. Because of the small value of this concentration, this is usually quite difficult, but for a limited number of cases this has now been accomplished (see Van Dam & Jansen 1991 for review). This then will give a coefficient for the control of substrate concentration in the chemostat (denoted as [X]) by the growth rate: $C_D^{[X]}$. The inverse of this coefficient is numerically equal to the coefficient of control of growth rate by the substrate concentration, as measured in the non-chemostat experiment. The proof of the latter property is as follows. Let us consider a small increase in dilution rate of the chemostat. $C_D^{[X]}$ now is the relative change in concentration of the growth limiting substrate in the culture, divided by the relative change in dilution rate: $d\ln[\text{substrate}]/d\ln D$. Looking at it from the perspective of the cells, the substrate concentration has increased and

they have responded by proceeding to a new growth rate. For the cells the situation must, in principle, be quite the same as that in a batch culture where the substrate concentration, kept at a fixed magnitude, is now kept at a slightly increased value. They adjust their growth rate. Therefore, C_S^J is equal to $d\ln|J|/d\ln[\text{substrate}]$. Because D and J are equal, $C_S^J = 1/C_D^{[X]}$.

An alternative way of looking at the control of growth rate in a chemostat is to add the influx pump as a first reaction ('reaction 0') to the metabolic pathway (see Fig. 1B) with the property of having zero elasticity coefficients with respect to all metabolite concentrations. This first reaction then sets the constant flux (i.e., growth rate), and makes the pathway substrate an internal variable. Metabolic control analysis, including the calculation of control coefficients from elasticity coefficients, may now be performed. First one writes the matrix E:

$$\mathbf{E} = \begin{bmatrix} 1 & 0 & 0 & 0 \\ 1 & -\varepsilon_X^1 & -\varepsilon_Y^1 & -\varepsilon_Z^1 \\ 1 & -\varepsilon_X^2 & -\varepsilon_Y^2 & -\varepsilon_Z^2 \\ 1 & -\varepsilon_X^3 & -\varepsilon_Y^3 & -\varepsilon_Z^3 \end{bmatrix}$$

Inversion of this matrix gives the matrix that gives all control coefficients:

$$\mathbf{C} = \begin{bmatrix} C_D^J & C_1^J & C_2^J & C_3^J \\ C_D^X & C_1^X & C_2^X & C_3^X \\ C_D^Y & C_1^Y & C_2^Y & C_3^Y \\ C_D^Z & C_1^Z & C_2^Z & C_3^Z \end{bmatrix}$$

C_D^J quantifies the control exerted by the pump (set at dilution rate D) on the steady-state growth rate J. Because of the zeros in the matrix **E**, the control matrix bears zeros at the same positions. This implies that $C_D^J = 1$ and $C_i^J = 0$ for any i; all control on growth rate lies in the pump and no control resides in any of the enzymes in the bacteria. Indeed, if in a chemostat the activity of an 'important' enzyme in the bacteria was increased (e.g., by addition of IPTG in the case of an operon under the control of the *lac* promoter), the growth rate of the bacteria in the newly attained steady state would not change; it would still equal the dilution rate of the chemostat. What would change of course is the concentration of the growth substrate in the chemostat; C_i^X does *not* equal zero; in a chemostat the cellular enzymes do control the concentration of the growth substrate, as they continue to control the concentrations of metabolites. It should be noted that in the definition of control coefficients used here, the change in enzyme activity is supposed to be the same for all individual bacterial cells. If, in contrast, one of the cells were to mutate, then its growth rate would be enhanced and it could still out-compete the wild-type cells, although again, its ultimate steady-state growth rate (after finishing the competition) would return to the preset dilution rate of the chemostat. In a chemostat, all control on growth rate resides in the pump and cellular enzymes do not control the growth rate in a chemostat. Under comparable conditions in batch cultures the same enzymes would control growth rate. In either case, we note that the dynamics of these systems are sufficiently complex that it is doubtful that a 'true' steady state is attained (Kell et al. 1991).

The above is an illustration of a potentially powerful aspect of MCA: because its definitions are so akin to definitions used in technological process optimization, its analyses can be made congruent with analyses of the process optimization of bioreactors. Indeed, the technological processes around the bioprocess (in this simplest example represented by the dilution rate D) can be taken into account by describing them as additional 'metabolic' processes. For instance, one may define, measure, and come to understand a coefficient of control of the aeration apparatus on the intracellular ATP concentration or on the concentration of the relevant bioproduct.

In the biotechnological context it is relevant to ask if MCA may be used in studies meant to increase yields or efficiencies of biotechnological processes. Previously we have indicated a strategy for such optimizations (Kell & Westerhoff 1986 a, b; Westerhoff & Kell 1987). If the interest lies in obtaining an intermediary metabolite of a microbe, then one should first obtain an estimated map of the metabolic pathway of interest. Subsequently, one should obtain estimates of kinetic properties of the enzymes in the pathway of interest. Rather than extensive knowledge of all kinetic properties, it suffices to know the so-called elasticity coefficients of the enzymes with respect to the metabolites. These elasticities correspond to the kinetic orders by which the rates depend on the metabolite concentrations, i.e., for a normal, far-from equilibrium Michaelis-Menten enzyme the elasticity for the substrate is $K_m/([S]+K_m)$, i.e., 1 far below the K_m and approaching zero far above the K_m. One then puts this information into a matrix of elasticity coefficients (see above for an example), inverts this matrix, and obtains the coefficients for the control of pathway flux and metabolite concentrations by pathway enzymes. Above we demonstrated this procedure for the pathway of Fig. 1A. The chemostat may be analyzed by using the extended pathway of Fig. 1B, as also illustrated above. The enzymes that have the highest control coefficients with respect to the metabolite concentration of interest (or with respect to production rates of metabolites, though, in a chemostat, not with respect to growth rate, see above) are the candidates for genetic or other engineering, either by modifying their intracellular concentration or their elasticity coefficients (Kell & Westerhoff 1986 a).

Engineering approaches that affect more than a single enzyme at the same time can be dealt with similarly (Westerhoff & Kell 1987). The MCA as illustrated above for Fig. 1 yielded a matrix **C** of control coefficients as the inverse of a matrix **E**. For a manipulation that affects the activity of more than a single enzyme, one may write the change in enzyme activities as a column vector dln(e_1), dln(e_2), dln(e_3). The change in variables that occurs when that change in enzyme activities is implemented is then given by the matrix product of **C** and this column vector:

$$\begin{bmatrix} d\ln|J| \\ d\ln X \\ d\ln Y \end{bmatrix} = C \begin{bmatrix} d\ln e_1 \\ d\ln e_2 \\ d\ln e_3 \end{bmatrix}$$

This property may also be used in the reverse sense. If one wishes to change the microbes' metabolism, say the pathway flux by 10%, the concentration of X by -5%, leaving the concentration of Y unaffected, one may substitute these values for the vector on the left hand side of the above equation and then invert the equation to read:

$$\begin{bmatrix} \delta \ln e_1 \\ \delta \ln e_2 \\ \delta \ln e_3 \end{bmatrix} = \mathbf{E} \begin{bmatrix} 0.10 \\ -0.05 \\ 0.00 \end{bmatrix}$$

where we used the fact that $\mathbf{E}$ equals the inverse of the $\mathbf{C}$ matrix. The left hand side of the above equation gives the best first-order estimates of the changes in enzyme activities one should establish in order to obtain the desired change in the microbe's physiology. The above example demonstrates how one can in principle calculate how to manipulate the cell in order to produce a desired metabolic phenotype. Although, obviously, the calculation has its limitations, it seems too simple *not* to make it before one embarks on extensive projects of genetic engineering.

4. Limits to conventional metabolic control analysis

Enzyme activities regulated by other pathways are not considered

As described above, metabolic control analysis was developed for the purpose of the analysis of the control of pathways of intermediary metabolism. Such pathways were conceptualized as a set of enzymes present at fixed activities (concentrations), acting on metabolites whose concentrations were freely variable. The steady-state metabolite concentrations and the pathway flux(es) are then a function of the enzyme concentrations, and it is the latter functional dependence that is quantified by the control coefficients.

In microbial physiology, the conceptual framework in which the enzyme concentrations are constant is valid only for short-term phenomena. For time-windows exceeding say 5 minutes, enzyme concentrations are likely to change due to changes in their rates of transcription, translation or degradation, and such changes are often relevant for the regulation of metabolic pathways. Well-known examples of the latter are the induction of the *lac* operon by lactose and the repression of the *his* operon by histidine. The corresponding type of regulation has rarely been considered in MCA (but see Barthelmess et al. 1974; Westerhoff et al. 1990), though it has been considered in Biochemical Systems Analysis (e.g., Savageau 1976).

Regulation of the *amount* of enzyme is not the only phenomenon that interferes with the conventional concept of MCA. It is also assumed that activation or inactivation of an enzyme, e.g., through covalent modification in reactions catalyzed by other enzymes, do not occur as part of the internal variation of the system. However, very notable examples of this are found in the signal transduction pathways of both eukaryotes and prokaryotes.

The recognition that regulated gene expression and covalent enzyme modification is not yet part and parcel of MCA raises a number of questions:

- Is it possible to extend MCA so as to include these phenomena?
- Are the laws that govern metabolic regulation also applicable if regulation through variable gene expression or covalent enzyme modification occurs in parallel?
- Is it possible quantitatively to weigh the relative importance of regulation at the metabolic, gene-expression and covalent enzyme modification levels?

We believe that the answers to these questions are in the affirmative, and will shortly (section 5) seek to show this with respect to the glutamine synthetase regulatory cascade of *Escherichia coli*. There is an additional problem with MCA, however, and that is that in principle it considers only small changes in system parameters.

Large changes are not considered

As reviewed above, MCA discusses control of physiology in terms of control and elasticity coefficients. These are defined in mathematical terms as derivatives of effects with respect to causes. Such derivatives may be translated into magnitudes of effects divided by the magnitude of their causes for very (in fact infinitesimally) small effects. The advantage of definitions in terms of derivatives are that they are (i) unambiguous and (ii) they facilitate application of mathematics, such that control laws can be deduced that specify the connections between the control coefficients.

However, in many actual cases of regulation, changes are not truly small; they may well amount to 200% rather than to 1%. In such cases, MCA is only a first-order approximation of the actual status of the control of the pathway (still better than the conventional qualitative analyses). Second order extensions to MCA have been developed, but are fraught with complexity. The question therefore is: are there alternative methods to MCA that deal with substantial changes during regulatory transitions in a better-than-first-order approximation? Below we shall discuss two of these approaches, called Mosaic Non Equilibrium Thermodynamics and Biochemical Systems Analysis. First however, we shall deal with the possibility that enzyme activities and/or concentrations vary as a function of changes other than those made directly by the experimenter.

5. Modular metabolic control analysis

The control of the regulatory cascade of glutamine synthetase in E. coli as a model system

As we discussed in section 4, standard MCA discusses a physiological system as a single network in which all reactions are connected. In actual practice, cellular physiology is more organized than that. Several levels can be distinguished: the level of intermediary metabolism, the level of protein metabolism (synthesis, modification and degradation), the level of mRNA metabolism (transcription and decay). At the level of protein metabolism one can observe cascades of enzymes covalently modifying one another.

Conceptually, biochemists and cell physiologists have tended to separate the various levels of regulation of metabolism. Control is said to be at the level of transcription, at the level of translation, or 'just' metabolic. Conventional MCA did not acknowledge such a separation.

Recently MCA has been developed so as to analyze the contribution of the various controlling levels. Most explicitly this was done for systems with variable gene expression and for regulatory cascades: the modular metabolic control theory (Westerhoff et al. 1990; Kahn & Westerhoff 1991). The latter approach started from the general formalism of Reder (1988) and has the advantage of providing a better understanding of the control of the hierarchical levels of a cascade. Here, the modular control theory will be made explicit for the glutamine synthetase regulatory cascade, which is a complex and interesting system controlling the assimilation of NH_4^+ in enteric bacteria (review: Rhee et al. 1988).

Glutamine synthetase (GS) catalyses the incorporation of ammonium into glutamate resulting in glutamine. The activity of GS can be modified by the enzyme adenylyl-transferase (AT_a) to produce the less active form GS-AMP. The same enzyme catalyses both adenylylation and deadenylylation (AT_d) of GS. The transferase activity is stimulated by the regulatory protein P_{II}, while the deadenylylase activity is enhanced by the uridylyl form of P_{II} (P_{II}-UMP). The modification of P_{II} is catalyzed by uridylyl-transferase (UT_u), which has also, in an analogy to adenylyl-transferase, a deuridylylation activity (UT_d). UT_u and UT_d are regulated allosterically by α-ketoglutarate and by glutamine. Thus UT is a sensor for the nitrogen status of the cell. This cascade can be divided into three modules (see the boxes in Fig. 2). Each module contains a set of reactions which are connected with each other, but not with reactions from other modules. Therefore separate modules interact solely via effector-type interactions, i.e. substances from one module may affect reactions in another module without being a substrate or a product in this other module. Thus, in Fig. 2, both α-ketoglutarate and glutamine (which are in module 3) act as positive allosteric effectors of, respectively, UT_u and UT_d, whilst glutamine is a negative effector of UT_u, in module 1. The structure of this metabolic network may be summarized in terms of a stoichiometry matrix **N** (see e.g. Reder 1988; Holstein & Greenshaw 1991), which consists of the stoichiometric coefficients of the reactions of the network. To clarify the mean-

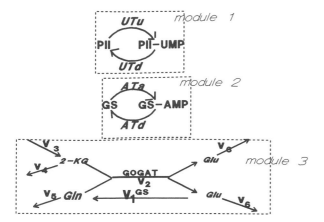

Fig. 2. Scheme of the pathways involved in ammonia assimilation in *E. coli.* 2-KG: α-ketoglutarate, Glu: glutamate, Gln: glutamine, GS: glutamine synthetase, PII: protein II, UTu: uridylyl transferase, UTd: deuridylylase, ATa: adenylyl transferase, ATd: deadenylylase, GOGAT: glutamine-2-ketoglutarate aminotransferase.

ing of this stoichiometry matrix for the glutamine synthetase regulatory cascade, let $\mathbf{x}$ be the column-vector of the molarities $[x_i]$ and $\mathbf{v}$ the column vector of the rates $[v_i]$; the time-dependent evolution of the system will be determined by the matrix product:

$$d\mathbf{x}/dt = \mathbf{N} \cdot \mathbf{v}$$

This can be expressed for the glutamine synthetase regulatory cascade (see Fig. 2). For the purpose of this article, we will simplify the treatment and consider that the GS/GOGAT cycle is the only route of ammonium assimilation, ignoring the glutamate dehydrogenase (GDH) route (as we would do with a glutamate dehydrogenase negative mutant). We obtain:

In the stoichiometry matrix, the modular structure of the metabolic network is indicated by the dashed lines. In shorthand, we may write:

$$\mathbf{N} = \begin{bmatrix} \mathbf{N}_1 & \mathbf{0} & \mathbf{0} \\ \mathbf{0} & \mathbf{N}_2 & \mathbf{0} \\ \mathbf{0} & \mathbf{0} & \mathbf{N}_3 \end{bmatrix}$$

where $\mathbf{N}_i$ indicates the stoichiometry matrix of the module of interest. Now we may build an elasticity matrix containing all elasticity coefficients (see section 2) of the system:

$$\boldsymbol{\varepsilon} = \quad [\partial\ln|v_i|/\partial\ln x_j]$$

Here ∂ refers to partial differentiation. Each coefficient ε_{ij} of the elasticity matrix $\boldsymbol{\varepsilon}$ quantifies the effect of molecular species j (at molarity x_j) upon rate v_i. The elasticity matrix $\boldsymbol{\varepsilon}$ of the glutamine synthetase regulatory cascade is decomposed into blocks according to the modular structure of the system, as mentioned above:

$$\boldsymbol{\varepsilon} = \begin{bmatrix} \boldsymbol{\varepsilon}_1^1 & \mathbf{0} & \boldsymbol{\varepsilon}_3^1 \\ \boldsymbol{\varepsilon}_1^2 & \boldsymbol{\varepsilon}_2^2 & \mathbf{0} \\ \mathbf{0} & \boldsymbol{\varepsilon}_2^3 & \boldsymbol{\varepsilon}_3^3 \end{bmatrix}$$

where $\boldsymbol{\varepsilon}_1^1$ (which is itself a matrix, hence written bold face) refers to module *1* (above), and so on. The block $\boldsymbol{\varepsilon}_2^1$ is null because the state of the GS interconversion cycle (module *2*) does not directly influence the rates of the P_{II} interconversion cycle (module *1*). Similarly the block $\boldsymbol{\varepsilon}_1^3$ is null because P_{II} and P_{II}-UMP do not directly influence the metabo-

$$\begin{bmatrix} dP_{II}/dt \\ dP_{II}\text{-UMP}/dt \\ dGS/dt \\ dGS\text{-AMP}/dt \\ d\alpha\text{-KG}/dt \\ dGlu/dt \\ dGln/dt \end{bmatrix} = \begin{bmatrix} -1 & 1 & 0 & 0 & 0 & 0 & 0 & 0 & 0 & 0 \\ 1 & -1 & 0 & 0 & 0 & 0 & 0 & 0 & 0 & 0 \\ 0 & 0 & -1 & 1 & 0 & 0 & 0 & 0 & 0 & 0 \\ 0 & 0 & 1 & -1 & 0 & 0 & 0 & 0 & 0 & 0 \\ 0 & 0 & 0 & 0 & 0 & -1 & 1 & -1 & 0 & 0 \\ 0 & 0 & 0 & 0 & -1 & 2 & 0 & 0 & 0 & -1 \\ 0 & 0 & 0 & 0 & 1 & -1 & 0 & 0 & -1 & 0 \end{bmatrix} \cdot \begin{bmatrix} v_{UTu} \\ v_{UTd} \\ v_{ATa} \\ v_{ATd} \\ v_1 \\ v_2 \\ v_3 \\ v_4 \\ v_5 \\ v_6 \end{bmatrix}$$

lic rates of module 3. In principle, the block ε_3^2 differs from 0 because the GS interconversion cycle can also be directly influenced by the metabolic status. However, to simplify the presentation, this effect is neglected here. A more complete treatment will appear elsewhere (Kahn & Westerhoff 1991). The block ε_2^3 is special, in that it contains the elasticities of the glutamine synthetase reaction towards GS and GS-AMP. If, for the sake of simplicity, we assume that GS-AMP is fully inactive, we can write:

$$\varepsilon_2^3 = \begin{bmatrix} 1 & 0 \\ 0 & 0 \\ 0 & 0 \\ 0 & 0 \\ 0 & 0 \\ 0 & 0 \end{bmatrix}$$

This is because with the definitions for the rates given in Fig. 2:

$$\begin{bmatrix} d\ln|v_1| \\ d\ln|v_2| \\ d\ln|v_3| \\ d\ln|v_4| \\ d\ln|v_5| \\ d\ln|v_6| \end{bmatrix} = \begin{bmatrix} 1 & 0 \\ 0 & 0 \\ 0 & 0 \\ 0 & 0 \\ 0 & 0 \\ 0 & 0 \end{bmatrix} \cdot \begin{bmatrix} d\ln[GS] \\ d\ln[GS\text{-}AMP] \end{bmatrix}$$

Similarly we can construct a matrix of control coefficients $\mathbf{C}$ containing all the flux control coefficients of the system, and decompose it into blocks following the modular structure of the system:

$$\mathbf{C} = \begin{bmatrix} \mathbf{C}_1^1 & \mathbf{C}_2^1 & \mathbf{C}_3^1 \\ \mathbf{C}_1^2 & \mathbf{C}_2^2 & \mathbf{C}_3^2 \\ \mathbf{C}_1^3 & \mathbf{C}_2^3 & \mathbf{C}_3^3 \end{bmatrix}$$

where the submatrix $\mathbf{C}_j^i$ contains the flux control coefficients describing the sensitivities of the fluxes within module i to changes in the process-activities in module j.

Now suppose we have studied the metabolic part of this system (module 3) without operation of the regulatory cascade (that is, at a constant, clamped, level of GS adenylylation, equal to the steady-state level of GS adenylylation normally attained). We can analyze the control of such a subsystem and place the resulting control coefficients into an *intrinsic* control matrix which we will note $\mathbf{C}_3$. $\mathbf{C}_3$ will in general be different from the matrix $\mathbf{C}_3^3$ containing the control coefficients of module *3* when the regulatory cascade is left to operate (that is, the level of GS adenylylation is allowed to vary freely). Note, however, that the elasticity matrix of the metabolic module *3* (or of any other module) is an intrinsic property and therefore is ε_3^3 both in the clamped and in the non-clamped system. Recently, we have been able to demonstrate a relationship allowing to calculate the matrix $\mathbf{C}_3^3$ from the knowledge of the intrinsic *flux* control matrix $\mathbf{C}_3$, the intrinsic *concentration* control matrices $\mathbf{S}_1$, $\mathbf{S}_2$ and $\mathbf{S}_3$, and the elasticity matrices ε_j^i (Kahn & Westerhoff 1991, $\mathbf{I}$ is the identity matrix):

$$\mathbf{C}_3^3 = \mathbf{C}_3 \cdot (\mathbf{I} - \varepsilon_2^3.\mathbf{S}_2.\varepsilon_1^2.\mathbf{S}_1.\varepsilon_3^1.\mathbf{S}_3)^{-1}$$

Here it has been assumed that module 3 does not directly affect module 2. The first term of this product is the *intrinsic* control matrix of the metabolic module, whereas the second term describes the effect of the cyclic regulation on the control within the module. Thus the control matrix can be calculated as the product of an intrinsic control matrix and a regulatory term referring to the regulation through the other modules. Similarly, this can be done for the concentration control coefficients. The importance of these relationships is at least twofold: (i) the regulatory effects of a process on a flux in the same module are equal to the regulatory effects if the module were in isolation, divided by a term that measures the regulatory effects through all other modules (and that goes to 1 if those other effects are absent or incompletely connected), and (ii) it is possible to determine the control properties of the modules separately before constructing the control of the entire system. In this manner, modular metabolic control analysis marries reductionism and holism (see section 7).

Moreover, the above modular control analysis allows one to quantify the *regulatory strengths* (Kahn & Westerhoff 1992) of the cascade, which express quantitatively the sensitivities of nitrogen assimilation fluxes to fluctuations in the levels of P_{II}

uridylylation. These are the elements of the following *regulation matrix*:

$$\mathbf{R}_1^3 = \mathbf{C}_2^3 . \boldsymbol{\varepsilon}_1^2$$
$$= \mathbf{C}_3^3 . \boldsymbol{\varepsilon}_2^3 . \mathbf{S}_2 . \boldsymbol{\varepsilon}_1^2$$

It is the product of an elasticity matrix $\boldsymbol{\varepsilon}_1^2$ expressing the response of the GS adenylylation cycle to changes in P_{II} uridylylation, and of a control matrix $\mathbf{C}_2^3$ expressing the control of nitrogen assimilation fluxes by GS adenylylation (Kahn & Westerhoff 1991).

This brief account of how Modular MCA can be applied to the GS regulatory cascade indicates that the way is now open for the quantitative analysis of rather more complex systems than was heretofore possible. First, one decomposes the system into its relevant disconnected modules. Second, the control of each module is analyzed individually, by clamping the concentrations in the other modules. Third, the complex system is analyzed as a whole, both experimentally and by mathematical reconstruction. If experiment and mathematical reconstruction are consistent, one may consider that the model on which the latter was based is a sound quantitative model. Thus Modular MCA is a method for treating the control of complex metabolic systems by exploiting their structure, when the completely direct treatment is too complicated.

6. Larger changes: biochemical systems theory and mosaic non equilibrium thermodynamics

The elasticity and control coefficients of MCA are defined as derivatives, i.e., as the ratios of infinitely small changes. In practical applications, these are replaced by ratios between small changes. However, as discussed in section 4, for many cases of interest in microbial physiology, larger changes are important. Why not then use the same definitions and theory for larger changes? Thus one might be inclined to define the coefficient of the control of enzyme *i* exerted on flux J as the percentage decrease in flux divided by the percentage reduction in activity of that enzyme. However, the magnitude of such a control coefficient would al-

most always depend on the percentage change in enzyme activity. And, for a linear pathway, the flux control coefficient of each enzyme would tend to approach 1 when the inhibition of activity approached 100%, since complete elimination of that enzyme will reduce the pathway flux to zero. Indeed, the magnitude of the sum of all the control coefficients would depend on the magnitude of the percentage change in enzyme activity made in the determination of the control coefficients.

Up to this moment, no complete solution for this dilemma has been found. What is left for the analysis of larger changes is:

(i) just use MCA and accept the result as approximate for the description of the actual control and regulation
(ii) perform a complete integration of all the kinetic equations of the system
(iii) use approximating, but simpler descriptions
(iv) use methods of artificial intelligence (Kell & Davey 1991).

The disadvantage of approach (ii) is that it requires the precise knowledge of most kinetic characteristics of the system. In addition, because of the incongruence between standard integration subroutines and the structure of metabolism and physiology, this procedure tends to lose touch with biochemistry. Object-oriented programming approaches have recently been used in attempt to alleviate the latter problem (Stoffers et al. 1991).

The third method has been applied, at times with considerable success. This degree of success may seem somewhat surprising as it would seem impossible to approximate the richness of the kinetics of cellular reactions by simpler rate equations and still simulate cellular behavior. However, the success becomes more understandable if it is granted that we are generally less interested in understanding the *precise* quantitative behavior of physiological systems than in understanding the *essence* of such behavior, even though the latter may include quantitative aspects (such as synergism) (Savageau 1976).

Non Equilibrium Thermodynamics (NET) has been an approach, which, although its accuracy in describing all the kinetic features of biochemical kinetics is severely limited, could still make one understand the essence of phenomena such as free-energy transduction, coupling, and optimal states (Caplan & Essig 1983; Stucki 1980; Westerhoff & Van Dam 1987). For microbial growth this led to insights into why the efficiency of microbial growth may be as low as it is (Westerhoff et al. 1983; see however Heijnen 1991). A variant of NET, called Mosaic Non Equilibrium Thermodynamics (MNET) was developed so as to remove the most apparent inconsistencies between NET and biochemical kinetics (reviewed: Westerhoff & Van Dam 1987). It has been applied to enhance understanding of what underlies the phenomenon of growth rate-dependent and growth rate-independent maintenance metabolism (Hellingwerf et al. 1982), as well as the basis for the distinction between Carbon- and Energy-limited growth (Westerhoff & Van Dam 1987). The most recent review of this method may be found in Rutgers et al. (1991).

MNET may be considered an extension of a branch of MCA that focuses on the control aspects of free-energy metabolism (Westerhoff & Van Dam 1987). Biochemical Systems Analysis (BST) is a method that is parallel to MCA. Its basic approach may be rationalized as follows: If one describes the dependence of a reaction rate on its substrate concentration by an elasticity coefficient, and one assumes that the elasticity coefficient does not change much as the substrate concentration is increased, then one may integrate and describe a reaction rate by:

$$v = k \cdot [S]^{\varepsilon_s}$$

Transition to logarithmic space allows one to integrate systems with this type of rate equations in a simple way (Savageau 1976; Voit 1991). A problem arises when groups of reactions are aggregated or when reversible reactions are considered, but even for those cases the approximation has been shown to work reasonably well (Voit & Savageau 1987; Voit 1991). Up to this moment BST has been used to describe the general behaviour of biological systems and indeed, qualitative conclusions of general

value have been attained. The application of the method in direct experimentation has remained limited, because the basic definitions in BST are somewhat remote from experimental observables, in contrast to the definitions of MCA and MNET (see however, Groen & Westerhoff 1990).

7. Concluding remarks

Methods to reduce complexity:
the rationalization of reductionism
In this paper we have discussed a number of modern approaches to the quantitative analysis of microbial physiology. This should be regarded as a parallel to the review of concrete applications of these methods by Van Dam & Jansen (1991).

It is often suggested that the sole aim of quantitative methods should be to describe, accurately to the second decimal place, rates and concentrations in systems. We take issue with this. The more important aim of these methods is to realize a basic tenet of biochemistry and biophysics, i.e., that, in principle biology should be explicable in terms of physical and chemical principles. And, 'explicable' should mean 'explicable in the physical chemical sense', i.e., in principle including the quantitative detail.

Too often, the latter tenet has been subject to immaterial debate between holists and reductionists, the latter emphasizing studies of single mechanisms that occur in cells, the former stressing that doing so destroys the essence of cellular, organismal and ecological organization. The methods we have discussed here provide a scientific link between molecular mechanisms (elasticity coefficients) and properties of the cell as a whole (control coefficients). Indeed, laws such as the summation and the connectivity theorems (see above) are the very expression of the fact that the whole is more than the simple sum of its parts.

Of course, we note that, to date, most of these types of analyses take into account only limited aspects of cellular organization. Organization on the basis of metabolic channelling (e.g. Welch & Keleti 1990) is not usually considered (see however Kell & Westerhoff 1985, 1990; Westerhoff and Kell

1988; Kacser et al. 1990; Ovádi 1991), and neither are phenomena, often called 'self-organization', which invoke bistability and hysteresis (Nicolis & Prigogine 1977; see however Cortassa et al. 1991). In truth, it is not always easy, and may not be possible in principle, to put Humpty Dumpty back together again (Kell & Welch 1991).

Acknowledgements

HVW & DBK thank Dr. Stouthamer for continued interest in and discussions of the application of quantitative methods to microbial physiology. DBK thanks the Wellcome Trust, the SERC and the SERC Biotechnology Directorate for financial support. DK thanks the INRA (France) and EC (BRIDGE) for support. HVW thanks the Netherlands Organisation for Scientific Research (NWO) for much of the same.

References

Acerenza L, Sauro HM & Kacser H (1989) Control analysis of time-dependent metabolic systems. J. Theor. Biol. 137: 423–444

Barthelmess IB, Curtis CF & Kacser H (1974) Control of flux to arginine in *Neurospora crassa*. J. Mol. Biol. 87: 303–316

Brand MD & Murphy MP (1997) Control of electron flux through the respiratory chain in mitochondria and cells. Biolog. Rev. 62: 141–193

Burns JA, Cornish-Bowden A, Groen AK, Heinrich R, Kacser H, Porteous JW, Rapoport SM, Rapoport TA, Stucki JW, Tager JM, Wanders RJA & Westerhoff HV (1985) Control analysis of metabolic systems. Trends Biochem. Sci. 10: 16

Caplan SR & Essig A (1983) Bioenergetics and Linear Nonequilibrium Thermodynamics. The Steady State. Cambridge, Massachusetts: Harvard University Press

Cascante M, Franco R & Canela EI (1989) Use of implicit methods from general sensitivity theory to develop a systematic approach to metabolic control I. Unbranched pathways. Mathem. Biosci. 94: 271–288

Cornish-Bowden A & Cárdenas M-L (Eds) (1990) Control of Metabolic Processes. Plenum Press, New York

Cornish A, Greenwood JA & Jones CW (1988) The relationship between glucose transport and the production of succinoglucan exopolysaccharide by *Agrobacterium radiobacter*. J. Gen. Microbiol. 134: 3111–3122

Cortassa S, Aon MA & Westerhoff HV (1991) Linear Non Equilibrium Thermodynamics describes the Dynamics of an Autocatalytic System. Biophys. J. 60: 794–803

Fell DA & Sauro HM (1985) Metabolic control and its analysis. Additional relationships between elasticities and control coefficients. Eur. J. Biochem. 148: 555–561

Flint HJ, Tateson RW, Barthelmess IB, Porteous DJ, Donachie WD & Kacser H (1981) Control of the flux in the arginine pathway of *N. crassa*. Biochem. J. 200: 231–246

Groen AK & Westerhoff HV (1990) Modern Control Theories: a Consumers Test. In: Cornish-Bowden A & Cárdenas M-L (Eds) Control of Metabolic Processes (pp 101–118). Plenum Press, New York

Groen AK, Van der Meer R, Westerhoff HV, Wanders RJA, Akerboom TPM & Tager JM (1982) Control of metabolic fluxes. In: Sies H (Ed) Metabolic Compartmentation (pp 9–37). Academic Press, New York

Heijnen JJ (1991) A new thermodynamically based correlation of chemotrophic biomass yields. A. van Leeuwenhoek 60: (this issue)

Heinisch J (1986) Isolation and characterization of the two structural genes coding for phosphofructokinase in yeast. Mol. Gen. Genet. 202: 75–82

Hellingwerf KJ, Lolkema JS, Otto R, Neijssel OM, Stouthamer AH, Harder W, Van Dam K & Westerhoff HV (1982) Energetics of microbial growth: an analysis of the relationship between growth and its mechanistic basis by mosaic non-equilibrium thermodynamics. FEMS Microbiol. Lett. 15: 7–17

Holstein H & Greenshaw CP (1991) A numerical treatment of metabolic control models. In: Westerhoff HV (Ed) Biothermokinetics Intercept, Andover, UK (in press)

Kacser H (1983) The control of enzyme systems in vivo: elasticity analysis of the steady state. Biochem. Soc. Trans. 11: 35–40

Kacser H & Porteous JW (1987) Control of metabolism: what do we have to measure? Trends Biochem. Sci. 12: 5–14

Kacser H, Sauro HM & Acerenza L (1990) Control analysis of systems with enzyme-enzyme interactions. In: Cornish-Bowden A & Cárdenas M-L (Eds) Control of Metabolic Processes (pp 251–257). Plenum Press, New York

Kahn D & Westerhoff HV (1991) Control theory of regulatory cascades. J. Theor. Biol. 153: 255–285

— (1992) The regulatory strength: how to be precise about regulation and homeostasis. Biotheor. Acta (in press)

Keleti T & Ovádi J (1988) Control of metabolism by dynamic macromolecular interactions. Curr. Top. Cellul. Regul. 29: 1–33

Kell DB & Davey CL (1991) On fitting dielectric spectra using artificial neural networks. Bioelectrochem. Bioenerg. (in press)

Kell DB, Ryder HM, Kaprelyants AS & Westerhoff HV (1991) Quantifying heterogeneity: flow cytometry of bacterial cultures. A. van Leeuwenhoek 60: (this issue)

Kell DB & Welch GR (1991) No turning back: reductionism and complexity in molecular biology. The Times Higher Education Supplement 9–8–91, p 15

Kell DB & Westerhoff HV (1985) Catalytic facilitation and membrane bioenergetics. In: Welch GR (Ed) Organized Multienzyme Systems. Catalytic Properties (pp 63–138). Academic Press, New York

— (1986a) Metabolic Control Theory: its role in microbiology and biotechnology. FEMS Microbiol. Rev. 39: 305–320

— (1986b) Towards a rational approach to the optimisation offlux in microbial biotransformations. Trends Biotechnol. 4: 137–142

— (1990) Control analysis of organised multienzyme systems. In: Srere PA, Jones ME & Mathews C (Eds) Structural and Organizational Aspects of Metabolic Regulation (pp 273–289). Wiley-Liss, New York

Kell DB, Van Dam K & Westerhoff HV (1989) Control analysis of microbial growth and productivity. In: Banmberg S, Hunter I & Rhodes M (Eds) Microbial products: New Approaches. Soc. Gen. Microbiol. Symp. 44 (pp 61–93). Cambridge University Press

Middleton RJ & Kacser H (1983) Enzyme variation, metabolic flux, and fitness. Alcohol dehydrogenase in *Drosophila melanogaster*. Genetics 105: 633–650

Monod J (1942) Recherches sur la croissance des cultures bactériennes, Herman et Cie, Paris

Nicolis G & Prigogine I (1977) Self-Organisation in Nonequilibrium Systems. John Wiley & Sons, New York

Ovádi J (1991) On the physiological significance of metabolic channeling. J. Theor. Biol. 152: 1–22

Reder C (1988) Metabolic control theory: a structural approach. J. Theor. Biol. 135: 175–202

Rhee SG, Bang WG, Koo JH, Min KH & Park SCV (1988) Regulation of glutamine synthetase activity and its biosynthesis in *Escherichia coli*: mediation by three cycles of covalent modification. In: Boon Chock P, Huang CY, Tsou CL & Wang JH (Eds) Enzyme Dynamics and Regulation (pp 136–145). Springer, Berlin

Rutgers M, Van Dam K & Westerhoff HV (1991) Control and thermodynamics of microbial growth. Rational tools for bioengineering. CRC Crit Rev. Biotechnol. 11:367–395

Sauro HM, Small JR & Fell DA (1987) Metabolic control and its analysis. Extensions to the theory and the matrix method. Eur. J. Biochem. 165: 215–221

Savageau MA (1976) Biochemical Systems Analysis: A Study of Function and Design in Molecular Biology. Addison-Wesley, Reading, MA

Savageau MA & Voit EO (1982) Power-law approach to modelling biological systems. I. Theory. J. Ferment. Technol. 60: 221–228

Schaaff I, Heinisch J & Zimmermann FK (1989) Overproduction of glycolytic enzymes in yeast. Yeast 5: 285–290

Small JR (1988) Theoretical aspects of metabolic control. Ph. D. thesis, Oxford Polytechnic

Small JR & Fell DA (1989) The matrix method of metabolic control analysis: its validity for complex pathway structures.

J. Theor. Biol. 136: 181–197

— (1990) Metabolic control analysis: sensitivity of control coefficients to elasticities. Eur. J. Biochem. 191: 413–420

Stoffers HJ, Sonnhammer ELL, Blommestijn GJF, Raat HNJ & Westerhoff HV (1991) METASIM: object oriented modelling of cell regulation. CABIOS (in press)

Stucki JW (1980) The optimal efficiency and the economic degrees of coupling of oxidative phosphorylation. Eur. J. Biochem. 109: 269–283

Torres NV, Mateo F, Melendez-Hevia E & Kacser H (1986) Kinetics in metabolic pathways. A system *in vitro* to study the control of flux. Biochem. J. 234: 169–174

Van Dam K & Jansen N (1991) Quantification of control of microbial metabolism by substrates and enzymes. A. van Leeuwenhoek 60: (this issue)

Voit EO (Ed) (1991) Canonical Nonlinear Modeling. S-System Approach to Understanding Complexity. Van Nostrand Reinhold, New York

Voit EO & Savageau MA (1987) Accuracy of alternative representation for integrated biochemical systems. Biochemistry 26: 6869–6880

Walter RP, Kell DB & Morris JG (1987) The roles of osmotic stress and water activity in the inhibition of the growth, glycolysis and glucose phosphotransferase system of *Clostridium pasteurianum*. J. Gen. Microbiol. 133: 259–266

Walsh K & Koshland DE, Jr (1985) Characterisation of the rate-controlling steps *in vivo* by the use of an adjustable expression vector. Proc. Natl. Acad. Sci. USA 82: 3577–3581

Welch GR & Keleti T (1990) In: P Srere, ME Jones & C Mathews (Eds) Structural and organizational aspects of metabolic regulation, UCLA Symposia on Molecular and Cellular Biology, New Series, Vol 134 (pp 321–330). Alan R. Liss, New York

Westerhoff HV & Kell DB (1987) Matrix method for determining the steps most rate-limiting to metabolic fluxes in biotechnological processes. Biotechnol. Bioengin. 30: 101–107

— (1988) A control theoretical analysis of inhibitor titration assays of metabolic channelling. Comm. Molec. Cellul. Biophys. 5: 57–107

Westerhoff HV & van Dam K (1987) Thermodynamics and Control of Biological Free Energy Transduction. Elsevier, Amsterdam

Westerhoff HV, Hellingwerf KJ & van Dam K (1983) Efficiency of microbial growth is low, but optimal for maximum growth rate. Proc. Natl. Acad. Sci. 80: 305–309

Westerhoff HV, Koster JG, van Workum M & Rudd KE (1990) On the control of gene expression. In: Cornish-Bowden A & Cárdenas M-L (Eds) Control of Metabolic Processes (pp 399–412). Plenum Press, New York

Antonie van Leeuwenhoek **60**: 209–223, 1991.
© 1991 *Kluwer Academic Publishers. Printed in the Netherlands.*

Quantification of control of microbial metabolism by substrates and enzymes

K. van Dam & N. Jansen
E.C. Slater Institute for Biochemical Research, Biotechnological Center, University of Amsterdam, Plantage Muidergracht 12, 1018 TV Amsterdam, The Netherlands

Key words: control, enzyme, limitation, microbial growth, substrate

Abstract

The control of substrates or enzymes on metabolic processes can be expressed in quantitative terms. Most of the experimental material found in the literature, however, has been obtained under non-standardized conditions, precluding definite conclusions concerning the magnitude of control. A number of representative examples is discussed and it is concluded that a quantitative analysis of the factors that control metabolism is essential for understanding the microbial behaviour.

Introduction

The metabolism of a microorganism is a complex network of reactions, catalyzed by numerous enzymes. When one uses, for instance, the expression: 'a microbe's metabolism is limited by energy' it is evident that this must be a gross simplification of a subtle combination of causes and effects. It is a basic law of nature that any change in conditions will cause compensatory mechanisms in the metabolic network to counteract that change. As a result, the concentration of intermediates in the metabolic pathways, and in the longer run also the concentration of enzymes may change, resulting in a microorganism of a makeup that is different from the original one.

As a consequence of this complexity, understanding microbial metabolism requires the evaluation of the relative importance of a large number of regulatory effects. This necessitates the use of precise and quantitative statements when discussing the factors that 'limit' or 'control' a microorganism's metabolism. Fortunately, a number of coherent theories of quantitation of control have been developed over the past years (see review by Westerhoff et al. in this issue). In this review we will make most extensive use of the control theory as developed originally by Heinrich & Rapoport (1973) and Kacser & Burns (1973), later extended by others (see Westerhoff & van Dam 1987). In essence, this theory describes control by any factor on a metabolic flux as a differential; the 'control coefficient of effector P on flux J' is defined as the relative change in the flux, caused by a small relative change in the effector:

$$C_P^J = (dJ/J)/(dP/P)$$

Even beyond implementation of this quantitative definition, a statement like 'the microorganism is growing under energy-limited conditions' lacks precision as to what of the following is meant: Is the microorganism not producing sufficient ATP to allow maximal growth? Would an increase in the enzymes involved in ATP production increase the rate of growth? Or would an increase in the substrates, required for ATP production stimulate growth? To what extent is each of these factors responsible for the sub-optimal performance? Questions of this type have to be formulated in a quantitative sense to be amenable to any answer. Thus, one can ask: how large is the control of a particular enzyme on a particular flux *or* how large is the control of a substrate on a particular flux. By extension, one may ask the same question for a group of enzymes (or even a group of substrates) in a defined concentration ratio or even of a whole segment of metabolism, provided that segment is well-defined and its activity can be modulated in a controlled way (Westerhoff and Van Dam, 1987). Thus, for instance, theoretically one could ask: what is the control of the catabolic machinery ('energy generation') on the growth of a microorganism? To answer such a question, one would have to specifically modulate the activity of the catabolic machinery without affecting the input concentrations of the substrates of the machinery. The control coefficient would then be defined as the ratio between the change in activity of the catabolic machinery and the change in growth.

In this review, we will discuss experimental data concerning the control coefficients on microbial metabolism of two groups of effectors: the substrates and the enzymes. Unfortunately, quantitative data that can be used in the framework of the modern control theories are rather scarce. Most of the available experimental material has been obtained under non-standardized conditions. Therefore, we will have to limit ourselves to the discussion of a number of representative examples.

Some relevant issues in control theory

It must be stressed that a control coefficient as defined in the control theory is independent of any model of describing metabolism: it is a ratio of two (in principle) measurable quantities. Increasing detail in a metabolic model should serve to make the relation between the measured quantities and the underlying biochemical detail increasingly understandable. It should be further noted that in its simplest form this particular control coefficient is unequivocal only for steady states where the network itself is not changing, i.e. the composition of

the microorganism is a constant and the input and output of the metabolism is also assumed constant. In those cases where the relaxation to the new steady state is accompanied by a change in the enzymatic composition of the microorganism, two such control coefficients can be defined: one which does not and one which does take network changes into account (see Westerhoff et al. 1990; Westerhoff et al. 1992). We will mainly address the former type of control coefficient in this review, but want to emphasize already at this point that the assumption of constant cellular composition is an often unwarranted simplification that has to be experimentally verified where possible.

It is easy to confuse the variables and the parameters in an experiment. For instance, one may wish to determine the effect of a substrate on the growth rate, for which a chemostat would seem the most appropriate experimental setup. However, in a chemostat experiment one fixes the growth rate by setting the dilution rate. Thus, the variable which one wants to measure and the parameter which one dictates have been interchanged. To obtain the answer in this particular example, one may perform the measurement of the substrate at a number of growth rates (= dilution rates), interpolate to the desired substrate concentration and determine the relative change in substrate concentration divided by the relative change in dilution rate. The mathematical inverse of this control coefficient equals the control exerted by substrate concentration on growth rate (Westerhoff et al. 1992). A technical development which presents a solution to this problem was reported by Kleman et al. (1991). These authors constructed a 'glucosestat', in which the substrate concentration is kept constant through feedback and feedforward control of the dilution rate. In this way the substrate concentration (the parameter) is fixed and the growth rate (the variable) is allowed to develop to its steady-state value. As presented, the system responds to glucose concentrations in the mM range. It would need to be adapted for those cases where the actual steady-state concentration of the limiting substrate is in the μM range, as is often the case for sugars and amino acids.

In any metabolic network the control on any flux

of all the enzymes involved adds up to 1; this is the so-called 'summation theorem'. This is an important practical tool in verifying whether all the controlling steps have been pinpointed: as long as the sum is much less than 1, there is certainly an important controlling factor missing. However, the converse is not true. This is partly due to the fact that there are also enzymes with a negative control on a pathway. A simple example is the control of enzymes that hydrolyse ATP on growth: increase in their activity will decrease growth. Another complicating factor are so-called 'channelled pathways'. In such sequences the enzymes operate functionally as a unit; decrease in any of the enzymes in such a pathway has the effect of decreasing the total flux through the pathway. In extreme cases, the control on the flux of *each* of the enzymes in a channelled pathway may become 1.

For the control of substrates on the flux through a pathway there is not such a simple summation theorem. However, one can obtain a useful relation in which the control by all substrates is quantitatively expressed, by making use of the 'connectivity theorem' between the elasticity (epsilon) of the velocity (v) through an enzyme (e) towards its effectors (Z) and the control of those effectors on the flux through a pathway (Rutgers 1990):

$$C_Z^J = C_e^J \cdot \varepsilon_Z^v$$

Control of substrates on microbial metabolism

One might expect that determination of the control exerted by substrates (and products) on microbial growth and product formation would have been completed for most cases of interest. The only thing one has to do is measure the growth of the microorganism as a function of the relevant substrate concentrations, just like one would do in determining the kinetics of an isolated enzyme. The control exerted by the particular substrate on the growth rate would then be defined as the relative change in flux J (= growth rate) divided by the relative change in substrate (S) concentration:

$$C_S^J = (dJ/J)/(dS/S)$$

Unfortunately, in practice this is not so simple. For most substrates, the apparent affinity towards substrates is so high that the rate of growth will change only at very low concentrations of substrate; at higher concentration the maximal rate of growth with respect to that substrate will be reached. As a result of this, in the relevant range of substrate concentrations in batch cultures no steady state can be reached, because relatively large changes in substrate concentration are accompanied by insignificant growth. It was probably this fact that inspired Monod (1942) to develop the theory and practice of the so-called 'chemostat'. It is generally assumed that in the chemostat *one* of the nutrients is limiting the rate of growth, so that the supply of fresh medium just balances the disappearance of that particular nutrient through its conversion to biomass and other products. Monod assumed that the uptake of the limiting nutrient by the microorganism followed saturable, Michaelis-Menten type, kinetics and that the subsequent metabolic reactions just balanced this uptake. Thus, in the steady state there should be a low concentration of the limiting substrate in the chemostat and the dependence of the growth rate on that concentration should be hyperbolic.

The proposal of Monod remained a working hypothesis for a long time because of the inherent technical difficulties in measuring the low steady-state concentrations of the limiting substrate. During the sampling procedure, the cells will take up the residual substrate from the medium, which will lead to an underestimation of the substrate concentration present in the culture. In order to avoid this kind of problem as much as possible it is necessary to use cultures with a very low dry weight (Rutgers et al. 1990), and to separate cells from culture fluid within less than a second (depending on the growth rate). Another problem is the variation of the measured substrate concentration with the actual moment of sampling, especially if a series of samples taken on different days are compared. Rutgers et al. (1987) have shown that, even though the steady-state concentration of biomass becomes constant within less than ten generations (i.e. the yield on the limiting substrate has become constant; Stouthamer 1973), it may still take 40–60 generations

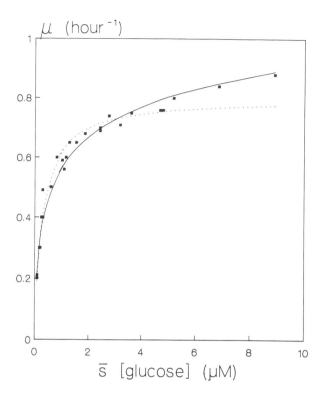

Fig. 1. Relation between the steady-state concentration of glucose and the specific growth rate of *Escherichia coli* ML30 in glucose-limited chemostat culture. The data are fitted with a hyperbolic (. . .) or with a logarithmic (——) function. Reproduced from Senn (1989).

before the concentration of the limiting substrate has become constant (see also Hartl et al. 1985). Thus, one either has to take samples after a fixed number of generations after the disturbance of the old steady state, or one has to wait until the limiting substrate concentration has become constant. Several different sampling techniques have been developed by different groups. Mulder (1988) centrifuged samples for potassium determinations through silicone oil, and Postma et al. (1988) dropped samples for glucose determinations into liquid nitrogen, thawed them, and centrifuged them at 4° C. Of all methods published in this context the quickest way to separate cells from fluid is sampling by making use of a vacuum filtering technique (Rutgers et al. 1987; Senn 1989). With the above methods Monod's hypothesis could be tested with accuracy. It turned out that the predicted hyperbolic relationship did not always give the best fit to the

experimental data (Schulze & Lipe 1964; Owens & Legan 1987; Rutgers et al. 1987; Senn 1989; Rutgers et al. 1989). For instance, Senn (1989) concluded that a logarithmic dependence of growth on limiting substrate concentration was a better description of the findings than a hyperbolic (Fig 1). This logarithmic dependence had been presumed by the Mosaic Non-Equilibrium Thermodynamic (MNET) description of microbial growth (Westerhoff et al. 1982; Westerhoff & Van Dam 1987). According to this model, the growth rate depends on the free energy change in the metabolic reactions. For instance, the free energy of anabolism (growth) at constant concentration of biomass is a logarithmic function of the (limiting) anabolic substrate. It is in practice, however, not easy to distinguish definitely between a hyperbolic and a logarithmic curve, because the regions of very low and very high substrate concentrations are the most important in comparing these two curves, and unfortunately these points are experimentally the most difficult to determine accurately.

The simple metabolic control theory considers the microorganism as a defined entity of constant composition. If, for instance, the uptake system of the limiting substrate would be inducible by that substrate, it is clear that the degree of induction would depend on the growth rate because the steady-state concentration of that substrate increases with the growth rate. Therefore, in the simple Monod approach a distortion of the dependence of growth rate on steady-state substrate concentration may be expected. An example of the opposite direction of regulation is found in the changes in the activity of the PEP: sugar phosphotransferase system with growth rate; in this case higher growth rates seem to be associated with lower activity of the uptake system (Neijssel et al. 1977; O'Brien et al. 1980). Up to this point, the effect of one single substrate on the growth rate was discussed. However, just like in enzyme kinetics each of the substrates will affect the reaction rate, it may be expected that the growth rate of a microorganism depends on the concentration of each of the nutrients. For simplicity one usually assumes that all nutrients except one are present in 'saturating' concentration. In practice, the experimenter will try to arrange the composition of the medium such that changes in all but one of the components do not affect the growth rate. However, there are many possible situations where more than one substrate has a significant control on the growth rate.

The situation of 'multiple-substrate limitation' is experimentally even more difficult to handle than that of the common single-substrate limitation in the chemostat. Some studies have been reported in which a transition between different substrates in the chemostat was effected. It is useful to distinguish between several types of dual-substrate limitation. First we have substrates that are metabolized and (partially) incorporated into cell material. To this class belong the transitions between either the usual carbon or the usual nitrogen sources. In a transition between two different carbon source limitations, we may have a situation where the one can replace the other with respect to catabolic and anabolic purposes. An example of such a transition is given in the work of Brinkmann et al. (1990). These authors cultivated *Hansenula polymorpha* in the chemostat and varied the ratio of methanol and xylose in the input medium. Comparing the results with those obtained on methanol or xylose as single carbon substrate, they concluded that the capacity of the linear methanol dissimilatory pathway is not rate-limiting and that methanol oxidase itself is the rate-limiting enzyme for growth on methanol. Such a conclusion is, however, not warranted for several reasons. Firstly, there are many different changes in the cells during these experiments and, secondly, the enzyme activities held responsible for the changes in growth rate were not actually determined.

The second type of transition is that between a carbon and a nitrogen source limitation; in this case a more complex change probably occurs. In any case, it is most likely that the composition of the cells will be different in the two limiting situations. Minkevich et al. (1988) studied the transition between glucose and ammonia limitation in the chemostat by changing their relative concentration in the input medium. These authors found that there was a transition region where the concentrations of the two substrates in the chemostat were below the limit of detection of their methods and concluded that under those conditions both substrates were limiting growth. However, the fact that a substrate is (almost) completely consumed does not prove that it has a high control on growth. For the latter conclusion, it has to be shown that the relative growth rate depends strongly on the relative change in substrate concentration.

Rutgers (Rutgers et al. 1990) performed a similar study, but with the required extra quantitative detail. Specifically, he determined the steady-state concentrations of both glucose and ammonia in a chemostat culture of *Klebsiella pneumoniae*, growing at different rates and with different input concentrations of these substrates. Thus, he was able to construct a three-dimensional plot in which the steady-state concentrations of glucose and ammonia were related to the growth rate (Fig. 2). From such a plot, one can read off the variation of the growth rate with either of the two substrate concentrations at constant concentration of the other substrate. Through this procedure, Rutgers derived the control of each of the two substrates on growth at constant concentration of the other substrate. It was found that there is a large region of conditions where glucose and ammonia exert *simultaneously* significant control on the growth of *K. pneumoniae*.

As already discussed above, most theories describing bacterial growth assume a constant cell composition. For the experiments as done by Minkevich (1988) and Rutgers (1990), however, it can be shown that the cell composition is changing when the input ratio of the limiting substrates is changed. When going from ammonium-limited conditions to glucose-limited conditions, the nitrogen yield decreases, because at first ammonia is completely incorporated into biomass while at higher concentrations it is excreted (or not taken up). This means that in the region where glucose and ammonium are both almost completely consumed, when one alters the input ratio of these substrates in the feed medium, either the cell composition has to change or the composition of the metabolic products has to change and probably even both occur simultaneously. Thus, in order to go through such a region of double substrate limited growth the cell has to adapt its metabolism in a

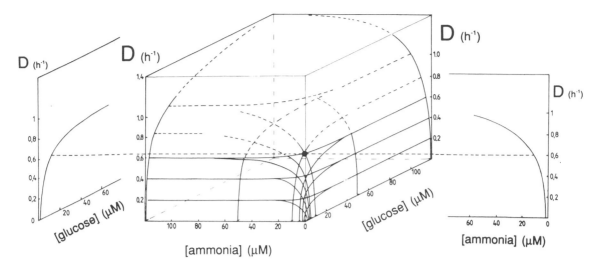

Fig. 2. Three-dimensional surface, relating the steady-state concentrations of glucose and ammonia to the rate of growth of *Klebsiella pneumoniae* in chemostat culture. Reproduced with permission from Rutgers et al. (1990). © 1990, Springer-Verlag.

flexible way, and it is quite possible that the cells at the beginning of such an experiment are rather different from the ones at the end of it. From the point of view of the simple metabolic control theory this is a problem since the biomass composition is not constant at all compositions of the medium. If possible, one should compare the rates of growth under such conditions that there has not yet occurred a significant change in cell composition. An extended form of the metabolic control theory can, however, cope with the problem of changes in cell composition (Westerhoff et al. 1990; Westerhoff et al. 1992). In this version two types of control coefficients are introduced, the simple one used in this review and another one allowing for the changes in biomass composition as the result of changes in gene expression.

A third class of substrates involves those that are required for growth and that are incorporated in the cell in the form in which they are taken up. An example is the cation potassium. Even though it might be considered an 'inert' solute, the intracellular concentration of potassium varies with the growth rate (Aiking & Tempest 1976; Mulder 1988). In fact, it was suggested that the growth rate of *Escherichia coli* does not depend on the extracellular but on the intracellular concentration of the ion (Mulder 1988). This emphasizes again the ge-

neral point that one should be precise in defining which parameter determines which variable: if the intracellular concentration of potassium should be considered the parameter, it is another example of a parameter that is experimentally difficult to set.

The transition between two limiting substrates can also be used to determine experimentally to what extent these substrates are mutually exchangeable. We (Jansen et al. unpublished) performed such a study to examine the claim (Buurman et al. 1989) that potassium and ammonium ions may substitute for each other under certain conditions. To this end, we determined the steady-state concentration of potassium and ammonium in chemostat cultures of *Klebsiella* at different input ratios of these ions. At low pH a region of control by ammonium and a region of control by potassium were separated by a region where both ions simultaneously exerted a significant control. In contrast, at high pH there was no condition where only potassium had a high control on the growth rate. These results confirm in a quantitative sense that under the latter conditions ammonium can substitute for potassium.

Looking back at the experiments that are relevant for quantitating the control of substrates on the growth rate of microorganisms, we must conclude that there are only few reports that can be used in the ultimate quantitative sense of control. What experimental evidence there is, complies with the notion that often most substrates are present in 'saturating' concentration for growth. One can, however, find conditions where more than one substrate has a significant control coefficient with respect to growth rate.

In nature, the evolutionary importance of control of growth by substrates may be less than that by enzymes; most natural conditions probably resemble a pulsed batch culture where periods of maximal growth rate alternate with periods of starvation. Only during the periods close to starvation, control by substrate may become an important competitive criterion.

Therefore, the control of enzymes on growth, discussed in the next section, is probably more important for understanding why microbial cells are constituted in the way we isolate them.

Control of enzymes on microbial metabolism

According to the original model of Monod (1942) *one* enzymatic step of metabolism, for instance the uptake of the substrate, is assumed to be limiting. The modern theories of control show that we have to be more exact, because in principle the control of a certain flux may be distributed over all the participating enzymes. To be precise, we would have to measure the effect on the flux of a small change in one of the participating enzymes to determine the control by that enzyme on a particular flux. This is experimentally a difficult task, because it is not possible to simply add small amounts of an enzyme to the system within the intracellular compartment.

Technically it is most easy to change the activity of an enzyme (or an ensemble of enzymes) by adding an irreversible inhibitor. This approach has been used to advantage in the study of the distribution of control of mitochondrial oxidative phosphorylation over the different participating enzymes (Groen 1984). An important prerequisite for this kind of study is that the inhibitor reacts specifically with *one* of the components of the system and that there is an independent way to ascertain which fraction of that enzyme has bound the inhibitor.

An example of this is given in the study of the control of glycolysis in *Streptococcus cremoris* by glyceraldehyde-3-phosphate dehydrogenase (Poolman et al. 1987). These authors determined the inhibitory effect of the relatively unspecific inhibitor iodoacetate on glycolysis and at the same time established to what extent the enzymes in an extract of the same cells were inhibited. They found that specifically the activity of glyceraldehyde-3-phosphate dehydrogenase was inhibited and the control coefficient of this enzyme on the glycolytic flux was close to 1. It should be noted, however, that one always has to keep the possibility open that in the whole cells other enzymes were inhibited by the rather unspecific inhibitor iodoacetate as well.

Poolman & Konings (1988) also established a correlation between the growth rate of *Streptococcus lactis* and *Streptococcus cremoris* and the rate of amino acid uptake, when growth at different pH values was studied. At high pH the maximal rate of uptake of glutamate appears to be just sufficient to account for the rate of growth. This finding does strictly speaking not prove that the uptake process has a high control on growth, since growth at different pH values may also cause changes in the activity of other enzymes.

Another example of apparent control of the uptake process on metabolism is given by the work of Birkenhead et al. (1988). These authors showed a correlation between an increase in transport of dicarboxylic acids and the rate of growth or nitrogen fixation in *Bradyrhizobium japonicum* cells. The increase in dicarboxylic acid uptake was effected by introduction of relevant genes from *Rhizobium meliloti*. That such a correlation does not necessarily prove that there is a causal relation, is shown by the fact that the rate of succinate uptake is approximately doubled by the presence of the plasmid, independent of the growth substrate. However, the specific growth rate is only increased with dicarboxylic acids as carbon source. Iwami & Yamada (1985) compared the rate of glycolysis between *Streptococcus sanguis* cells grown in the chemostat under glucose-limited or glucose-excess conditions. The difference in growth condition led to differences in enzyme composition of the cells. The rate of glycolysis in the first type of cells was higher. It turned out that of the measured enzyme activities glyceraldehyde-3-phosphate dehydrogenase and the PEP : glucose phosphotransferase uptake system were strongly increased. The authors concluded that these increased activities were the cause of the increase in glycolytic flux. This interpretation of the data is not allowed, because several enzymatic activities had changed at the same time. Any quantitative conclusion concerning the control of any one of the enzymes on the glycolytic flux would require variation of its activity at constant activity of all other enzymes.

A similar study was reported by Galazzo & Bailey (1990), who determined the control of different enzymes in yeast on the glycolytic flux, by comparing suspended and immobilized *Saccharomyces cerevisiae* cells. The kinetics of the enzymes involved were determined from in vivo ^{31}P-NMR measurements and inserted into an overall kinetic model of glycolysis. The authors concluded that entrapment of yeast cells in an alginate matrix leads to shift in control from the glucose uptake step to the fructose-6-phosphate kinase step. The problem with these experiments is again that the comparison is not clear-cut, because more than one factor changes simultaneously, prohibiting derivation of a control coefficient. Moreover, the matrix that was inverted to estimate control coefficients from elasticity coefficients did not properly correspond to the experimental system (Bailey & Galazzo, personal communication, 1991).

More global experiments to determine with the same inhibitor whether catabolism or anabolism controls microbial growth in an unrestricted batch culture (Kell et al. 1989) are less satisfying, because in this case it was not proven that iodoacetate affected only enzymes of catabolism. In fact, it is even difficult to make a strict distinction between catabolism and anabolism, since the two have enzymes in common.

An elegant way to vary the intracellular activity of single enzymes is to modulate the expression of their genes. An early example of the determination of control by an enzyme by genetic means on microbial metabolism was reported Flint et al. (1980,

1981). These authors modulated the activity of enzymes of the pathway leading to synthesis of arginine in *Neurospora crassa*. This was done by a combination of mutations in the relevant genes and the use of heterokaryons, carrying genes with different activity. In this way, a large range of activities of the enzymes in the arginine synthesis pathway could be covered. The conclusion from these experiments was that at the wild-type level of activity none of these enzymes had a high control on the flux to arginine.

The group of Niederberger (Valinger et al. 1989) performed a similar study on the control by the enzymes of amino acid synthesis on the flux through that pathway in yeast. They used a combination of controllable plasmids and variation of the gene dosage by constructing nuclear hybrids with one to four copies in a diploid strain. From this study the conclusion follows that also in yeast the enzymes involved in amino acid biosynthesis at wild-type activity have little control on the rate of this process.

An even more finely tuned approach to the determination of the control of metabolism by an enzyme was taken by Walsh & Koshland (1985), by making use of inducible expression of the relevant enzyme as coded on a plasmid. These authors investigated the control of citrate synthase on metabolic fluxes and growth of *E. coli*. By use of an adjustable expression vector, they could vary the activity of citrate synthase over a large range around the wild type level. From the results, they concluded that citrate synthase at wild-type level of activity has little control on growth with glucose as the substrate, but that the same enzyme has significant control on growth on acetate (although the control is still less than 1). These findings illustrate that the control by an enzyme depends on the pathway and the conditions considered.

The authors rationalize their findings by considering the difference in metabolic pathways with the two growth substrates. With acetate as the growth substrate (in contrast to the case of glucose), the cell is critically dependent on the functioning of the glyoxylate bypass of the citric acid cycle to produce building blocks for biosynthesis.

However, this does not necessarily imply that enzymes of this pathway must have a high control on the flux through that pathway.

Later work by Stueland et al. (1988) has identified isocitrate dehydrogenase as another enzyme with important control on the flux through the glyoxylate cycle. Isocitrate dehydrogenase activity can be modulated by phosphorylation/dephosphorylation by a bifunctional enzyme: isocitrate dehydrogenase kinase/phosphatase. It turns out that the rate of pyruvate-induced dephosphorylation (i.e. activation) of isocitrate dehydrogenase is almost proportional to the level of this bifunctional enzyme, suggesting that the bifunctional enzyme has a control of 1 on this dephosphorylation rate. This is, of course, not unexpected: an enzyme should have a control of 1 on the reaction it catalyzes. However, the in vivo situation is complex as revealed by the fact that in the same cells the rate of glucose-induced dephosphorylation of isocitrate dehydrogenase is almost not dependent on the level of the bifunctional enzyme. This is probably because glucose and pyruvate induce different levels of substrates and effectors for the bifunctional enzyme. Since the control is defined for a pathway at constant input and output, this illustrates that any conclusion about control holds only for very strictly defined boundary conditions.

Some other recent examples on the use of controlled expression of enzymes to determine their control of fluxes can be given. Ruijter (Ruijter et al. 1991) determined the control of Enzyme II^{Glc} of the PEP:glucose phosphotransferase system (PTS) on fluxes in *Escherichia coli*. By using an adjustable expression vector, he could modulate the activity of Enzyme II^{Glc} over a large range. He found that the magnitude of the control of this enzyme depended on the pathway considered, as expected. Enzyme II^{Glc} at wild-type level has a relatively high control (0.6) on the uptake of glucose via the PTS, but a low control on the complete pathway of oxidation of glucose or on growth on glucose (Fig.3). As a point of warning, it should be mentioned that the control coefficients were derived from measurements of the rate of each process at apparent saturating concentrations of the substrates. Since

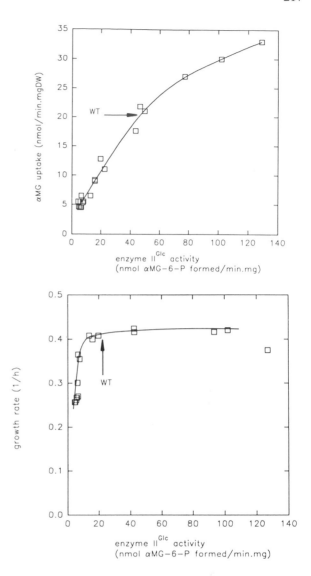

Fig. 3. Relation between sugar uptake or growth rate and Enzyme II^{Glc} activity in *Escherichia coli*. Data from Ruijter et al. (1991).

the cell composition may depend on the rate of growth, it may be that (part of) the effect of changes in Enzyme II^{Glc} is modulated by the fact that the measurements were done at different growth rate.

This finding is interesting in view of the regulatory role of the components of the PTS in sugar metabolism in bacteria in general. It has been made plausible (Postma & Lengeler 1985) that the degree of phosphorylation of Enzyme III^{Glc} determines whether other uptake systems are inhibited (inducer exclusion) or whether adenylate cyclase is activated. On the basis of such a mechanism, one may expect that the relative activities of the enzymes adjacent to III^{Glc}, i.e. HPr and II^{Glc}, are tuned in such a way that changes in flux through the PTS result in strong changes in the level of phosphorylation of Enzyme III^{Glc}. However, the extent to which this will be the case depends not only on the control by these adjacent enzymes, but also on their elasticity toward the substrates, i.e. how strongly they respond to changes in concentration of substrate and product. The finding that Enzyme II^{Glc} has a high control on glucose uptake does therefore not necessarily imply that initiation of glucose uptake via the PTS could not result in significant dephosphorylation of Enzyme III^{Glc}.

Zimmermann and co-workers (Schaaff et al. 1989) looked at the control of the enzymes of glycolysis in yeast on the metabolic fluxes. They found that over a rather wide range of activities above wild-type level there was relatively little effect of varying enzyme activity on the glycolytic flux (or on the concentration of key glycolytic intermediates). This might indicate that the uptake process is a limiting factor in the flux. It would be interesting to test this by using permeabilised cells. One may contrast the findings of Zimmermann and co-workers concerning the absence of significant control of any of the enzymes on glycolysis in yeast with those discussed above for the same process in *S.cremoris* (Poolman et al. 1987), where glyceraldehyde-3-phosphate dehydrogenase appeared to have a high control. It should, however, be kept in mind that the latter conclusion is open to criticism (see above). Yet, it is entirely feasible that the distribution of control in the major metabolic pathways is very different between cells.

In a different approach, Dean et al. (1986) investigated the control of lactose permease and β-galactosidase on the 'metabolic fitness' (expressed by the relative growth rate) of cells with lactose as substrate. This was done by constructing mutants with different kinetic properties of these enzymes. By selection of spontaneous revertants of a nonsense mutation in the lacZ gene, β-galactosidase enzymes with moderately different activity were obtained. Strains with such slightly different en-

zymes could hardly have been picked up by traditional screening procedures. After transduction of the mutation, the properties of isogenic *E. coli* strains with different β-galactosidase activity were compared. To this end, they were competed against each other in lactose-limited chemostat cultures. To minimize problems with spontaneous mutations, leading to constitutive expression of the lacZ gene, an excess of an inducer (IPTG) was also present. It was found that the maximal growth rate of *E. coli* depends on the activity of β-galactosidase only at activities less than 5% of wild-type. At wild-type level of activity β-galactosidase apparently has virtually no control on the flux from lactose to biomass.

In a subsequent study Dykhuizen et al. (1987) quantitated the combined effect of changes in the lactose translocator and β-galactosidase on the metabolic fluxes in *E. coli*, again by competition between strains in lactose-limited chemostats (see Fig. 4). From these measurements, they concluded (again) that β-galactosidase at wild-type level has a low control on the flux from lactose to biomass. On the other hand, the control of the lactose translocator at wild-type level on the same process was rather high (C = 0.55). The authors go on to conclude from this that the selective pressure on the lactose translocator must be higher than on the β-galactosidase. Note, however, that all the above conlusions are based on experiments in lactose-limited chemostats; the conclusions are generally not valid for other conditions. To come to more firm conclusions, it would be worthwhile to investigate how the changes in enzyme activity in the chemostat are accompanied by changes in the steady-state level of the limiting substrate.

Portillo & Serrano (1989) performed a study to determine the control of the plasma membrane H$^+$-ATPase on the growth rate of yeast. They constructed mutants in which the ATPase activity varied between 10 and 100% of wild type and showed that the growth rate depended almost linearly on this activity. The conclusion that this enzyme has a high control on growth of yeast is justified for the lower growth rates, but remains speculative for the higher rates, since no values higher than wild type were obtained.

An interesting study concerning the control of growth (and other fluxes) by the ATP synthase in *E. coli* was reported by Jensen (1991). He studied the rate of growth, respiration rate and growth yield of *E. coli* as a function of the expression of ATP synthase under different conditions. To this end, the genes coding for ATP synthase were expressed from an inducible plasmid, just like in the study of the control coefficient of Enzyme II^{Glc} on metabolic processes (Ruijter et al. 1991). Using the non-fermentable substrate succinate, at low induction of the *atp* operon there is almost a proportional increase in growth rate with increase in ATP synthase activity; this implies that in that region the ATP synthase has a control of 1 on the growth rate. However, at wild-type level of ATP synthase activity, the control of this enzyme on the growth rate is almost zero. Interestingly, at yet higher levels of expression, the control of ATP synthase on the growth rate becomes negative. Perhaps, under these conditions the excess ATP synthase activity leads to excessive ATP breakdown.

Using the fermentable substrate glucose, the control of ATP synthase on the growth rate of *E. coli* is much smaller. This can be rationalized by saying that with glucose as substrate ATP production can be sustained by substrate-linked phosphorylation. Thus, even in the absence of ATP synthase the growth rate is still 80% of that at wild-type level of this enzyme. As a consequence, the control of ATP synthase on the growth rate is small (less than 0.04) at low activity of this enzyme. At wild-type level, the control of ATP synthase on growth rate is almost zero and at even higher levels it is negative, just as in the case of succinate as the substrate.

Interestingly, it seems that the maximal growth rate of *E. coli* is attained at wild-type level of ATP synthase, either with glucose or with succinate as growth substrate. This led Jensen (1991) to suggest that the level of this enzyme is adjusted to achieve maximal rate of growth of this organism, in line with the suggestion of Westerhoff et al. (1982, 1987), based on thermodynamic arguments.

In a recent review, Jensen and Pedersen (1990) analyze which factors control the maximal growth rate of *E. coli*. Their main conclusion is that this

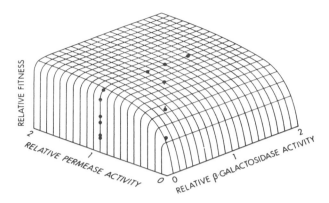

Fig. 4. Three-dimensional surface, relating the activity of β-galactosidase and lactose permease to the rate of growth (denoted here as 'relative fitness') of *Escherichia coli* in chemostat culture. Reproduced with permission from Dykhuizen et al. (1987). © 1987, Genetics Society of America.

control can primarily be attributed to subsaturation of the macromolecular biosynthetic apparatus with substrates and catalytic components, i.e. the concentration of the substrates is below the K_m of their enzymes (see also Marr 1991). In some respects this is a truism: in the steady state, the formation and breakdown of all intracellular components must be balanced. Nevertheless, the analysis is interesting, since it discloses where possible control points may be found.

Jensen and Pedersen (1990) emphasize that there is a close connection between the control of growth rate and the so-called stringent response, i.e. the cessation of stable RNA accumulation after a nutritional downshift. In contrast to earlier claims that chain elongation rates in *E. coli* are invariable, they adduce evidence that the maximal capacity of the cells to synthesize RNA and protein exceeds the capacity to generate building blocks for these processes. Also, the intracellular concentration of promoters is in excess of RNA polymerase molecules. As a consequence, there is competition between promoters for RNA polymerase. Furthermore, mRNA's are in excess of ribosomes. In fact, the rate of total protein synthesis correlates with the intracellular concentration of ribosomes. It is difficult to interpret this in terms of control: in an *E. coli* cell growing 4 times faster, the average ribosome incorporates 1.6 times as many amino acid residues per second. This means that in the slower growing cells the ribosomes function submaximally. The stringent response is accompanied by accumulation of ppGpp. This leads to an inhibition of RNA chain elongation, so that more RNA polymerase is occupied and less available for the initiation of new chains. The differential effect of the stringent response on expression of different RNA's can be easily explained in this scenario by differential affinity of promoters for RNA polymerase.

The importance of the stringent response for the metabolism at low growth rates has also been extensively discussed by Stouthamer et al. (1990). These authors conclude that at (very) low growth rates a redistribution of metabolic fluxes occurs, leading to an energetic growth yield that is different from that at higher growth rates.

The use of models of microbial growth

The MNET theory and Monod's theory are not the only ones for describing bacterial growth. Growth models can be roughly divided into two categories: the unstructured models and the structured models. In an unstructured model none of the cell's (bio)chemical reactions are taken into account. Monod's theory belongs to this category, but also some more complex theories as developed by Dabes et al. (1973), Powell (1967), Droop (1973), Roels (1980) and Erickson et al. (1978). What all these models have in common is that they give a very simplified picture of the real situation, although sometimes they can give very good fits with experimental data. However, none of these models gives insight into how the biochemical processes in the cell determine growth, and they can only be applied to steady-state situations and non-variable cell compositions. So if more complex situations are to be examined, also more complex models are needed in order to extract more detailed information. Models that do take into account all relevant variables, are called structured models. The advantage of such models is clear, but an equally clear disadvantage is their complexity. Shuler and coworkers (Shuler et al. 1979; Domach et al. 1984;

Ataai & Shuler 1987; Shu & Shuler 1989) presented a biochemically structured model for the growth of *E. coli*. They have tried to describe the properties of one single cell under ammonium or glucose limitation as the sum of its biochemical components. Joshi and Palsson (1988) simplified this model to a 'three pool model' by lumping together certain groups of reactions, and this model gave good fits with experimental data. The disadvantage of both models is of course that they can only be applied to the bacterium they have been developed for. Also they leave a number of aspects of the metabolism unclarified, like for instance ATP leakage processes.

A hybrid between a structured and an unstructured model is the already mentioned MNET theory as developed by Westerhoff & Van Dam (1987). This theory is an extension of near equilibrium thermodynamics which considers a bacterium as a black box with an input flow, an output flow, and some unspecified reactions in between that couple these two. In MNET this black box is opened and the bacterium is divided into catabolism, anabolism, and ATP flux to couple these two. The equations that come with this theory can be made as complicated as necessary for the conditions examined, and as a consequence this model can be used to predict the cell's behavior when some changes are being made.

All models mentioned above assume that the cell composition does not change with increasing growth rate, so in fact they assume that the macromolecular components, the enzyme content and the enzyme activities remain constant when the growth rate changes. That this is evidently not true can be illustrated by many examples.

It has been established long ago that the cell composition does vary with the growth rate (Herbert 1961; Bremer & Dennis 1987), so this would imply that for every new growth condition a new determination of the cell composition would be needed. Applying the MNET model the importance of this complication was evaluated by using an extended set of equations (Mulder et al. 1989), and it was shown that if the cell composition changed, one could no longer expect a linear relation between the flux through anabolism and catabolism. The equations allowed a quantitative assessment of the deviations from linearity as a result of changes in cell composition. It was shown that the dependence of growth on catabolism is rather insensitive to such changes in cell composition (Mulder 1988). For cultures grown under ammonium or glucose limitation the most extreme observed changes in cell composition would lead to less than 10% change in the relation between the flux through anabolism and catabolism. Under other conditions this change might conceivably be larger, specifically if there would be a high rate of lipid synthesis.

Concluding remarks

A solid theoretical basis for quantifying control of metabolic fluxes in microorganisms is now available. However, relatively little experimental material available in the literature can be used within this framework. To some extent this is the result of technical difficulties in obtaining the appropriate data: for the control of enzymes on fluxes it is difficult to satisfy the boundary condition of constant cell composition (apart from the relevant enzyme), for the control of substrates on fluxes it is difficult to maintain those substrates at the required (usually low) concentration while at the same time measurable fluxes occur. For the latter type of measurement, the chemostat is ideally suited, but one has to realize that in the chemostat the parameter (in this case the substrate concentration) and the variable (in this case the growth rate) have been practically interchanged.

More importantly, however, most experimentalists seem to be satisfied with non-quantitative answers. Control analysis suggests precise experiments, that give quantitative information about the role of enzymes and substrates in metabolism. For an understanding of metabolic fluxes and the factors that control them and for the rational manipulation of these fluxes, such knowledge is indispensable. The recent additions to the metabolic control theory (see Westerhoff et al. 1990; Westerhoff et al. 1992) have extended its applicability to

the more general cases, where changes in conditions are accompanied by changes in the enzymatic makeup of the cells. It is to be hoped that in this way the measured control coefficients can ultimately be related to the properties of the underlying biochemical machinery of the microbial cells.

Apart from any theory, measurement of the control coefficients of different factors on different processes gives a quantitative insight in the importance of each factor in determining the overall performance of that process. It is important to remember that in the measurement of such a control coefficient an exact definition of process and factor is required. Thus, in speaking of 'energy limitation' one might want to know what is the 'control coefficient of energy on the growth rate'; for this, one would have to define 'energy'. Theoretically it is possible to determine the control coefficient of ATP or even of the phosphate potential on growth; however, to determine that control coefficient one would have to be able to fix that (intracellular) parameter at the relevant values. Similarly, in speaking of 'carbon limitation' one might want to know the 'control coefficient of the carbon source on the growth rate'; again, one would have to define 'carbon' – most likely by defining the particular form of carbon that is limiting, for example the carbon-containing growth substrate.

The experimental material that *is* available supports some of the general notions that followed from the control theory. For instance, often the control of one particular enzyme on a flux is low. This follows from the fact that in complex pathways the control is distributed over all the constituting enzymes. Only in rare cases will the control of, for instance, the uptake system on growth be close to 1, as implied in the original Monod theory.

Furthermore, the control of a particular enzyme or substrate depends on the pathway and condition considered. The same factor that may have strong control in one condition may have no significant control in another condition.

Thus, in analyzing microbial metabolism one will have to determine many different control coefficients. It is, however, our conviction that only by such painstaking quantitative work real insight in the functioning of the complex metabolic network of cells will be gained.

Acknowledgement

We thank Drs Postma, Rutgers and Westerhoff for constructive criticism during the preparation of this manuscript.

References

Aiking H & Tempest DW (1976) Growth and physiology of *Candida utilis*NCYC321 in potassium-limited chemostat culture. Arch. Microbiol. 108: 117–124

Ataai MM & Shuler ML (1987) A mathematical model for prediction of plasmid copy number and genetic stability in *Escherichia coli*. Biotechnol. Bioeng. 30: 389–397

Birkenhead K, Manian SS & O'Gara F (1988) Dicarboxylic acid transport in *Bradyrhizobium japonicum*: use of *Rhizobium meliloti dct* gene(s) to enhance nitrogen fixation. J. Bact. 170: 184–189

Bremer H & Dennis PP (1987) Modulation of chemical composition and other parameters of the cell by growth rate. In: Neidhart FC (Ed) *Escherichia coli* and *Salmonella typhimurium*: Cellular and Molecular Biology, Vol 2 (pp 1527–1542). ASM, Washington DC

Brinkman U, Mueller RH & Babel W (1990) The growth rate-limiting reaction in methanol-assimilating yeasts. FEMS Microbiol. Rev. 87: 261–266

Buurman ET, Pennock J, Tempest DW, Teixeira de Mattos MJ & Neijssel OM (1989) Replacement of potassium ions by ammonium ions in different microorganisms grown in potassium-limited chemostat culture. Arch. Microbiol. 152: 58–63

Dabes JN, Finn RK & Wilke CR (1973) Equations of substrate-limited growth. Case for Blackman kinetics. Biotech. Bioeng. 15: 1159–1177

Dean AM, Dykhuizen DE & Hartl DL (1986) Fitness as a function of β-galactosidase activity in *Escherichia coli*. Genet. Res. Camb. 48: 1–8

Domach MM, Leung SK, Cahn RE, Cocks GG & Shuler ML (1984) Computer model for glucose-limited growth of a single cell of *Escherichia coli* B/r-A. Biotechnol. Bioeng. 26: 203–216

Droop MR (1973) Some thoughts on nutrient limitation in algae. J. Phycol. 9: 264–272

Dykhuizen DE, Dean AM & Hartl D (1987) Metabolic fluxes and fitness. Genetics 115: 25–31

Erickson LE, Minkevich IG & Eroshin VK (1978) Utilization of mass-energy balance regularities in the analysis of continuous-culture data. Biotech. Bioeng. 20: 1595–1621

Flint HJ, Porteous DJ & Kacser H (1980) Control of the flux in the arginine pathway of *Neurospora crassa*. Biochem. J. 190: 1–15

Flint HJ, Tateson RW, Barthelmess IB, Porteous DJ, Donachie WD & Kacser H (1981) Control of the flux in the arginine pathway of *Neurospora crassa*. Biochem. J. 200: 231–246

Galazzo JL & Bailey JE (1990) Fermentation pathway kinetics and metabolic flux control in suspended and immobilized *Saccharomyces cerevisiae*. Enzyme Microb. Technol. 12: 162–172

Groen AK (1984) Quantification of control studies on intermediary metabolism. PhD Thesis, University of Amsterdam

Hartl DL, Dykhuizen DE & Dean A (1985) Limits of adaption: the evolution of selective neutrality. Genetics 111: 655–674

Heinrich R & Rapoport TA (1973) Linear theory of enzymatic chains: its application for the catalysis of the crossover theorem and of the glycolysis of human erythrocytes. Acta Biol. Med. Germ. 31: 479–494

Herbert D (1961) The chemical composition of micro-organisms as a function of their environment. In: Meynell CG & Gooder H (Eds) Microbial Reaction to the Environment. Symp. Soc. Gen. Microbiol. 11: 391–416

Iwami Y & Yamada T (1985) Regulation of glycolytic rate in *Streptococcus sanguis* grown under glucose-limited and glucose-excess conditions in a chemostat. Infect. Immun. 50: 378–381

Jensen KF & Pedersen S (1990) Metabolic growth rate control in *Escherichia coli* may be a consequence of subsaturation of the macromolecular biosynthetic apparatus with substrates and catalytic components. Microbiol. Rev. 54: 89–100

Jensen PR (1991) Growth physiology of *Escherichia coli*strains with variable expression of the *atp* operon. PhD Thesis, Danmarks Tekniske Hojskole

Joshi A & Palsson BO (1988) *Escherichia coli* growth dynamics: a three-pool biochemically based description. Biotech. Bioeng. 31: 102–116

Kacser H & Burns JA (1973) Control of [enzyme] flux. In: Davies DD (Ed) Rate Control of Biological Processes (pp 65–104). Cambridge University Press

Kell DB, Westerhoff HV & van Dam K (1989) Control analysis of microbial growth and productivity. In: 44th Symp. Soc. Gen. Microbiol, Baumberg S, Hunter L & Rhodes M (Eds) (pp 61–93). Cambridge University Press

Kleman GL, Chalmers JJ, Luli GW & Strohl WR (1991) Glucose-stat, a glucose-controlled continuous culture. Appl. Environ. Microbiol. 57: 918–923

Marr AG (1991) Growth rate of *Escherichia coli*. Microbiol. Rev. 55: 316–333

Minkevich IG, Krinitzkaya AY & Eroshin VK (1988) A double substrate limitation zone of continuous microbial growth. In: Kyslic P, Dawes EA, Klumphanzl V & Novak M (Eds) Continuous Culture (pp 171–184). Academic Press, London

Monod J (1942) Recherches sur la croissance des cultures bacteriennes. Herman et Cie, Paris

Mulder MM (1988) Energetic aspects of bacterial growth; a mosaic non-equilibrium thermodynamic approach. PhD thesis, University of Amsterdam

Mulder MM, van der Gulden HML, Postma PW & van Dam K

(1989) Macromolecular composition of *Klebsiella aerogenes*NCTC 418 under glucose- and ammonia-limiteed conditions in continuous culture. Biochim. Biophys. Acta 936: 406–412

Neijssel OM, Hueting S & Tempest DW (1977) Glucose transport capacity is not the rate-limiting step in the growth of some wild-type strains of *Escherichia coli* and *Klebsiella aerogenes* in chemostat culture. FEMS Microbiol. Lett. 2: 1–3

O'Brien RW, Neijssel OM & Tempest DW (1980) Glucose: phosphoenolpyruvate phosphotransferase activity and glucose uptake rate of *Klebsiella aerogenes* growing in chemostat culture. J. Gen. Microbiol. 116: 305–314

Owens JD & Legan JD (1987) Determination of the Monod substrate saturation constant for microbial growth. FEMS Micobiol. Rev. 46: 419–432

Poolman B, Bosman B, Kiers J & Konings WN (1987) Control of glycolysis by glyceraldehyde-3-phosphate dehydrogenase in *Streptococcus cremoris* and *Streptococcus lactis*. J. Bact. 169: 5887–5890

Poolman B & Konings WN (1988) Relation of growth of *Streptococcus lactis* and *Streptococcus cremoris*to amino acid transport. J. Bact. 170: 700–707

Portillo F & Serrano R (1989) Growth control strength and active site of yeast plasma membrane ATPase studied by site-directed mutagenesis. Eur. J. Biochem. 186: 501–507

Postma E, Scheffers WA & van Dijken JP (1988) Adaptation of the kinetics of glucose transport to environmental conditions in the yeast *Candida utilis* CBS621: a continuous culture study. J. Gen. Microbiol. 134: 1109–1116

Postma PW & Lengeler J (1985) Phosphoenolpyruvate: carbohydrate phopsphotransferase system of bacteria. Microbiol. Rev. 49: 232–269

Powell EO (1967) Microbial physiology and continuous culture; In: Powell EO, Evans CGT, Strange RE & Tempest DW (Eds) Proceedings of the Third Int.Symp. (pp 34–55). HMSO, London

Roels JA (1980) Bioengineering report. Application of macroscopic principles to microbial metabolism. Biotech. Bioeng. 22: 2457–2514

Ruijter G, Postma PW & van Dam K (1991) Control on glucose metabolism by Enzyme II[Glc] of the phosphoenolpyruvate-dependent phosphotransferase system in *Escherichia coli*. J. Bacteriol. (in press)

Rutgers M (1990) Control and thermodynamics of microbial growth PhD Thesis, University of Amsterdam

Rutgers M, Teixeira de Mattos MJ, Postma PW & van Dam K (1987) Establishment of the steady state in glucose-limited chemostat cultures of *Klebsiella pneumoniae*. J. Gen. Microbiol. 133: 445–451

Rutgers M, Balk PA & van Dam K (1989) Effect of concentration of substrates and products on the growth of *Klebsiella pneumoniae* in chemostat cultures. Biochim. Biophys. Acta 977: 142–149

Rutgers M, Balk PA & van Dam K (1990) Quantification of multiple-substrate controlled growth. Simultaneous ammonium and glucose limitation in chemostat culturs of *Klebsiella pneumoniae*. Arch. Microbiol. 153: 478–484

Schaaff I, Heinisch J & Zimmermann FK (1989) Overproduction of glycolytic enzymes in yeast. Yeast 5: 285–290

Schulze KL & Lipe RS (1964) Relationship between substrate concentration, growth rate and respiration rate of *Escherichia coli* in continuous culture. Archiv. für Mikrobiol. 48: 1–20

Senn HP (1989) Kinetik und Regulation des Zuckerabbaus von *Escherichia coli* ML30 bei tiefen Zuckerkonzentrationen; PhD thesis, ETH Zürich

Shu J & Shuler ML (1989) A mathematical model for the growth of a single cell of *E.coli* on a glucose/glutamine/ammonium medium. Biotechnol. Bioeng. 33: 1117–1126

Shuler ML, Leung S & Dick CC (1979) A mathematical model for the growth of a single bacterial cell. Ann. N. Y. Acad. Sci. 326: 35–55

Stouthamer AH (1973) A theoretical study on the amount of ATP required for synthesis of microbial cell material. Antonie van Leeuwenhoek 39: 545–565

Stouthamer AH, Bulthuis B & Van Verseveld HW (1990) Energetics of growth at low growth rates and its relevance for the maintenance concept. In: Poole RK, Bazin MJ & Keevil CW (Eds) Microbial Growth Dynamics, Vol 28 (pp 85–102). SGM Publications

Stueland CS, Gorden K & LaPorte DC (1988) The isocitrate dehydrogenase phosphorylation cycle; identification of the primary rate-limiting step. J. Biol. Chem. 263: 19475–19479

Valinger R, Braus G, Niederberger P, Künzler M, Paravicini G, Schmidheini T & Hütter R (1989) Cloning of the *LEU2* gene of *Saccharomyces cerevisiae* by in vivo recombination. Arch. Microbiol. 152: 263–268

Walsh K & Koshland D (1985) Characterization of rate-controlling steps in vivo by use of an adjustable expression vector. J. Biol. Chem. 82: 3577–3581

Westerhoff HV, Lolkema JS, Otto R & Hellingwerf KJ (1982) Thermodynamics of bacterial growth. The phenomenological and the mosaic approach. Biochim. Biophys. Acta 683: 181–220

Westerhoff HV & van Dam K (1987) Thermodynamics and Control of Biological Free-energy Transduction. Elsevier, Amsterdam

Westerhoff HV, van Heeswijk W, Kahn D & Kell DB (1992) Antonie van Leeuwenhoek 60 (this issue)

Westerhoff HV, Koster JG, van Workum M & Rudd KE (1990) On the control of gene expression. In: Cornish-Bowden A & Luz Cardenas M (Eds) Control of Metabolic Processes. NATO ASI Series A: Life Sciences, Vol 190 (pp 399–412)

Antonie van Leeuwenhoek **60**: 225–234, 1991.
© 1991 *Kluwer Academic Publishers. Printed in the Netherlands.*

On multiple-nutrient-limited growth of microorganisms, with special reference to dual limitation by carbon and nitrogen substrates

Thomas Egli
Swiss Federal Institute for Water Resources and Water Pollution Control (EAWAG), Überlandstrasse 133, CH-8600 Dübendorf, Switzerland

Key words: chemostat, growth yield, growth limitation, multiple nutrient limitation, carbon, nitrogen

Abstract

Simultaneous limitation of microbial growth by two or more nutrients is discussed for dual carbon/nitrogen-limited growth in continuous culture. The boundaries of the zone where double-limited growth occurs can be clearly defined from both cultivation data and cellular composition and they can be also predicted from growth yield data measured under single-substrate-limited conditions. It is demonstrated that for the two nutrients carbon and nitrogen the zone of double nutrient limitation is dependent on both the C : N ratio of the growth medium and the growth (dilution) rate. The concept on double-(carbon/nitrogen)-limited growth presented here can be extended to other binary and multiple combinations of nutrients.

Introduction

The use of defined mineral media for the growth of microorganisms was a major development in microbiology. In combination with advances in cultivation techniques, e.g. the chemostat culture, it enabled the reproducible growth of microbes in selected environments under defined conditions and to study the effects of specific nutrients on various physiological phenomena. As a general rule, defined mineral media have been formulated so as to allow microbes to synthesize their cellular components from single sources of carbon, nitrogen, phosphorus, etc. Additionally, it became common practice for the relative concentrations of the individual nutrients to be adjusted such that, as formulated in the 'Law of the minimum' by Justus von Liebig (1840), only one of them, usually the carbon source, restricted the maximum quantity of biomass that could be produced, with all other nutrients in excess (Monod 1942; Stephenson 1949).

In contrast to the conditions which microbes are exposed to in laboratory cultures, the environments encountered in Nature are entirely different. There, microbes have to cope with both the simultaneous presence and low concentrations of a multiplicity of homologous nutrients, i.e., compounds that can satisfy the same physiological function. This is especially true for carbon, nitrogen and phosphorus sources. In fact, it has been pointed out that the presumption that limitation of growth in Nature is limited by either a single substrate or nutrient is probably an erroneous assumption (Veldkamp & Jannasch 1972; Pearl 1977). Although, considerable progress has been made in recent years with respect to understanding the simultaneous growth on either two carbon sources or two electron acceptors (Harder & Dijkhuizen 1976; Harder & Dijkhuizen 1982; Egli et al. 1986; Robertson & Kuenen 1990), there have been few reports concerning the simultaneous limitation by multiple non-homologous nutrients, as discussed by Harrison (1972) and Baltzis & Fredrickson

(1988). Consequently, most microbiologists still support the view that only one compound can be growth-limiting at any particular time.

In this contribution evidence supporting the existence of well-defined multiple-substrate-limited growth regimes will be evaluated. Due to the paucity of data, examples of the simultaneous limitation of growth by two macro-nutrients, carbon and nitrogen, will be emphasized, but it will become clear that the principle of multiple-nutrient-limited growth is equally applicable to a wide range of other nutrient limitations. The influence that simultaneous growth limitation by carbon and nitrogen has on cell composition and cell physiology will be examined and a working hypothesis for multiple-nutrient-limited growth of microbes will be presented.

The phenomenon of multiple-nutrient-limited growth

First hints that growth of phytoplankton may be limited by several factors simultaneously were reported for oligotrophic freshwater systems (summarized in Pearl 1977). With respect to microorganisms, preliminary experimental data suggesting the occurrence of simultaneous limitation of growth by two heterologous nutrients was reported by Cooney et al. (1976) for nitrogen and phosphorus, Hueting & Tempest (1979) for carbon and nitrogen and for carbon and potassium, Harrison (1972) and Hamer et al. (1975) for gaseous nutrients, and by Egli (1982) for carbon and nitrogen. However, it was only recently, that the existence of a double carbon/nitrogen-limited growth regime under chemostat culture conditions was clearly demonstrated for both bacteria and yeast on the basis of both the culture parameters and the physiological characteristics of the growing cells (Egli & Quayle 1984; Egli & Quayle 1986; Gräzer-Lampert et al. 1986; Minkevich et al. 1988; Duchars & Attwood 1989; Rutgers et al. 1990).

A typical example for the response generally observed as a function of the C : N ratio in the feed medium is illustrated in Fig. 1 for a culture of *Hyphomicrobium* ZV620 growing at a constant di-

lution rate of $0.054\,h^{-1}$ in a chemostat with methanol/NH_4^+ as the carbon/nitrogen sources. In this experiment the concentration of the nitrogen source (NH_4^+) in the medium reservoir was kept constant at $223\,mg\,l^{-1}\,NH_4^+ - N$, whilst the concentration of methanol was increased stepwise from $0.95\,g\,l^{-1}$ (C : N = 1.6) to $12.4\,g\,l^{-1}$ (C : N = 20.85). Judging from the residual concentrations of methanol and NH_4^+, the cellular composition and the synthesis of NH_4^+-assimilating enzymes, three distinct growth regimes were recognized:

- methanol limitation where ammonia was in excess (C : N < 7.1)
- ammonia limitation were methanol was in excess (C : N > 12.6)
- a transition regime where both methanol and NH_4^+ were below the detection limit (7.1 < C : N < 12.6)

It is striking that, although apparently growing nitrogen-limited at C : N > 7.1, where no residual nitrogen was detectable, the dry weight of the culture still increased linearly when additional carbon substrate was added to the growth medium (Fig. 1A). When growth was limited by a single nutrient only, the cells exhibited essentially a constant overall cellular composition with respect to N, protein and storage product contents (Fig. 1B). Within the transition growth regime cell composition was extremely dependent on the ratio of the two limiting substrates and it was adjusted according to the actual availability of the two nutrients which restricted the synthesis of cell material. It was calculated that the storage material poly-β-hydroxybutyrate (PHB) synthesized under these conditions accounted for up to 90% of the additional biomass produced (Gräzer-Lampert et al. 1986). Generalizing, one can conclude from the data that during carbon-limited growth, the content of protein and probably also of nucleic acids is high and the storage product content is low, whereas the opposite is true for cells grown under nitrogen limitation. This finding is not new and has been reported for many microbes although, as far as the author is aware, overall cell composition was always implied rather

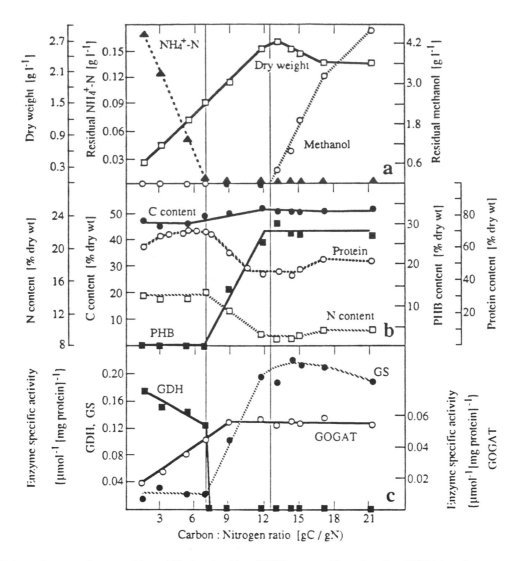

Fig. 1. Cellular and enzymatic composition of *Hyphomicrobium* ZV620 grown with methanol and NH_4^+ in a chemostat at a constant dilution rate, $D = 0.054\,h^{-1}$, as a function of the carbon : nitrogen ratio in the feed [$S_0(NH_4^+ - N)$ was held constant, S_0(methanol) was increased]. Abbreviations: GS, glutamine synthetase, GOGAT, glutamate synthase; GDH, glutamate dehydrogenase; PHB, poly-β-hydroxybutyrate. Data from Gräzer-Lampert et al. (1986).

than shown to be independent of the medium C : N ratio.

As a response to the growth conditions described above, the protein composition of cells of *Hyphomicrobium* ZV620 was not only affected quantitatively, but also qualitatively. The data in Fig. 1C demonstrate that in this bacterium, ammonia was assimilated mainly via the low affinity glutamate dehydrogenase system under nitrogen excess growth conditions, whereas, as soon as the availability of ammonia became restricted, the high af-

finity glutamine synthetase/glutamate synthase system took over. It was surprising to see how differently synthesis of the three enzymes involved was regulated. Gräzer-Lampert et al. (1986) also showed that not only the concentration of the total glutamine synthetase protein increased during the change from C- to N-limited conditions, but also the fraction of deadenylated, i.e., active glutamine synthetase increased.

A very similar behaviour with respect to cell composition and enzyme regulation has been reported for the methylotrophic yeast *Hansenula polymorpha* during chemostat growth at a constant dilution rate of $0.10\,h^{-1}$ with media containing different C : N ratios, where a C/N-limited transition growth regime was observed between $12 < C : N < 31$ (Egli & Quayle 1986). Recently, Minkevich et al. (1988), using the same experimental approach, reported the existence of a double carbon/nitrogen-limited growth regime for the yeast *Candida valida*, Duchars & Attwood (1989) for *Hyphomicrobium* X, Al-Awadhi et al. (1990) for the growth of a thermotolerant methylotrophic *Bacillus* sp., and Rutgers et al. (1990) for *Klebsiella pneumoniae*. All these authors observed a virtually identical pattern of behaviour for the cellular composition of these organisms as that described above for *Hyphomicrobium*. From all this experimental evidence for different microorganisms, one can conclude that the existence of a transition, i.e., a double-substrate-limited, growth regime between two distinct single-nutrient-limited zones for growth is a general phenomenon.

Prediction of range of double-substrate-limited growth

Accepting the existence of the phenomenon prompts the question of whether the range where growth is limited by restrictions in the availability of two (or more) nutrients simultaneously can be predicted. From their experimental data, Egli & Quayle (1986) demonstrated that the bounderies of the three growth regimes at a certain fixed growth rate could be predicted when the growth yields for the individual substrates are known.

The empirical equation for the concentration of biomass in the culture (x), as first proposed by Egli & Quayle (1986) for the two nutrients carbon and nitrogen, can be deduced from the conceptual scheme shown in Fig. 2, i.e.,

$$x = (c_0 - c) \cdot Y_{X/C} = (n_0 - n) \cdot Y_{X/N} \tag{1}$$

where c_0 and n_0 are the concentrations of carbon and nitrogen in the chemostat feed, c and n are the corresponding residual concentrations of the two

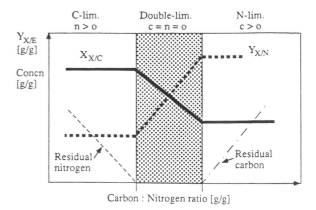

Fig. 2. Conceptual diagram showing the dependence of the growth yield for carbon ($Y_{X/C}$) and nitrogen ($Y_{X/N}$) on the carbon : nitrogen ratio in the growth medium supplied. It is assumed that growth takes place in a chemostat at a constant dilution rate. The residual concentrations of the carbon (c) and the nitrogen (n) source, respectively, in the culture are shown. The yield is given as dry biomass formed per g of element ($Y_{X/E}$).

nutrients and $Y_{X/C}$ and $Y_{X/N}$ are the biomass growth yield coefficients based on carbon and nitrogen. Because both c and n approach zero under carbon/nitrogen-limited growth, equation (1) can be rewritten in terms of the C : N ratio in the inflowing medium, i.e.,

$$c_0/n_0 = Y_{X/N}/Y_{X/C} \tag{2}$$

Assuming constant yield coefficients under single-substrate-limited growth conditions, the boundary between carbon limitation and carbon/nitrogen limitation can be calculated by substituting values for $Y_{X/N}$ and $Y_{X/C}$ values measured under carbon-limited growth conditions into equation (2). Similarly, substituting $Y_{X/N}$ and $Y_{X/C}$ measured for nitro-

Table 1. Predicted and experimentally observed boundaries for single and double-substrate(carbon/nitrogen)-limited growth of *Hyphomicrobium* ZV620 with methanol and NH_4^+ at a constant dilution rate of $0.054\,h^{-1}$ in the chemostat.

	$Y_{X/C}$	$Y_{X/N}$	C : N predicted	C : N observed
C-limited	0.99	7.19	7.2	7.1
N-limited	0.80	10.30	12.8	12.6

Experimental data from Gräzer-Lampert et al. (1986).

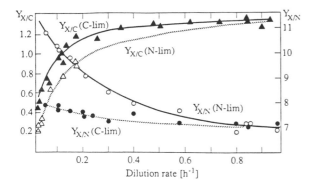

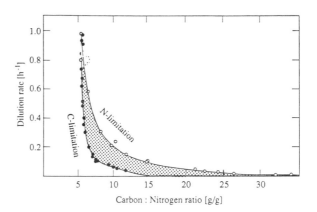

Fig. 3. Growth yield coefficients for carbon ($Y_{X/C}$) and nitrogen ($Y_{X/N}$) for a culture of *Klebsiella pneumoniae* growing in the chemostat either carbon- or nitrogen-limited. Glycerol and NH_4^+ were used as sole sources of carbon and nitrogen, respectively. Each yield coefficient is given in g dry biomass produced per g of element. Data from Tempest et al. (1967), Herbert (1976) and Tempest & Neijssel (1978).

Fig. 4. Dependence of the nature of growth limitation on the carbon : nitrogen ratio of the growth medium supplied and the growth rate for a culture of *Klebsiella pneumoniae* growing in a chemostat with glycerol and NH_4^+ as the sole sources of carbon and nitrogen.

gen-limited grown cells would result in the medium C : N ratio where the culture switches from carbon/nitrogen-limited to nitrogen-limited growth. Applying equation (2) to the growth of *Hyphomicrobium* ZV620 with methanol and NH_4^+ as carbon and nitrogen sources, the values calculated for the two boundaries (Table 1) correspond closely with those found in the experiment shown in Fig. 1.

Equation (1) implies that in order to obtain a zone of double-substrate-limited growth with substrates S_1 and S_2, the growth yield coefficients Y_{X/S_1} and/or Y_{X/S_2} have to be different under S_1-limited and S_2-limited growth conditions, respectively. Additionally, the equation also indicates that the wider the range within which the growth yield coefficients for the two substrates can vary, the more extended the double-substrate-limited growth regime will be. Egli & Quayle (1986) also predicted that the transition growth regime would become narrower with increasing growth rates because differences in cellular composition and growth yield are markedly more pronounced in slow growing cells (Herbert 1976). This prediction was recently confirmed for a culture of *C. valida* growing at three different dilution rates with ethanol and NH_4^+ as carbon and nitrogen sources (Minkevich et al. 1988).

Applying the concept to literature data

The simplicity of the method discussed for calculating the different boundaries of growth limitation regimes prompted further examination of the literature. If the growth yield coefficients for two nutrients S_1 and S_2 are known for S_1- and S_2-limited growth at different dilution rates in chemostat culture, it would be possible to predict the zone of double S_1/S_2-limited growth for microbes as a function of their growth rates.

One of the best documented examples found in the literature was that for *K. pneumoniae* growing in a synthetic medium with glycerol and ammonia as the only sources of carbon and nitrogen, respectively (Fig. 3). Although the data were derived from different sources (Tempest et al. 1967; Herbert 1976; Tempest & Neijssel 1978) and, in some cases, required recalculation and transformation, they clearly demonstrate the influence of growth rate and the nature of growth limitation on biomass yield coefficients based on carbon and nitrogen. Data for $Y_{X/C}$(glycerol) during NH_4^+-limited growth were scarce, especially at high growth rates and showed considerable scatter. Nevertheless, because at the maximum specific growth rate $Y_{X/C}$(N-lim) is theoretically identical to $Y_{X/C}$(C-lim), the values of $Y_{X/C}$ for nitrogen-limited growth can be interpolated for dilution rates higher than $0.2\,h^{-1}$.

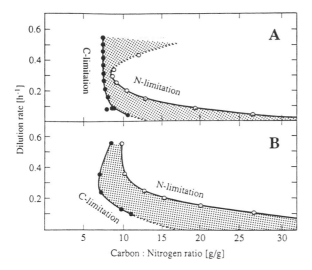

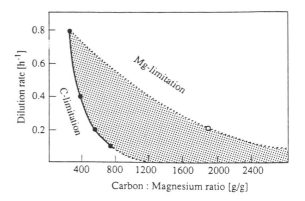

Fig. 6. Predicted dependence of the nature of growth limitation on the carbon : magnesium ratio of the growth medium supplied and the dilution rate for chemostat growth of *Klebsiella pneumoniae* with glucose and magnesium. Calculation of boundaries is based on data from Strange & Hunter (1976), Tempest & Dicks (1967) and Tempest & Neijssel (1978).

Fig. 5. Dependence of the nature of growth limitation on the carbon : nitrogen ratio of the growth medium supplied and the growth rate for chemostat growth of (A) *Candida utilis* with glucose and NH_4^+ and (B) *Candida valida* with ethanol and NH_4^+. Calculation of boundaries for the growth of *C. utilis* is based on data from Herbert (1976) and Aiking & Tempest (1977). Boundaries for *C. valida* are adapted from Minkevich et al. 1988.

From this set of data c_0/n_0 ratios were calculated and the corresponding zone boundaries for growth limitation obtained are shown as a function of growth rate in Fig. 4. As predicted, the extension of the carbon/nitrogen-limited growth regime for a culture of *K. pneumoniae* is strongly dependent on the imposed growth rate.

For example, $D = 0.8\,h^{-1}$, which is close to the maximum specific growth rate, simultaneous limitation by carbon and nitrogen should be found during growth with media of C : N ratios in between $5.4 - 5.6$ and would, therefore, be difficult to detect experimentally. However, at $D = 0.05\,h^{-1}$ double limitation should extend over a much wider range from a C : N ratio of 10.6 to 20.6. The zone of double limitation not only becomes wider with decreasing growth rates but is, at the same time, moved towards higher C : N ratios. As a consequence, if the bacterium was grown in a medium with a C : N ratio higher than approximately 5.5, growth should become nitrogen-limited at higher dilution rates. With the same medium, the

culture would be carbon-limited at growth rates lower than approximately $0.5\,h^{-1}$.

Similar boundaries for carbon/nitrogen-limited growth of the two yeasts *C. utilis* with glucose/NH_4^+ and *C. valida* with ethanol/NH_4^+ (Fig. 5) were calculated from chemostat data reported in the literature (Herbert 1976; Aiking 1977; Minkevich et al. 1988). In the case of *C. utilis* (Fig. 5A) the data reported for growth yield coefficients under nitrogen-limited conditions at dilution rates $> 0.3\,h^{-1}$ were scarce and they suggest that at high growth rates the zone of C/N-limited growth would become wider again. This could be explained by the extensive excretion of carbonaceous products under such conditions. In contrast, for the growth of *C. valida* with ethanol and ammonia the calculations predicted a zone of double limitation comparable to that for *K. pneumoniae*.

From the evidence presented, it can be envisaged that zones of double-substrate-limited growth will also occur for chemostat growth of microorganisms with substrate combinations other than carbon and nitrogen. Unfortunately, few reliable data were found in the literature on growth yield coefficients at different growth rates for substrate combinations such as, e.g., carbon/magnesium, carbon/phosphorus or carbon/sulfur. Especially data on growth yield coefficients for carbon under carbon excess conditions were rarely reported

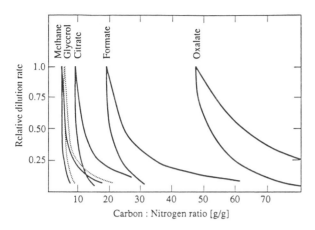

Fig. 7. Influence of growth yield (or degree of reduction of substrates) on the position of boundaries of the double-C/N-limited growth regime for a hypothetical microorganism. For calculation, maximum growth yield factors were taken from the literature (Linton & Stephenson 1978) and the same pattern for cell composition and maintenance as shown in Fig. 3 for *Klebsiella pneumoniae* was applied. Maximum $Y_{X/C}$ values used for calculation were for methane 1.50, for glycerol 1.35, for citrate 0.80, for formate 0.38 and for oxalate 0.15.

in the literature examined. The best documented example is that of growth of *K. pneumoniae* with glycerol and magnesium under both carbon and magnesium limitation (Strange & Hunter 1967; Tempest & Dicks 1967; Tempest & Neijssel 1982) and the boundaries of the zone for double-(carbon/magnesium)-limited growth predicted from the data are shown in Fig. 6. Unfortunately, in this example the boundary from carbon/magnesium-limited to magnesium-limited growth conditions is documented at one single dilution rate only and more data are clearly needed. Nevertheless, the emerging outline of the shape of the double-substrate-limited zone can be be predicted to be even more extended than that for carbon/nitrogen-limited growth of this bacterium.

Some conclusions and outlook

With respect to the two nutrients under consideration, one can conclude from the information presented here that the position and shape of the carbon/nitrogen-limited growth regime is mainly influenced by two parameters: growth yield coeffi-

cients and the plasticity of cell composition with respect to carbonaceous and nitrogenous cell constituents. Obviously, the influence of growth yield implies that for highly oxidized substrates, i.e., substrates giving low growth yield coefficients (Linton & Stephenson 1978) this regime will be shifted towards higher C : N medium ratios. Furthermore, at low dilution rates endogenous activity (maintenance) together with death/lysis and cryptic growth are the parameters which will determine the shape of the zone of double-nutrient limitation and the shift towards higher medium C : N ratios with decreasing growth rates. The effect of growth with substrates of different degrees of reduction is demonstrated in Fig. 7.

The results reported here with respect to the two macro-nutrients, carbon and nitrogen, clearly demonstrate that growth of microbes in continuous culture is frequently not only limited by one single nutrient at any particular time, but can also be limited by two (or more) nutrients simultaneously. The zone of double nutrient limitation is dependent on the C : N ratio of the growth medium and the boundaries of the two independent single-substrate-limited growth regimes can be predicted from the growth yield coefficients based on carbon and nitrogen, respectively, measured during single-substrate-limited growth. It is envisaged that the concept of double-nutrient-limited growth presented here for carbon and nitrogen can be extended to other combinations of nutrients. Cells growing within such double-nutrient-limited zones are characterized by a potentially variable cell composition with respect to both the structural components and the enzymes expressed, which results in increased metabolic flexibility and versatility. The data reported by Wanner & Egli (1990) on the patterns of growth and changes of cellular composition of *K. pneumoniae* during limitation by different nutrients in batch culture suggests that extensively extended zones of double-substrate-limited growth for nutrient combinations such as carbon/phosphorus and carbon/potassium can be expected at low growth rates.

From the data presented it can be assumed that microbial growth in the chemostat can be limited by more than two nutrients simultaneously. As an

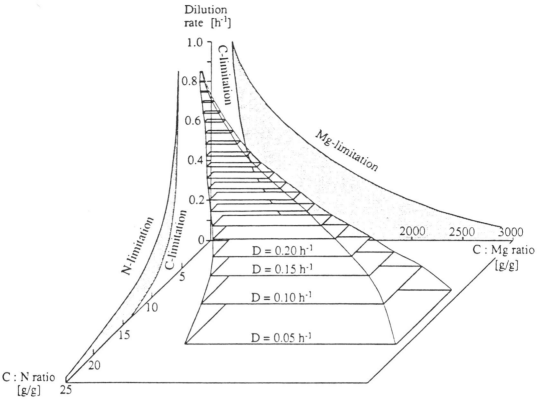

Fig. 8. Predicted zone of triple nutrient (carbon, nitrogen and magnesium) limitation for growth of *Klebsiella pneumoniae* in chemostat culture, as a function of dilution rate. Calculation of boundaries is based on data from Strange & Hunter (1967), Tempest & Dicks (1967) and Tempest & Neijssel (1978).

example, the theoretical position of the zone of triple-substrate-limited growth by carbon, nitrogen and magnesium at different dilution rates has been calculated for *K. aerogenes* from literature data (Fig. 8). The data clearly predict an extended zone of simultaneous limitation by the three nutrients at low growth rates.

The concept presented here on double-substrate-limited growth allows several conclusions to be drawn concerning growth of microorganisms under both laboratory and environmental conditions. First, with respect to growth with carbon and nitrogen sources, one can predict that studies on nitrogen-limited growth at low growth rates may require rather higher ratios of C : N in the medium than are commonly used, i.e., the lower the dilution rate, the higher the medium C : N ratio needed to ensure distinctly nitrogen-limited growth conditions. Additionally, it can be concluded that tests to

confirm that growth is carbon-limited should be performed at high dilution rates, whereas low dilution rates should be employed when testing for nitrogen limitation. Also, the fact that in the example presented in Fig. 1A the dry weight in the culture still increased linearly in the carbon/nitrogen-limited zone indicates that the traditional procedure of assessing the limiting factor in a growth medium via the amount of biomass produced can lead to ambiguous results.

From a microbial ecologist's viewpoint, the shape of the limitation zones in Fig. 4 indicate the crucial importance of the growth yield coefficient on the regime of growth limitation under natural environmental conditions. Considering that growth rates of bacteria in ecosystems are low, i.e., frequently in the range of $0.001–0.01\,h^{-1}$, for extended periods of time (Moriarty 1986), then in the case of the two macro-nutrients carbon and nitrogen, environmental C : N ratios have to be rather high before microbial growth becomes fully nitrogen-limited.

With respect to wastewater treatment, the existence of a double-substrate-limited zone provokes the speculation that simultaneous removal of, e.g., carbonaceous and nitrogenous compounds in the double-limited zone should be possible. Such an application can be envisaged for the treatment of industrial wastewaters where it is occasionally possible to adjust the ratio of different nutrients by mixing wastewaters from different processes.

Although the concept of double nutrient limitation still needs further examination, it is hoped that this contribution will add a shade of grey to our black and white thinking concerning the nutrient-limited growth of microbes.

Acknowledgements

The author is indepted to G. Hamer for his continued interest in this work and the many stimulating discussions, to C.A. Mason for his help during the preparation of the manuscript and to H. Bolliger for drawing the figures. A preliminary version of this article was published recently (Egli & Schmidt 1990).

References

Aiking H & Tempest DW (1977) Rubidium as a probe for function and transport of potassium in the yeast *Candida utilis* NCYC 321, grown in chemostat culture. Arch. Microbiol. 115: 215–221

Al-Awadhi N, Egli T, Hamer G & Mason CA (1990) The process utility of thermotolerant methylotrophic bacteria: I. An evaluation in chemostat culture. Biotechnol. Bioeng. 36: 816–820

Baltzis BC & Fredrickson AG (1988) Limitation of growth rate by two complementary nutrients: some elementary but neglected considerations. Biotechnol. Bioeng. 31: 75–86

Cooney C, Wang DIC & Mateles RI (1976) Growth of *Enterobacter aerogenes* in a chemostat with double nutrient limitations. Appl. Environ. Microbiol. 31: 91–98

Duchars MG & Attwood MM (1989) The influence of the carbon : nitrogen ratio of the growth medium on the cellular

composition and regulation of enzyme activity in *Hyphomicrobioum* X. J. Gen. Microbiol. 135: 787–793

Egli T (1982) Regulation of protein synthesis in methylotrophic yeasts: repression of methanol dissimilating enzymes by nitrogen limitation. Arch. Microbiol. 131: 95–101

Egli T & Quayle JR (1984) Influence of the carbon : nitrogen ratio on the utilization of mixed carbon substrates by the methylotrophic yeast *Hansenula polymorpha*. Proceedings of the 100th Annual Meeting of the Society for General Microbiology, Warwick (p M13)

— (1986) Influence of the carbon : nitrogen ratio of the growth medium on the cellular composition and the ability of the methylotrophic yeast *Hansenula polymorpha* to utilize mixed carbon sources. J. Gen. Microbiol. 132: 1779–1788

Egli T & Schmidt CR (1990) Dual-nutrient-limited growth of microbes, with special reference to carbon and nitrogen substrates. In: Hamer G, Egli T & Snozzi M (Eds) Mixed and Multiple Substrates and Feedstocks (pp 165–167). Hartung-Gorre, Constance

Egli, T, Bosshard C & Hamer G (1986) Simultaneous utilization of methanol-glucose mixtures by *Hansenula polymorpha* in chemostat: influence of dilution rate and mixture composition on utilization pattern. Biotechnol. Bioeng. 28: 1735–1741

Gräzer-Lampert SD, Egli T & Hamer G (1986) Growth of *Hyphomicrobium* ZV620 in the chemostat: regulation of NH_4^+-assimilating enzymes and cellular composition. J. Gen. Microbiol. 132: 3337–3347

Hamer G, Harrison DEF, Harwood JH, Topiwala HH (1975) SCP production from methane. In: Tannenbaum SR & Wang DI (Eds) Single Cell Protein II (pp 357–369). MIT Press, Cambridge, Mass.

Harder W & Dijkhuizen L (1976) Mixed substrate utilization. In: Dean ACR, Ellwood DC, Evans GGT, Melling J (Eds) Continuous Culture 6: Applications and New Fields (pp 297–314). Ellis Horwood, Chichester

— (1982) Strategies of mixed substrate utilization in microorganisms. Phil. Trans. R. Soc. Lond. B 297: 459–480

Harrison DEF (1972) Physiological effects of dissolved oxygen tension and redox potential on growing populations of microorganisms. J. Appl. Chem. Biotechnol. 22: 417–440

Herbert D (1976) Stoicheiometric aspects of microbial growth. In: Dean ACR, Ellwood DC, Evans CGT & Melling J (Eds) Continuous Culture 6: Applications and New Fields (pp 1–30). Ellis Horwood, Chichester

Hueting S & Tempest DW (1979) Influence of the glucose input concentration on the kinetics of metabolite production by *Klebsiella aerogenes* NCTC 418: growing in chemostat culture in potassium- or ammonia-limited environments. Arch. Microbiol. 123: 189–199

von Liebig J (1840) Organic Chemistry in its Application to Agriculture and Physiology [English translation by L. Playfair]. Taylor and Walton, London

Linton JD & Stephenson RJ (1978) A preliminary study on growth-yields in relation to the carbon and energy-content of various organic growth substrates. FEMS Microbiol. Lett. 3: 95–98

Minkevich IG, Krynitskaya AY & Eroshin VK (1988) A double substrate limitation zone of continuous microbial growth. In: Kyslik P, Dawes EA, Krumphanzl V & Novak M (Eds) Continuous Culture (pp 171–189). Academic Press, London

Monod J (1942) La croissance des cultures bactériennes. Hermann, Paris

Moriarty DJW (1986) Measurement of bacterial growth rates in aquatic systems from rates of nucleic acid synthesis. Adv. Microb. Ecol. 9: 245–292

Pearl HP (1977) Factors limiting productivity of freshwater ecosystems. Adv. Microb. Ecol. 6: 75–110

Robertson LA & Kuenen JG (1990) Mixed terminal electron acceptors (oxygen and nitrate). In: Hamer G, Egli T & Snozzi M (Eds) Mixed and Multiple Substrates and Feedstocks (pp 97–106). Hartung-Gorre, Constance

Rutgers M, Balk PA & van Dam K (1990). Quantification of multiple-substrate controlled growth. Simultaneous ammonium and glucose limitation in chemostat cultures of Klebsiella pneumoniae. Arch. Microbiol. 153: 478–484

Stephenson M (1949) Growth and nutrition. In: Bacterial Metabolism, 3rd ed (pp 159–178). Longmans, Green & Co., London

Strange RE & Hunter JR (1967) Effect of magnesium on the survival of bacteria in aqueous suspension. In: Powell EO, Evans CGT, Strange RE & Tempest DW (Eds) Microbial Physiology and Continuous Culture (pp 102–123). H.M.S.O., London

Tempest DW, Herbert D & Philipps PJ (1967) Studies on the growth of Aerobacter aerogenes at low dilution rates in a chemostat. In: Powell EO, Evans CGT, Strange RE & Tempest DW (Eds) Microbial Physiology and Continuous Culture (pp 240–257). H.M.S.O., London

Tempest DW & Dicks JW (1967) Interrelationships between potassium, magnesium, phosphorus and ribonucleic acid in the growth of Aerobacter aerogenes in a chemostat. In: Powell EO, Evans CGT, Strange RE & Tempest DW (Eds) Microbial Physiology and Continuous Culture (pp 140–145). H.M.S.O., London

Tempest DW & Neijssel OM (1978) Eco-physiological aspects of microbial growth in aerobic nutrient-limited environments. Adv. Microb. Ecol. 2: 105–153

Veldkamp H & Jannasch HW (1972) Mixed culture studies with the chemostat. J. App. Chem. Biotechnol. 22: 105–123

Wanner U & Egli T (1990) Dynamics of microbial growth and cell composition in batch culture. FEMS Microbiol. Rev. 75: 19–44

Antonie van Leeuwenhoek **60**: 235–256, 1991.

A new thermodynamically based correlation of chemotrophic biomass yields

J.J. Heijnen

Delft University of Technology, Department of Biochemical Engineering, Delft, The Netherlands

Key words: biomass yield, chemotrophic growth, Gibbsenergy dissipation, thermodynamic efficiencies

Abstract

A new, generally applicable, thermodynamically based method is proposed to provide an estimation of the biomass yield on arbitrary organic and inorganic substrates. Aerobic, anaerobic, denitrifying growth systems with and without reversed electrontransport are covered. The biomass yield can be estimated with only 15% error in a very wide range of microbial growth systems and biomass yields (0.01–0.80 C-mol/(C)-mol). This method is based on the use of 'Gibbs energy dissipated per C-mol produced biomass' (designated as D_S^{01}/r_{Ax}) as the central parameter. Moreover the insufficiency of other methods based on Y_{ATP}, Y_{Ave}, η_o, Y_C and enthalpy or Gibbs energy efficiencies is shortly discussed. Also it appeared to be possible to understand the obtained correlation of D_S^{01}/r_{Ax} in general biochemical terms.

Introduction

Microbial growth occurs on a wide variety of organic and inorganic compounds, using either chemical or light energy. For biotechnological processes of industrial interest, chemotrophic growth is most relevant. One of the most important parameters in such processes is the yield (Y_{DX}) of biomass (X) on electron donor (D). Y_{DX} is conventionally defined as the amount of produced biomass (in C-mol) per amount of electron donor consumed (in C-mol for organic donors or in mol for inorganic donors). Because of its prime importance Y_{DX} has been studied extensively for many microbiological growth systems. It has been found that Y_{DX} can vary widely in the range of 0.01–1.0 C-mol biomass/(C)-mol electron donor, and depends strongly on the microorganism and its growth substrates. In practice, an estimate of Y_{DX} is frequently required before the biochemical capabilities and properties of microorganisms are known. The only information available, known as black box information, would deal with the composition of C-source, electron donor, electron acceptor, N-source and the biomass.

This black box information is conveniently represented in the form of a macro-chemical equation which describes the material balance for the production of *1 C-mol* of biomass from electron donor, acceptor, N-source, H_2O, HCO_3^-, H^+.

Figure 1 provides the principle and 2 examples of such a macro-chemical equation. The examples consider aerobic growth of *Pseudomonas oxalaticus* on oxalate and the anaerobic growth of *Saccharomyces cerevisiae* on glucose with ethanol production.

The coefficients can be calculated straightforward from the known Y_{DX} and the required conservation of C,H,O,N and electric charge (see Appendix 1).

If Y_{DX} is unknown then all the stoichiometric coefficients, except the coefficient of + 1 for biomass, become functions of Y_{DX}. These functions are obtained from the mentioned conservation relations. Considering the wide variety of potentially useful microbial growth systems the estimation of Y_{DX}, based on only black box information, poses a serious problem.

The solution lies in the selection of a parameter

General Macro-chemical equation

$$-(...)\,electron\,donor - (...)\,N\text{-}source - (...)\,electron\,acceptor$$

$$+1\,C\text{-}mol\,Biomass + (...)\,H_2O + (...)\,HCO_3^- + (...)\,H^+ = 0$$

Example 1

Aerobic growth of Pseudomonas oxalaticus on oxalate; $Y_{DX} = 0.086$ C-mol/C-mol

$$-5.815\,C_2O_4^{2-} - 0.2\,NH_4^+ - 1.857\,O_2 - 0.8\,H^+ - 5.415\,H_2O + CH_{1.8}O_{0.5}N_{0.2} + 10.63\,HCO_3^- = 0$$

Example 2

Anaerobic growth of Saccharomyces cerevisiae on glucose; $Y_{DX} = 0.14$ C-mol/C-mol

$$-1.1904\,C_6H_{12}O_6 - 0.2\,NH_4^+ - 1.63\,H_2O + CH_{1.8}O_{0.5}N_{0.2} + 2.0308\,C_2H_6O + 2.081\,HCO_3^- + 2.28\,H^+ = 0$$

Fig. 1. The black box representation of microbial growth as a macro-chemical equation and 2 examples for the aerobic growth of *Pseud. oxal* on oxalate. $Y_{DX} = 0.086$, and the anaerobic growth of *Sacch. cer.* on glucose $Y_{DX} = 0.14$.

which can be quantified from an established correlation, and from which Y_{DX} can be calculated.

In this paper a new parameter is presented. Before doing so it is relevant to survey and critically evaluate past proposals.

A critical evaluation of published parameters to correlate Y_{DX}-values

In the past decades several parameters have been proposed (Table 1) to serve as a basis for the correlation of Y_{DX}-values. Recently (Heijnen & Van Dijken 1991) an extensive evaluation of these parameters has been provided. The properties of said parameters were evaluated with respect to:
– general applicability,
– an intrinsic maximum or minimum parameter value derived from the 2nd Law of thermodynamics,
– the need of only black box information,
– the absence of intrinsic problems.
It is not the intention to discuss here the relative success of these parameters to correlate Y_{DX}-values (see Roels 1983; Stouthamer 1979; Westerhoff & van Dam 1987 for overviews). Only the 4 mentioned aspects will be compared. The results of this evaluation are therefore only shortly summarised (Table 1).

Y_{ATP} has no intrinsic 2nd Law based limit, and for its application the black box information (Fig. 1) is not sufficient. One needs a detailed knowledge about ATP generating and consuming pathways. Y_{Ave}, η_o, Y_c have not an intrinsic 2nd law based limit. Furthermore the application of η_o is limited to aerobic, and of Y_C is limited to heterotrophic growth systems ($Y_C = 1$ by definition in autotrophic growth). The following parameters (η_H, η^{BB}, η^C and η^{EC}) are all energetic efficiencies. An important aspect of such efficiencies is that these, as well as being correlating parameters, are generally considered as a measure of thermodynamic process performance. A high value, e.g. 80%, is thought to indicate a good performance, 10% is considered bad. However energetic efficiencies are troubled with intrinsic problems which makes such on intuitive interpretation improper.

These problems are different for black box efficiencies of Gibbs energy η^{BB} and enthalpy η_H (Roels 1983), for the conservation efficiency η^C of Gibbs energy (Battley 1960), and for the Gibbs energy convertor efficiency η^{EC} (Westerhoff 1987). In the following paragraphs attention will be paid

Table 1. Evaluation of proposed parameters to correlate biomass yield.

Ref	Parameter	Symbol	General applicable	2nd law limit	Black box information sufficient	Absence intrinsic problems
1	Yield of biomass on ATP	Y_{ATP}	yes	no	no	yes
2	Yield on available electrons	Y_{Ave}	yes	no	yes	yes
3	Oxygen efficiency	η_o	no	no	yes	yes
4	Carbon efficiency	Y_C	no	no	yes	yes
3,5	Black box enthalpy efficiency	η_H	yes	no	yes	no
5	Black box Gibbs energy efficiency	η^{BB}	yes	yes	yes	no
7	Conservation Gibbs energy efficiency	η^C	yes	yes	no	no
5,6	Energy Converter Gibbs energy efficiency	η^{EC}	yes	yes	no	no

1) Stouthamer 1979; 2) Mayberry et al. 1967; 3) Minkevich & Eroshin 1973; 4) Linton & Stephenson 1978; 5) Roels 1983; 6) Westerhoff & Van Dam 1987; 7) Battley 1960, 1987.

to these problems because these have largely been neglected.

The application of the black box efficiencies for enthalpy, η_H, and Gibbs energy, η^{BB}, requires values of enthalpy and Gibbs energy of formation of the involved chemicals. These values are tabulated and are based on an assumed frame of reference (zero enthalpy and Gibbs energy of formation for the chemical elements at standard temperature pressure and concentration). This chosen frame of reference is completely independent of the specific process studied. Therefore other frames of reference are possible. It has been shown (Roels 1983) that the 'combustion' frame of reference (where enthalpy and Gibbs energy of O_2, H_2O, HCO_3^-, H^+ (pH = 7) are 0) is convenient. The enthalpy and Gibbs energy of formation of chemical compounds become then identical to the combustion energies. Table 2 provides the Gibbs energy of formation of chemical compounds for the thermodynamic and the combustion frame of reference.

The basic intrinsic problem of η_H and η^{BB} is now that, for the same microbial growth system, the value of η_H or η^{BB} changes if the frame of reference is changed. Figure 2A, B show an aerobic and an anaerobic growth example (taken from Fig. 1) for the calculation of the black box Gibbs energy efficiency η^{BB} for 2 frames of reference (taken from Table 2). The black box efficiency is defined (Roels 1983) as the smaller than 1 ratio of the sum of the Gibbs energy associated with all consumed chemicals and the sum of Gibbs energy associated with all produced chemicals. The produced and consumed chemicals are easily found from the macro-chemical equation. It follows clearly from Fig. 2A, B that a different frame of reference for the Gibbs energy of formation of the chemical compounds has a large effect on the calculated η^{BB}. It must therefore be concluded that black box efficiencies of enthalpy or Gibbs energy are not meaningful parameters to characterize the thermodynamic performance of the process of microbial growth. Moreover, for the black box enthalpy efficiency η_H there is also no intrinsic limit from the 2nd law. In the situation where heat uptake occurs (as perhaps during methanogenic growth on acetate, Heijnen & Van Dijken 1991), $\eta_N > 1$. Also it is obvious that η^{BB} is maximally equal to 1 due to the 2nd law of thermodynamics, and that it is generally applicable to microbial systems.

The conservation efficiency η^C (Battley 1960, 1987) is calculated from 2 chemical reactions, called the non-conservative and conservative reactions. The 'non-conservative' reaction describes the process which occurs if the micro-organism would convert *all* the substrate through its catabolic pathway. E.g. for anaerobic growth of *Saccharomyces cerevisiae* on glucose this would be the conversion of glucose to ethanol and CO_2; for the aerobic growth of *Pseudomonas oxalaticus* on oxalate this would be the combustion of oxalate to CO_2 and

Table 2. Gibbs energy of formation values for two different frames of reference (pH = 7). (Reproduced from Heijnen & van Dijken 1991. © 1991 John Wiley & Sons, Inc.)

Compound	Composition	Degree of reduction	Thermodyn. reference (pH = 7)	Combustion reference (pH = 7)	
			kJ/mol	kJ/mol	kJ/(C-)mol
biomass	$CH_{1.8}O_{0.5}N_{0.2}$	4.2	$-$ 67	$+$ 474.6	$+$ 474.6
H_2O	H_2O	0	$-$ 238	0	0
HCO_3^-	HCO_3^-	0	$-$ 588	0	0
NH_4^+	NH_4^+	0	$-$ 80	0	0
H^+	H^+	0	$-$ 40	0	0
O_2	O_2	$-$ 4	0	0	0
oxalate^{2-}	$C_2O_4^{2-}$	1	$-$ 676	$+$ 262	$+$ 131
carbon monoxide	CO	2	$-$ 137	$+$ 254	$+$ 254
formate$^-$	CHO_2^-	2	$-$ 335	$+$ 253	$+$ 253
glyoxylate$^-$	$C_2O_3H^-$	2	$-$ 458	$+$ 520	$+$ 260
tartrate^{2-}	$C_4H_4O_6^{2-}$	2.5	$-$ 1010	$+$ 1184	$+$ 296
malonate^{2-}	$C_3H_2O_4^{2-}$	2.66	$-$ 700	$+$ 867	$+$ 289
fumarate^{2-}	$C_4H_2O_4^{2-}$	3.0	$-$ 602	$+$ 1356	$+$ 339
malate^{2-}	$C_4H_4O_5^{2-}$	3.0	$-$ 845	$+$ 1352	$+$ 338
citrate^{3-}	$C_6H_5O_7^{3-}$	3.0	$-$ 1170	$+$ 2004	$+$ 334
pyruvate$^-$	$C_3H_3O_3^-$	3.33	$-$ 475	$+$ 1131	$+$ 377
succinate^{2-}	$C_4H_4O_4^{2-}$	3.50	$-$ 688	$+$ 1504	$+$ 376
gluconate$^-$	$C_6H_{11}O_7^-$	3.66	$-$ 1154	$+$ 2580	$+$ 430
formaldehyde	CH_2O	4	$-$ 130	$+$ 498	$+$ 498
acetate$^-$	$C_2H_3O_2^-$	4	$-$ 372	$+$ 844	$+$ 422
dihydroxyaceton	$C_3H_6O_3$	4	$-$ 450	$+$ 1437	$+$ 479
lactate	$C_3H_5O_3^-$	4	$-$ 519	$+$ 1326	$+$ 442
glucose	$C_6H_{12}O_6$	4	$-$ 918	$+$ 2856	$+$ 476
mannitol	$C_6H_{14}O_6$	4.33	$-$ 944	$+$ 3066	$+$ 511
glycerol	$C_3H_8O_3$	4.66	$-$ 489	$+$ 1638	$+$ 546
propionate$^-$	$C_3H_5O^-_2$	4.66	$-$ 361	$+$ 1481	$+$ 493
ethylene glycol	$C_2H_6O_2$	5.0	$-$ 323	$+$ 1172	$+$ 586
acetoin	$C_4H_8O_2$	5.0	$-$ 280	$+$ 2236	$+$ 559
butyrate$^-$	$C_4H_7O^-_2$	5.0	$-$ 378	$+$ 2096	$+$ 524
propanediol	$C_3H_8O_2$	5.33	$-$ 327	$+$ 1797	$+$ 599
butanediol	$C_4H_{10}O_2$	5.50	$-$ 322	$+$ 2432	$+$ 608
methanol	CH_4O	6	$-$ 175	$+$ 692	$+$ 692
ethanol	C_2H_6O	6	$-$ 182	$+$ 1314	$+$ 657
propanol	C_3H_8O	6	$-$ 176	$+$ 1950	$+$ 650
n-alkanes	$C_{15}H_{32}$	6.13	$+$ 60	$+$ 9720	$+$ 648
propane	C_3H_8	6.66	$-$ 24	$+$ 2100	$+$ 700
ethane	C_2H_6	7.0	$-$ 32	$+$ 1464	$+$ 732
methane	CH_4	8.0	$-$ 51	$+$ 816	$+$ 816
H_2	H_2	2	0	$+$ 238	$+$ 238

The effect of different frames of reference on η^{BB} for <u>aerobic</u> growth of <u>Pseudomonas oxalaticus</u> on oxalate; $Y_{DX} = 0.086$

Macro chemical equation

$$-5.815\,C_2O_4^{2-} -0.2\,NH_4^+ -1.857\,O_2 -0.8\,H^+ -5.415\,H_2O + CH_{1.8}O_{0.5}N_{0.2} + 10.63\,HCO_3^- = 0$$

Thermodynamic reference black box Gibbs energy efficiency

$$\eta^{BB} = \frac{5.815(-676)+0.2(-80)+0.8(-40)+1.857(0)+5.415(-238)}{1(-67)+10.63(-588)} = \frac{-5269}{-6317} = 0.83$$

Combustion reference black box Gibbs energy efficiency

$$\eta^{BB} = \frac{1(474.6)+10.63(0)}{5.815(262)+0.2(0)+0.8(0)+1.857(0)+5.415(0)} = \frac{474.6}{1523.5} = 0.31$$

a

The effect of different frames of reference on η^{BB} for <u>anaerobic</u> growth of <u>Saccharomyces cerevisiae</u> on glucose; $Y_{DX} = 0.14$

Macrochemical equation

$$-1.1904\,C_6H_{12}O_6 -0.2\,NH_4^+ -1.63\,H_2O + 1\,CH_{1.8}O_{0.5}N_{0.2} + 2.0308\,C_2H_6O + 2.081\,HCO_3^- + 2.28\,H^+ = 0$$

Thermodynamic reference Gibbs energy

$$\eta^{BB} = \frac{1.1904(-918)+0.2(-80)+1.63(-238)}{1(-67)+2.0308(-182)+2.081(-558)+2.28(-40)} = \frac{-1497}{-1751} = 0.855$$

Combustion reference Gibbs energy

$$\eta^{BB} = \frac{1(474.6)+2.0308(1314)+2.081(0)+2.28(0)}{1.1904(2856)+0.2(0)+1.63(0)} = \frac{3144}{3399} = 0.925$$

b

Fig. 2. Calculation of η^{BB} using 2 frames of reference for Gibbs energy of formation for aerobic or anaerobic growth. (a) aerobic growth *Pseudomonas oxalaticus*. (b) anaerobic growth *Saccharomyces cerevisiae*.

Gibbs energy conservation efficiency η^c according to Battley

Aerobic growth <u>Pseudomonas oxalaticus</u> on oxalate; $Y_{DX} = 0.086$

$$\eta^c = \frac{-1523 - (-1048)}{-1523} = 0.31$$

<u>Conservative reaction</u>

$$-5.815C_2O_4^{2-} - 0.2NH_4^+ - 0.8H^+ - 1.857O_2 - 5.415H_2O + 1CH_{1.8}O_{0.5}N_{0.2} + 10.63HCO_3^- = 0$$
$$\Delta G_C = -1048 \ kJ$$

<u>Non-conservative reaction</u>

$$-5.815C_2O_4^{2-} - 5.815H_2O - 2.908O_2 + 11.63HCO_3^- = 0$$
$$\Delta G_{NC} = -1523 \ kJ$$

a

Gibbs energy conservation efficiency η^c according to Battley

Anaerobic growth <u>Saccharomyces cerevisiae</u> on glucose; $Y_{DX} = 0.14$

$$\eta^c = \frac{-272 - (-256)}{-272} = 0.06$$

<u>Conservative reaction</u>

$$-1.1904C_6H_{12}O_6 - 0.2NH_4^+ - 1.63H_2O + CH_{1.8}O_{0.5}N_{0.2} + 2.0308C_2H_6O + 2.081HCO_3^- + 2.28H^+ = 0$$
$$\Delta G_C = -256 \ kJ$$

<u>Non-conservative reaction</u>

$$-1.1904C_6H_{12}O_6 - 2.381H_2O + 2.381C_2H_6O + 2.381HCO_3^- + 2.381H^+ = 0$$
$$\Delta GNC = -272 \ kJ$$

b

Fig. 3. Calculation of η^c with the non-conservative and conservative reaction according to Battley (a) aerobic growth *Pseudomonas oxalaticus* (b) Anaerobic growth *Saccharomyces cerevisiae.*

H_2O. The non-conservative reaction has a Gibbs energy of reaction of $\triangle G_{NC}$. Clearly $(- \triangle G_{NC})$ is then the maximal available Gibbs energy which could possibly be generated in the absence of growth. The 'conservative' reaction is in fact the macro-chemical reaction, with a reaction gibbs energy $\triangle G_C$. Clearly $(- \triangle G_C)$ is the dissipated Gibbs energy during growth. In relation to the maximal available Gibbs energy $(- \triangle G_{NC} - (- \triangle G_C)) = - \triangle G_{NC} + \triangle G_C$ represents then a measure of the Gibbs energy which has not been dissipated, but apparently has been used to produce biomass. Hence $\eta^C = (- \triangle G_{NC} + \triangle G_C)/(- \triangle G_{NC})$ is the fraction of the maximal available Gibbs energy which is conserved for growth. It is also clear that in the hypothetical case of thermodynamic equilibrium $\triangle G_C = o$, and therewith η^C has a maximal value of 1. Also η^C can generally be applied to any microbial growth system, and is not influenced by different frames of reference of Gibbs energy for chemical compounds. Using this approach Battley (1960) obtained a useful linear correlation between η^C and the maximal available Gibbs energy per c-mol of substrate for aerobic and anaerobic growth.

Figure 3A, B shows some sample calculations of η^C. It is clear that in order to apply η^C one must possess biochemical information about the catabolic route in order to establish the non-conservative reaction. Furthermore an intrinsic problem appears to be that $(- \triangle G_{NC})$ represents the maximal Gibbs energy if *all* substrate is catabolized. However, actually a part of the substrate is assimilated into biomass and in reality only a fraction is used to generate energy. Therefore η^C appears not to represent the actual energy metabolism, but must be considered as an operational definition. Nevertheless, from an empirical point of view the η^C-concept results in an interesting correlation.

Another efficiency proposal, η^{EC}, is based on the basic notion that microbial growth is the result of 2 coupled processes. Catabolism generates Gibbs energy and part of this Gibbs energy is used to produce biomass in an anabolic process. Hence anabolism is coupled to catabolism. η^{EC} is now the ratio of the Gibbs energy taken up in anabolism to the Gibbs energy released in catabolism. It is obvious that η^{EC} is maximally 1, due to the 2nd Law of thermodynamics. The calculation of η^{EC} requires the specification of an anabolic and a catabolic process. This means that the macrochemical equation must be split into 2 parts. Necessarily this leads to the restriction that the sum of proposed anabolic and catabolic process *must* equal the macrochemical equation. As a consequence a proposal for a catabolic process implicitly means that also an anabolic process has been defined. This anabolic process it then obtained as the difference between the known macro-chemical equation and the proposed catabolic equation. Along the some reasoning an anabolic proposal leads to an implicitly defined catabolic process. It is obvious that both the anabolic and catabolic proposals should be in agreement with general biochemical knowledge. The intrinsic problem of η^{EC} resides in the fact that, without detailed biochemical information about anabolism and catabolism, the required anabolic/catabolic split must be based on general biochemical arguments. However these general biochemical arguments still allow many splits, and in literature different proposals have been published (Roels 1983; Westerhoff & Van Dam 1987). Moreover, it is easily shown that these proposals are not in accordance with general accepted biochemical knowledge. This aspect has apparently been neglected to a large extent and therefore extra attention will be given to this point. It should however be realised that η^{EC} is not influenced by a change in frame of reference of Gibbs energy of formation of the chemical compounds. The same aerobic and anaerobic growth systems as in Figs 1 and 2 will be used to 2nd illustrate the different η^{EC}-proposals and their intrinsic problems.

It has been proposed (Roels 1983) to use the aerobic combustion Gibbs energy of biomass as the Gibbs energy conserved in the anabolism. This means that η^{EC} from Roels is based on the proposal of an *anabolic*process where biomass is produced from CO_2 under production of O_2. Figure 4A shows the resulting catabolic processes for the 2 examples. Although the proposed anabolic reaction of Roels is biochemically meaningless, the catabolic process for aerobic growth is acceptable. However for anaerobic growth the catabolic process, where O_2 is involved, is certainly meaningless.

Gibbs energy convertor efficiency η^{EC} according to Roels

Aerobic growth Pseudomonas oxalaticus on oxalate; $Y_{DX} = 0.086$

$$\eta^{EC} = \frac{474.6}{1523} = 0.31$$

Macro-chemical equation

$$-5.815C_2O_4^{2-} - 0.2NH_4^+ - 0.8H^+ - 1.857O_2 - 5.415H_2O + 1CH_{1.8}O_{0.5}N_{0.2} + 10.63HCO_3^-$$
$$\Delta G_R^{01} = -1048 \; kJ$$

Catabolism

$$-5.815C_2O_4^{2-} - 5.815H_2O - 2.908O_2 + 11.63HCO_3^-$$
$$\Delta G_R^{01} = -1523 \; kJ$$

Anabolism

$$\underline{-HCO_3^-} - 0.2NH_4^+ - 0.8H^+ + CH_{1.8}O_{0.5}N_{0.2} + \underline{1.05O_2} + 0.4H_2O$$
$$\Delta G_R^{01} = +474.6 \; kJ$$

Anaerobic growth Saccharomyces cerevisiae on glucose; $Y_{DX} = 0.14$

$$\eta^{EC} = \frac{474.6}{723} = 0.65$$

Macro-chemical equation

$$-1.1904C_6H_{12}O_6 - 0.2NH_4^+ - 1.63H_2O + CH_{1.8}O_{0.5}N_{0.2} + 2.0308C_4H_6O + 2.081HCO_3^- + 2.28H^+$$
$$\Delta G_R^{01} = -256 \; kJ$$

Catabolism

$$-1.1904C_6H_{12}O_6 - 2.03H_2O - \underline{1.05O_2} + 2.308C_2H_6O + 3.081HCO_3^- + 3.08H^+$$
$$\Delta G_R^{01} = -723 \; kJ$$

Anabolism

$$\underline{-HCO_3^-} - 0.2NH_4^+ - 0.8H^+ + CH_{1.8}O_{0.5}N_{0.2} + \underline{1.05O_2} + 0.4H_2O$$
$$\Delta G_R^{01} = +474.6 \; kJ$$

a

Gibbs energy convertor efficiency η^{EC} according to Westerhoff

Aerobic growth Pseudomonas oxalaticus on oxalate; $Y_{DX} = 0.086$

$$\eta^{EC} = \frac{343}{1391} = 0.246$$

Macro-chemical equation

$$-5.815C_2O_4^{2-} - 0.2NH_4^+ - 0.8H^+ - 1.857O_2 - 5.415H_2O + 1CH_{1.8}O_{0.5}N_{0.2} + 10.63HCO_3^- = 0$$
$$\Delta G_R^{01} = -1048 \; kJ$$

Catabolism

$$-5.315C_2O_4^{2-} - 5.315H_2O - 2.657O_2 + 10.63HCO_3^- = 0$$
$$\Delta G_R^{01} = -1391 \; kJ$$

Anabolism

$$-0.5C_2O_4^{2-} - 0.2NH_4^+ - 0.1H_2O - 0.8H^+ + 1CH_{1.8}O_{0.5}N_{0.2} + \underline{0.8O_2} = 0$$
$$\Delta G_R^{01} = +343 \; kJ$$

b

Fig. 4. Calculation of η^{EC} with the proposed anabolic and catabolic processes belonging to different η^{EC} proposals, using aerobic (*P. oxalaticus*) and anaerobic (*S. cerevisiae*) growth examples. (a) Roels, (b) Westerhoff (only aerobic).

An alternative proposal (Westerhoff & Van Dam 1987) is a catabolic process, where the organic substrate, diminished with 1 C-mol which is needed to deliver the carbon for 1 C-mol biomass, is oxidized with O_2 to HCO_3^- and H_2O. This proposal is only valid for aerobic growth, and some alternatives for anaerobic growth have been discussed (Rutgers 1990). Figure 4B shows the resulting anabolic/catabolic processes for the aerobic example only. Here again it must be concluded that an anabolic process where O_2 is produced is biochemically not sound.

This short discussion of these η^{EC} proposals, based on different splits clearly shows that the interpretation of η^{EC} as a meaningful measure of thermodynamic process performance of growth is not really valid because η^{EC} is based on anabolic/catabolic process definitions which are biochemically not realistic. One might ask whether a biochemical realistic definition of anabolism is possible. Indeed such anabolic equations have been provided (Bruinenberg e.a., 1983 and Frankena e.a. 1988). Frankena e.a. provide the following anabolic equation for citrate as C-source and NH_4^+ as N-source:

$$- 0.333 \ C_6H_5O_7^{3-} - 0.2 \ NH_4^+ - 0.3083 \ FAD$$

$$- 0.5917 \ NAD^+ - 1.166 \ H_2O + CH_{1.8} \ O_{.5} \ N_{0.2}$$

$$+ 0.7917 \ H^+ + HCO_3^- + 0.3083 \ FADH_2 +$$

$$0.5917 \ NADH = 0$$

Calculation of the $\triangle G_R$ leads to $- 1.90$ kJ which is very close to zero.

In general it will be found (see also Battley 1987) that the Gibbs energy of such anabolic processes using an organic C-source is always close to zero. Hence, based on the best available biochemical evidence, one might conclude that η^{EC} is close to zero for all heterotrophic growth systems. This also means that the hypothesis (Westerhoff & Van Dam 1987) that growth has evolved to a system which maintains a maximal rate, at a theoretical optimal $\eta^{EC} = 0.24$, seems highly questionable because the experimentally found $\eta^{EC} = 0.24$ is based on a biochemically unrealistic split. (Fig. 4B.) Before clos-

ing this section on energetic efficiencies it is interesting to remark that for aerobic heterotrophic growth, η^{BB} (combustion reference), η^C and η^{EC} (Roels) all give identical values. Hence totally different concepts can lead to the same η-values.

Summarizing it can be concluded (Table 1) that none off the discussed parameters possesses all the required properties. The range of application is to limited (Y_C, η_o), most have no intrinsic relation to the 2nd Law of thermodynamics ($Y_{ATP}, Y_C, \eta_o, Y_{Ave}, \eta_H$), some can't be applied if there is only black box information ($Y_{ATP}, \eta^C, \eta^{EC}$) and intrinsic problems due to frames of reference (η_H, η^{BB}), the need to define a split (η^{EC}) or being not representative of the energy metabolism (η^C) occur. Especially for η^{EC} it is easily shown that all proposed splits are biochemically not realistic.

The next section presents a new parameter which fulfills all requirements and which leads to a single correlation for a wide range of growth systems.

Gibbs energy dissipation per C-mol biomass produced as a predictive parameter for chemotrophic biomass yields

In the previous section it has been shown that the proposed parameters of Table 1 do fail to meet the mentioned requirements. However, the Gibbs energy dissipation per C-mol produced biomass, has been indicated (Roels 1983) as an interesting alternative parameter, which however has not been elaborated further. If we define D_s^{01} as the rate of Gibbs energy dissipated (kJ/m^{3}h) and r_{Ax} as the rate of biomass production (C-mol/m^{3}h), then D_s^{01}/r_{Ax} is equal to the Gibbs energy dissipated per C-mol produced biomass (kJ/C-mol). The calculation of this parameter follows straightforward from the macrochemical equation (Fig. 1). The Gibbs energy of reaction of the macrochemical equation is designated as $\triangle G_R^{01}$, (under biochemical standard conditions). Because the macro-chemical equation produces by definition 1 C-mol of biomass and the dissipated Gibbs energy is then $- \triangle G_R^{01}$, it's clear that eq. (1) holds

$$D_s^{01}/r_{Ax} = - \triangle G_R^{01} \tag{1}$$

If D_S^{01}/r_{Ax} is considered with respect to the requirements mentioned in the previous section it appears that all are fulfilled. D_S^{01}/r_{Ax} can be calculated for any microbial growth system based on only black box information, which is represented as the macro-chemical equation, by using eq. (1). D_S^{01}/r_{Ax} has a minimal value of 0, based on the 2nd Law of thermodynamics. Clearly D_S^{01}/r_{Ax} is independent of choices of frames of reference of Gibbs energy of formation, because $\triangle G_R^{01}$ is not influenced by different frames of reference. Also there is no need to specify input or output processes; D_S^{01}/r_{Ax} follows directly from the macro-chemical equation. The only problem resides in the fact that for a correct calculation of Gibbs energy dissipation one needs to consider the actual concentration of the reactants in the macro-chemical equation, (eq. 2).

$$\triangle G_{fi} = \triangle G_{fi}^0 + RT\ln(C_i) \qquad (2)$$

This information about concentration is often not available. As a compromise the Gibbs energy of the macrochemical reaction $\triangle G_R$ is calculated for pH = 7 and otherwise standard conditions (1 M or 1 bar at 25°C). This Gibbs energy is indicated by the superscript$_{01}$. It can be shown that deviations from these standard conditions generally exert a limited effect on D_S^{01}/r_{Ax}. Exceptions are to be found in anaerobic cultures with low ($10^{-2} - 10^{-4}$ ata) H_2-pressures or cultures with low pH.

From literature a set of Y_{DX} data for chemotrophic growth has been collected. In these experiments batch or continuous culture was applied with electron donor limitation in well defined mineral media. Unknown product formation was excluded by showing that carbon and redox balances were satisfied. Reliable macro-chemical equations could thus be calculated, giving meaningful values of D_S^{01}/r_{Ax}. A typical calculation of D_S^{01}/r_{Ax} is shown in Appendix 1.

Table 3 provides an overview of the *heterotrophic* microbial growth systems, and gives the electron acceptor, the electron donor (which is, for heterotrophic growth also the C-source), the degree of reduction of the electron donor, Y_{DX} and D_S^{01}/r_{Ax}. Table 4 gives the autotrophic systems.

It can be seen that a very wide range of conditions is covered comprising

- different micro-organisms,
- different C-sources with 1–6 C-atoms and degree of reduction from 0–8 (Roels 1983),
- O_2, NO_3^- or fermentation as electron acceptor,
- microbial systems where the same micro-organism can grow on a wide variety of substrates.

From each Y_{DX} value and the known C-source, N-source (ammonia), electron acceptor and the biomass composition ($CH_{1.8}O_{0.5}N_{0.2}$) the macro-chemical equation was established and D_S^{01}/r_{Ax} was calculated (Appendix 1).

If one studies the resulting dissipation values for heterotrophic growth (Table 3), a very simple correlation appears if D_S^{01}/r_{Ax} is plotted as a function of degree of reduction and C-chain length of the C-source (Fig. 5). It appears that D_S^{01}/r_{Ax} is mainly dependent on the C-source, notably its degree of reduction, (γ_D) and its carbon chain length (C), and is not much influenced by the electron acceptor applied. The correlation is described by eq. (3).

Heterotrophic growth

$$D_s^{01}/r_{Ax} = 200 + 18(6 - C)^{1.8} +$$
$$\exp[\{(3.8 - \gamma_D)^2\}^{0.16} \star (3.6 + 0.4C)] \qquad (3)$$

In this equation (3), γ_D is the degree of reduction and C the number of C-atoms of the organic substrate (which is carbon source and electron donor). The line in Fig. 5 is eq. (3), which is seen to describe the experimental data with 25–30% error. This eq. (3) can be used to calculate D_S^{01}/r_{Ax} for non-listed carbon-sources. Table 5 contains the average values of D_S^{01}/r_{Ax} for a number of C-sources obtained from Table 3 and 4.

From the available *autotrophic* growth data (Table 4) an even simpler correlation is obtained. If reversed electron transport (RET) is not needed (as for H_2 or CO as electron donor) then the dissipation is in line with Fig. 5 and eq. (3). (For CO_2 as a C-source ($\gamma_D = 0$, C + 1).

Table 3. Biomass yield and Gibbs energy dissipation for chemoheterotrophic growth. (Reproduced from Heijnen & van Dijken 1991. © 1991 John Wiley & Sons, Inc.)

Ref[1]	Micro-org.	Type of electron Acceptor[2]	Electron donor	Composition	γ_D	Y_{DX} C-mol /C-mol	D_S^{01}/r_{Ax} kJ/ C-mol
1	Pseudomonas oxalaticus	Aer	oxalate[2-]	$C_2O_4^{2-}$	1	0.086	1048
1	,,	,,	formate[-]	CHO_2^-	2	0.162	1089
1	,,	,,	glyoxylate[-]	$C_2O_3H^-$	2	0.220	709
1	,,	,,	tartrate[2-]	$C_4H_4O_6^{2-}$	2.5	0.280	584
1	,,	,,	malonate[2-]	$C_3H_2O_4^{2-}$	2.66	0.238	757
1	,,	,,	citrate[3-]	$C_6H_5O_7^{3-}$	3.0	0.390	383
1	,,	,,	succinate[2-]	$C_4H_4O_4^{2-}$	3.50	0.385	504
1	,,	,,	acetate[-]	$C_2H_3O_2^-$	4	0.406	567
1	,,	,,	fructose	$C_6H_{12}O_6$	4	0.505	470
1	,,	,,	glycerol	$C_3H_8O_3$	4.66	0.569	473
1	,,	,,	ethanol	C_2H_6O	6	0.558	702
2	Candida utilus	Aer	citrate[3-]	$C_6H_5O_7^{3-}$	3.0	0.411	340
2	,,	,,	pyruvate[-]	$C_3H_3O_3^-$	3.33	0.434	396
2	,,	,,	succinate[2-]	$C_4H_4O_4^{2-}$	3.50	0.448	366
2	,,	,,	gluconate[-]	$C_6H_{11}O_7^-$	3.66	0.559	296
2	,,	,,	glucose	$C_6H_{12}O_6$	4	0.595	327
2	,,	,,	xylose	$C_5H_{10}O_5$	4	0.490	497
2	,,	,,	acetate[-]	$C_2H_3O_2^-$	4	0.455	455
2	,,	,,	glycerol	$C_3H_8O_3$	4.666	0.692	316
2	,,	,,	acetoin	$C_4H_8O_2$	5.0	0.424	845
2	,,	,,	2-3 butanediol	$C_4H_{10}O_2$	5.5	0.446	890
2	Candida utilis	Aer	ethanol	C_2H_6O	6	0.617	592
3,4	Paracoccus denitrificans	,,	formate[-]	CHO_2^-	2	0.12	1636
3,4	,,	,,	malate[2-]	$C_4H_4O_5^{2-}$	3.0	0.42	333
3,4	,,	,,	succinate[2-]	$C_4H_4O_4^{2-}$	3.50	0.48	311

Table 3. Continued

3,4	,,	,,	gluconate⁻	$C_6H_{11}O_7^-$	3.66	0.51	371
3,4	,,	,,	mannitol	$C_6H_{14}O_6$	4.33	0.62	345
3,4	,,	,,	methanol	CH_4O	6	0.54	809
5	Thiobacillus acidophilus	Aer	formate⁻	CHO_2^-	2	0.10	2058
5	,,	,,	L-malate²⁻	$C_4H_4O_5^{2-}$	3.0	0.25	880
5	,,	,,	pyruvate⁻	$C_3H_3O_3^-$	3.33	0.32	704
5	,,	,,	glucose	$C_6H_{12}O_6$	4	0.40	717
5	,,	,,	glycerol	$C_3H_8O_3$	4.666	0.55	512
6,7,8	Heterotrophic microorganisms	Aer	oxalate²⁻	$C_2O_4^{2-}$	1	0.07	1399
6,7,8	,,	,,	formate⁻	CHO_2^-	2	0.18	933
6,7,8	,,	,,	malate²⁻	$C_4H_4O_5^{2-}$	3.0	0.375	429
6,7,8	,,	,,	citrate³⁻	$C_6H_5O_7^{3-}$	3.0	0.365	442
6,7,8	,,	,,	succinate²⁻	$C_4H_4O_4^{2-}$	3.50	0.400	467
6,7,8	,,	,,	gluconate⁻	$C_6H_{11}O_7^-$	3.666	0.51	371
6,7,8	,,	,,	glucose	$C_6H_{12}O_6$	4	0.61	308
6,7,8	,,	,,	lactate⁻	$C_3H_5O_3^-$	4	0.51	394
6,7,8	,,	,,	acetate⁻	$C_2H_3O_2^-$	4	0.41	557
6,7,8	,,	,,	formaldehyde	CH_2O	4	0.47	587
6,7,8	,,	,,	mannitol	$C_6H_{14}O_6$	4.333	0.56	433
6,7,8	,,	,,	glycerol	$C_3H_8O_3$	4.666	0.67	335
6,7,8	,,	,,	propionate⁻	$C_3H_5O_2^-$	4.666	0.480	556
6,7,8	,,	,,	acetone	C_3H_6O	5.33	0.445	813
6,7,8	,,	,,	ethanol	C_2H_6O	6	0.53	765
6,7,8	Heterotrophic microorganisms	Aer	methanol	CH_4O	6	0.54	809
6,7,8	,,	,,	propanol	C_3H_8O	6	0.575	658
6,7,8	,,	,,	n-alkanes	$C_{15}H_{32}$	6.13	0.57	662
6,7,8	,,	,,	butane	C_4H_{10}	6.5	0.445	1061
6,7,8	,,	,,	methane	CH_4	8.0	0.55	1011
4	Campylobacter sputum	Den	formate⁻	CHO_2^-	2	0.166	999
4	Paracoccus denitrificans	Den	succinate²⁻	$C_4H_4O_4^{2-}$	3.50	0.387	466
4	Paracoccus denitrificans	Den	gluconate⁻	$C_6H_{11}O_7^-$	3.666	0.505	358

Table 3. Continued

4	Campylobacter sputum	Den	lactate⁻	$C_3H_5O_3^-$	4	0.274	1064
4	Paracoccus denitrificans	Den	mannitol	$C_6H_{14}O_6$	4.333	0.506	500
1	Klebsiella pneumoniae	Ana	citrate³⁻	$C_6H_5O_7^{3-}$	3.0	0.073	185
1	”	”	pyruvate⁻	$C_3H_3O_3^-$	3.33	0.083	236
1	”	”	gluconate⁻	$C_6H_{11}O_7^-$	3.666	0.121	237
1	”	”	fructose	$C_6H_{12}O_6$	4	0.173	210
1	”	”	glucose	$C_6H_{12}O_6$	4	0.176	236
1	”	”	dihyd.acetone	$C_3H_6O_3$	4	0.150	257
1	”	”	mannitol	$C_6H_{14}O_6$	4.33	0.154	191
1	”	”	glycerol	$C_3H_8O_3$	4.666	0.093	254
1	Clostridium butyricum	Ana	gluconate⁻	$C_6H_{11}O_7^-$	3.666	0.143	219
1	”	”	glucose	$C_6H_{12}O_6$	4	0.176	229
1	”	”	mannitol	$C_6H_{14}O_6$	4.33	0.151	222
9	Methanobacterium formicicum	Ana	formate⁻	CHO_2^-	2	0.053	880
10	Methanobacterium soehngenii	Ana	acetate⁻	$C_2H_3O_2^-$	4	0.024	539
10	Methanosarcina barkeri	Ana	methanol	CH_4O	6	0.13	570
11	Butyribacterium methylotropicum	Ana	glucose	$C_6H_{12}O_6$	4	0.250	110
11	”	”	methanol	CH_4O	6	0.30	584
12	Pelobacter propionicus	Ana	lactate⁻	$C_3H_5O_3^-$	4	0.085	197
12	”	”	acetoin	$C_4H_8O_2$	5.0	0.08	358
12	”	”	butanediol	$C_4H_{10}O_2$	5.5	0.063	390
12	”	”	ethanol	C_2H_6O	6	0.028	785
12	”	”	propanol	C_3H_8O	6	0.019	792
12	Pelobacter carbinolicus	Ana	ethyl.glycol	$C_2H_6O_2$	5	0.073	617
12	”	”	acetoin	$C_4H_8O_2$	5.0	0.070	259
12	”	”	butanediol	$C_4H_{10}O_2$	5.5	0.036	244
13	Clostridium magnum	Ana	citrate³⁻	$C_6H_5O_7^{3-}$	3.0	0.03	551
13	”	”	glucose	$C_6H_{12}O_6$	4	0.32	139
13	”	”	acetoin	$C_4H_8O_2$	5.0	0.08	364
13	”	”	butanediol	$C_4H_{10}O_2$	5.5	0.072	353
14	Saccharomyces cerevisiae	Ana	glucose	$C_6H_{12}O_6$	4	0.14	255

[1] 1) Rutgers 1990; 2) Verduyn 1991; 3) Verseveld 1979; 4) Stouthamer 1988; 5) Pronk e.a. 1990; 6) Linton & Stephenson 1978; 7) Heijnen & Roels 1981; 8) Blevins and Perry 1971; 9) Aswell and Ferry 1978; 10) Zehnder 1989; 11) Lynd and Zeikus 1983; 12) Schink 1984a; 13) Schink 1984b; 14) Von Stockar & Birou 1989. [2] Aer = O_2, Den = NO_3^-, Ana = fermentation

If reversed electron transport occurs (as for $S_2O_3^{2-}$, NH_4^+ etc) then a more or less constant, but very high, dissipation of about 3500 kJ/C-mol is needed; irrespective of the nature of the electron donor.

Table 4. Biomass yield and Gibbs energy dissipation for chemoautotrophic growth. (Reproduced from Heijnen & van Dijken 1991. © 1991 John Wiley & Sons, Inc.)

Ref[1]	Micro-org.	Type of electron Acceptor[2]	Electron donor	Compo-sition[3]	Y_D	Y_{DX} C-mol /C-mol	D_S^{01}/r_{Ax} kJ/ C-mol
No reversed electron transport							
1	Methanobacterium arborophilum	Ana	hydrogen	$H_2(10^{-3})$	2	0.015	1076
2,3	Methanobacterium AZ	Ana	hydrogen	$H_2(10^{-3})$	2	0.019	840
4	Butyribacterium methylotrophicum	Ana	H_2/CO_2	H_2	2	0.056	440
4	,,	Ana	carbonmon.	CO	2	0.11	350
5	Alcaligenes eutrophus	Aer	hydrogen	$H_2(10^{-2})$	2	0.13	1267
6	Carboxydotrophic bact.	Aer	carbonmon.	CO	2	0.16	1105
With reversed electron transport							
7	Thiosphaera pantotropha	Aer	thiosulf.	$S_2O_3^{2-}$	8	0.16	4627
7	Thiobacillus neapolitanus	Aer	thiosulf.	$S_2O_3^{2-}$	8	0.16	4627
8	Thiobacillus ferrooxidans	Aer	thiosulf.	$S_2O_3^{2-}$	8	0.22	3237
9	Thiobacillus acidophilus	Aer	thiosulf.	$S_2O_3^{2-}$	8	0.23	3076
8	Thiobacillus ferrooxidans	Aer	tetrathion.	$S_4O_6^{2-}$	14	0.41	2761
10	Thiobacillus denitrificans	Aer	Sulfide	HS^-	8	0.30	2186
11	Thiobacillus ferrooxidans	Aer	iron(II)	Fe^{2+} (pH=1.6)	1	0.010	2927
12	Nitrosomonas europaea	Aer	ammonium	NH_4^+	6	0.06	4117
12	Nitrobacter sp.	Aer	nitrite	NO_2^-	2	0.017	3892

[1] 1) Morii e.a. 1987; 2) Taylor & Pirt 1977; 3) Fuchs e.a. 1979; 4) Lynd & Zeikus 1983; 5) Siegel & Ollis 1984; 6) Meijer & Schlegel 1983; 7) Robertson 1988; 8) Hazeu e.a. 1986; 9) Pronk e.a. 1990; 10) Sublette 1987; 11) Hazeu e.a. 1984; 12) Van Niel 1991
[2] Aer = O_2, Den = NO_3^-, Ana = fermentation
[3] () = pressure of H_2 in bar

Number of carbon atoms of carbon source

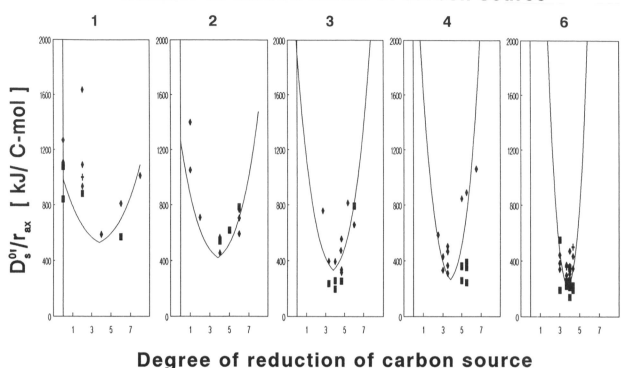

Fig. 5. The effect of carbon chain length and degree of reduction of the applied C-source (= electron donor) on the required Gibbs energy dissipation per C-mol biomass produced (data taken from Table 3 and 4). The drawn line corresponds to eq. (3). ◆ Aerobic, + denitrifying, ■ anaerobic growth system. (Reproduced from Heijnen & van Dijken 1991. ©1991 John Wiley & Sons, Inc.)

Autotrophic growth

$$- RET \ D_s^{01}/r_{Ax} = \ eq(3)$$
$$+ RET \ D_s^{01}/r_{Ax} = 3500 \ kJ/C\text{-}molbiomass \quad (4)$$

It is obvious that eq. (3) and eq. (4) can be used quite generally to calculate an estimate of the unknown Y_{DX} if the C-source (= organic electron donor or CO_2), N-source, electron acceptor are known. This calculation is a simple reversal of the procedure to calculate D_S^{01}/r_{Ax} from a known Y_{DX} using the macro-chemical equation. Appendix 1 shows an example of the reversed procedure.

Figure 6A shows a comparison of measured (Tables 3, 4) and calculated Y_{DX}-values, where the calculation is based on the average values of D_S^{01}/r_{Ax} listed in Table 5. Figure 6B shows the same comparison, but now D_S^{01}/r_{Ax} is taken from eq. (3) and

eq. (4) for the various C-sources. It is seen that Y_{DX} can be predicted with 13–19% error for an Y_{DX} range of 0.010–0.80 C-mol biomass/(C)-mol electron donor over a wide range of micro-organisms, electron acceptors and organic or inorganic electron donors. It is noted that the success of this approach is quite remarkable. In the following section attention will be given to understand and explain to some extent the found correlation of D_S^{01}/r_{Ax}.

Explanation of the obtained D_S^{01}/r_{Ax} correlation

It has been found that for *heterotrophic growth* (Table 3) D_S^{01}/r_{Ax}, to be called shortly 'dissipation', is mainly determined by the composition of the C-source (which is also the electron donor). Espe-

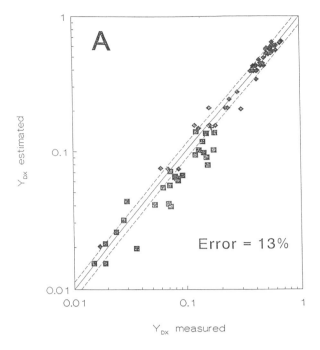

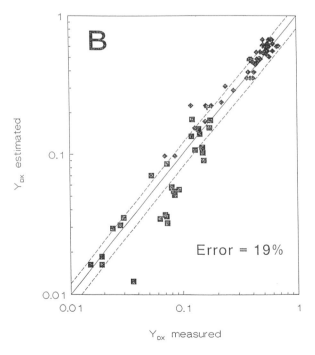

Fig. 6. Comparison between measured (Table 3 and 4) and calculated Y_{DX}-values. (a) Y_{DX} calculated from dissipation values taken from Table 5; (b) Y_{DX} calculated from dissipation values taken from eq. (3) and eq. (4). ◆ Aerobic + denitrifying ■ anaerobic growth system. (Reproduced from Heijnen & van Dijken 1991. © 1991 John Wiley & Sons, Inc.)

cially the degree of reduction (γ_D) and the number of C-atoms in the organic electron donor appear to be important. For a wide spectrum of micro-organisms, organic substrates (= electron donors) and electron acceptors, the same pattern is observed. Dissipation is minimal around $\gamma_D \approx 3.5$–4.5 and rises for both lower and higher γ_D-values. Consider for example organic electron donors with 2 carbon

Table 5. Average Gibbs energy dissipation values (D_s^{01}/r_{Ax}) for different C-sources (Reproduced from Heijnen & van Dijken 1991. © 1991 John Wiley & Sons, Inc.)

C-source compound	Degree of reduction	Carbon chain length	number of observations	D_s^{01}/r_{Ax} kJ/C-mol biomass	s.d. (%)
CO_2^1	0	1	9	3494	(23)
CO_2^2	0	1	3	1061	(16)
oxalate^{2-}	1.0	2	2	1224	(14)
CO	2.0	1	1	1105	(–)
formate$^-$	2.0	1	5	1107	(25)
glyoxylate$^-$	2.0	2	1	709	(–)
tartrate^{2-}	2.5	4	1	584	(–)
malonate^{2-}	2.67	3	1	757	(–)
malate^{2-}	3.0	4	2	380	(13)
citrate^{3-}	3.0	6	5	381	(31)
pyruvate$^-$	3.33	3	2	316	(25)
succinate^{2-}	3.5	4	5	422	(17)
gluconate$^-$	3.66	6	6	311	(19)
formaldehyde	4	1	1	587	(–)
acetate$^-$	4	2	4	529	(8)
lactate$^-$	4	3	2	296	(33)
dihydroxyacetone	4	3	1	257	(–)
glucose	4	6	8	284	(26)
glycerol	4.66	3	4	345	(23)
mannitol	4.33	6	5	338	(35)
propionate	4.66	3	1	556	(–)
ethyleneglycol	5.0	2	1	617	(–)
acetoin	5.0	4	3	457	(50)
butanediol	5.5	4	3	469	(53)
aceton	5.33	3	1	813	(–)
methanol	6	1	3	729	(15)
ethanol	6	2	4	712	(11)
propanol	6	3	2	725	(9)
n-alkanes	6.13	6	1	662	(–)
butan	6.5	4	1	662	(–)
methane	8	1	1	1011	(–)

^{1} for electron donor with reversed electron transfer; ^{2} for electron donor without reversed electron transfer.

atoms and different degree of reduction (see Table 5) like oxalate ($\gamma_D = 1$), glyoxylate ($\gamma_D = 2$), acetate ($\gamma_D = 4$), ethylene glycol ($\gamma_D = 5$) and ethanol ($\gamma_D = 6$), where the dissipation values are 1200, 700, 500, 600, 750 kJ/C-mol produced biomass.

Moreover there is the effect of the number of C-atoms in the organic C-source (= electron donor). Figure 5 clearly shows that dissipation increases for organic electron donors which have a smaller number of C-atoms. For example if one studies the organic electron donors with the same degree of reduction ($\gamma_D = 4$), but different number of C-atoms (formaldehyde, acetate, lactate, dihydroxyacetone, glucose, see Table 5), it appears clearly that D_S^{01}/r_{Ax} is halved when the number of C-atoms increases from 1 to 3. An analogous trend is observed for compounds with degree of reduction 2–2.5 (CO/formate, glyoxylate, tartrate), 2.7–3 (malonate, malate, citrate), 5–5.5 (ethylene glycol, acetoin). It also appears that the effect of carbon chain length is most pronounced between 1 and 4 C-atoms. The change of dissipation in the range 4–6 C-atoms is much less.

The effect of a change in electron acceptor, using the same organic electron donor, is rather limited, as can be seen if one compares O_2, NO_3^- and fermentation systems (Fig. 5, Table 3). There appears to be however, a tendency that in fermentative growth, where the electron transport chain does not operate, the dissipation is somewhat less (50–150 kJ/C-mol biomass) than in growth systems with

an electron transport chain (O_2, NO_3^-). The near independence of dissipation per C-mol produced biomass from the electron acceptor was already noted (Roels 1983).

A nice illustration is obtained with data (Von Stockar & Birou 1989) for the measured biomass yield of yeast growth on glucose under different regimes of O_2-supply and ethanol production (Table 6). It appears that Y_{DX} changes strongly but that D_S^{01}/r_{Ax} remains fairly constant if one changes gradually from full aerobic to full anaerobic growth. Furthermore it is noted that the measured heat production per C-mol produced biomass is not nearly as constant, (a change of a factor of 3.5), as the dissipation. For *chemoautotrophic* growth (Table 4), the dissipation appears to be mainly influenced by the absence (H_2, CO) or presence (Fe^{2+}, NO_2^-, NH_4^+, $S_2O_3^{2-}$ etc.) of reversed electron transport (R.E.T.). If R.E.T. is absent the dissipation is about 1000 kJ/C-mol biomass. If R.E.T. is required the dissipation is about 3500 kJ/C-mol biomass. The nature of the inorganic electron donor exerts no systematic influence. It is remarkable that for photoautotrophic growth of *Chlorella vulgaris* (where H_2O is the electron donor and where R.E.T. is known to occur) a dissipation of 3575 kJ/C-mol biomass is found (Iehana 1990).

One might wonder about the origins of the observed correlation for the dissipation per C-mol produced biomass (Fig. 5). A possible explanation can be based on the 'funnel' concept of microbial metabolism (Fig. 7) which is based on general biochemical knowledge. Here microbial metabolism is split in a primary part and a polymerisation part. The primary part converts the available carbon source (which is CO_2 or the organic electron donor) into central metabolites which are the building blocks for the production of biomass. In the polymerisation part said building blocks are converted into biomass. It is reasonable to assume that the polymerisation process is more or less the same for all micro-organisms and that analogous biochemical reactions are used. The reason for this is that in all micro-organisms the involved average building blocks are more or less the same (about 4–5 carbon atoms and degree of reduction, equal to that of biomass, of about 4.2) and that the principle of

Table 6. Biomass yield, heat production and Gibbs energy dissipation for growth of yeast under aerobic, partly anaerobic and anaerobic conditions. (Reproduced with modifications from Heijnen & van Dijken 1991. © 1991 John Wiley & Sons, Inc.)

Biomass yield	Ethanol yield	Measured heat production	Gibbs energy dissipation (D_S^{01}/r_{Ax})
C-mol /C-mol	C-mol /C-mol	kJ/C-mol biomass	kJ/C-mol biomass
0.57	0	339	332
0.52	0.082	313	307
0.40	0.228	250	306
0.23	0.440	160	312
0.19	0.512	114	230
0.14	0.566	95	255

252

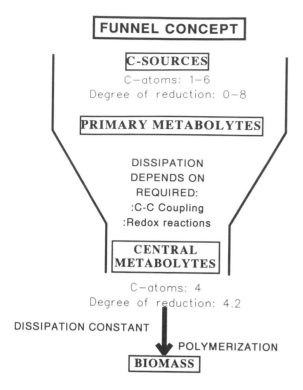

EXPLANATION BEHAVIOUR (D_S^{01}/r_{AX})

FUNNEL CONCEPT

C-SOURCES
C-atoms: 1–6
Degree of reduction: 0–8

PRIMARY METABOLYTES

DISSIPATION
DEPENDS ON
REQUIRED:
:C-C Coupling
:Redox reactions

CENTRAL METABOLYTES

C-atoms: 4
Degree of reduction: 4.2

DISSIPATION CONSTANT

POLYMERIZATION

BIOMASS

Fig. 7. The funnel concept of microbial metabolism.

unity of biochemistry is assumed. The primary part however depends highly on the available C-source (and probably also on the specific micro-organism). Depending on the number of C-atoms and γ_D, there must be performed very many or only few reactions in the primary part in order to convert the C-source into building blocks ($\gamma \approx 4.2$ and 4–5 C-atoms). For example, if CO_2 is the C-source, a micro-organism must carry out much more reduction reactions and carbon-carbon coupling reactions than a micro-organisms which uses glucose. Glucose is much closer to the redox level of about 4 and the number of C-atoms of about 4–5 of the building blocks than CO_2. Hence much more steps of redox and C-C coupling reactions are required for CO_2 as C-source. Because in principle each reaction adds to the Gibbs energy dissipation, it is clear that for CO_2 much more dissipation occurs than for glucose. Analogous reasoning applies to

e.g. formaldehyde and methanol. Similarly the presence of an electron transport chain (O_2, NO_3^-) results in additional reactions, compared to fermentation. Apparently this leads to some additional dissipation (50–150 kJ/C-mol extra). In the same way the presence of reversed electron transport during autotrophic growth leads apparently to an additional dissipation of + 2500 kJ/C-mol biomass in comparison to autotrophic growth without R.E.T.

Based on the above mentioned findings it appears that the dissipated Gibbs energy per C-mol produced biomass (D_S^{01}/r_{AX}) is a straightforward measure of the amount of chemical work which is required and spent to synthesize 1 C-mol of biomass from a given C-source, N-source, electron donor and electron acceptor. If more chemical processes (redox reactions, C-C coupling reactions, electron transport chain reactions, reversed electron transport reactions) are required to produce the biomass, the dissipation is higher. As such this provides a very simple explanation of the found correlation (Fig. 5) in general biochemical terms.

This explanation can be supported even more by looking to a well known comparable parameter, Y_{ATP}, which gives the grams of biomass which can be produced from 1 mol ATP. It is well known (Stouthamer 1973, 1979) that the theoretical Y_{ATP}, calculated from established biochemical knowledge, depends strongly on the applied C-source. For glucose, citrate, malate, acetate, ethanol, CO_2 without R.E.T. (Stouthamer 1973, Frankena e.a. 1988), CO_2 with R.E.T. (Kelly 1990), one has calculated theoretical Y_{ATP} values of 28.8, 20.9, 15.4, 10, 10, 6.6 and 2.5 g/mol ATP. These values can be converted to mol of ATP required per C-mol of biomass produced (using the fact that 1 C-mol of biomass ($CH_{1.8}O_{0.5}N_{0.2}$, is 24.6 g organic mass). If one compares this ATP-need with the found Gibbs energy dissipation per produced C-mol of biomass (taken from Table 5 for the mentioned compounds) one observes a very good correlation (Fig. 8).

This indicates that D_S^{01}/r_{AX} can be considered as the thermodynamic equivalent of the biochemically based ATP-parameter. Furthermore Fig. 8 supports the above mentioned explanation of the correlation in Fig. 5 in general biochemical terms. Figure 8 allows one further point of interest. From the slope one can calculate that per mol of theoretically spent ATP an overall Gibbs energy dissipation of about 280 kJ occurs. It is well known that under physiological conditions the Gibbs energy content of the ATP/ADP$^+$Pi couple is about 50 kJ/mol. Also it is known that in biosynthesis on organic substrates, ATP is mostly dissipated to provide irreversibility (Westerhoff & Van Dam 1987). Therefore it follows that during the generation of 1 mol ATP there occurs a Gibbs energy dissipation of about 230 kJ/mol ATP. This means that for the production of 1 ATP (= 50 kJ of Gibbs energy) an investment of 280 kJ of Gibbs energy is required. This would indicate an efficiency of Gibbs energy transformation during ATP synthesis of only about 50/280 = 18%. However, it has been reported (Stouthamer 1979) that the theoretical Y_{ATP} is not found in practice. Generally the actual Y_{ATP} is about 50% of the theoretically expected value, for reasons yet unknown. Referring to Fig. 8 this would mean an overall dissipation of 140 kJ per mol of actually spent ATP. This would lead to a dissipation of 90 kJ/mol ATP during ATP synthesis. The efficiency of Gibbs energy transformation during ATP synthesis becomes then 50/140 = 36%. For oxidative phosphorylation, energetic efficiencies of 25–45% have indeed been measured (Westerhoff & Van Dam 1987). It is interesting to note that the value-range of this efficiency is close to the calculated optimum value of energy convertors according to the proposed optimization strategies (Westerhoff & Van Dam 1987). Apparently the energy converter concept appears to be more suited to describe oxidative phosphorylation than to describe microbial growth.

Limits of applicability of the found correlation

Although the above found correlation is quite satisfying from the point of view of its range of applicability, the accuracy of the estimated Y_{DX} and its agreement with general biochemical knowledge, a word of caution seems appropriate. Equations (3, 4), or Table 5, will only give a first estimation of the biomass yield Y_{DX}. This value is bound to be of

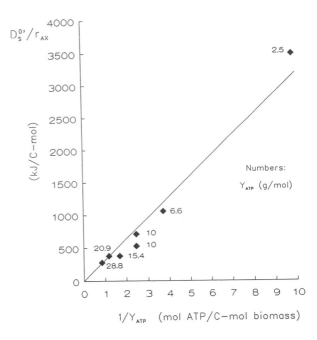

Fig. 8. Comparison between Gibbs energy dissipation and ATP-need for the production of biomass from several carbon-sources. (Reproduced from Heijnen & van Dijken 1991. © 1991 John Wiley & Sons, Inc.)

limited accuracy, because it is based on the average behaviour of micro-organisms. The actual biochemistry has not been taken into account. Of course, the absence of the need for biochemical details is the attractive feature of the present approach, but this also inherently limits the accuracy of yield predictions. This point is illustrated by *Thiobacillus acidophilus* and *Butyribacterium methylotrophicum* (Table 3, 4). These micro-organisms show a systematic deviation from the average behaviour. *T. acidophilus* has a much higher dissipation than most aerobic micro-organisms, presumably due to the low pH of 3.5. *B. methylotrophicum* has, for H_2 and CO, a much lower dissipation than other autotrophic anaerobic micro-organisms for reasons unknown. A second illustration of the limiting predictive capabilities is the anaerobic conversion of glucose to ethanol by micro-organisms. Using the present approach one obtains $Y_{DX} \approx 0.13$ which is correct for *S. cerevisiae*, but wrong for *Zymomonas mobilis*, where the experimentally found $Y_{DX} \approx 0.06$. The difference between the two micro-organisms, which have the same C-source, electron acceptor, N-source and biomass composition, is biochemical.

S. cerevisiae employs the glycolysis route which gives 2 mol ATP/mol glucose, while *Z. mobilis* uses the Entner-Doudoroff pathway which generates only 1 ATP/mol glucose. It can therefore be concluded that the present method can only provide a first estimate of Y_{DX}, but that a more accurate value can only be obtained by using detailed information about ATP generation/consumption or by performing actual measurements of Y_{DX}.

Conclusion

It has been shown that all published parameters which have been proposed to correlate biomass yields are subject to limitations and problems. Notably the Gibbs energy efficiencies suffer from intrinsic problems. However, it has been shown that a simple, and biochemically understandable, correlation can be based on D_S^{01}/r_{Ax}, being the Gibbs energy dissipation per C-mol produced biomass. This dissipation parameter is not subject to the above mentioned limitations and problems. It has been shown that, for a wide variety of microbial growth systems in which Y_{DX} varies between 0.01 and 0.80, this method provides an estimated biomass yield with an error of about 15%. Finally the limits of the applicability of this new method have been discussed.

Acknowledgements

The author wishes to acknowledge that J.P. van Dijken, K. van Dam and H.V. Westerhoff helped with stimulating discussion, L. Robertson helped with correcting the English text.

Appendix 1

Calculation of D_S^{01}/r_{Ax} and its use for the calculation of Y_{DX}

The dissipation of Gibbs energy per C-mol produced biomass is calculated most simply by first establishing the macro-chemical equation as introduced in Fig. 1. The macrochemical equation follows directly from a measured Y_{DX}-value. Suppose that one studies the aerobic growth of a microorganism on oxalate. One has checked the absence of products or other substrates and the N-source is NH_4^+. The biomass yield Y_{DX} been measured to be 0.086 C-mol biomass per C-mol oxalate. The biomass composition is taken as $CH_{1.8}O_{0.5}N_{0.2}$ (Roels 1983). Then one can formally write the macro-chemical equation as:

$$\frac{-1}{2*0.086}C_2O_4^{2-} + aNH_4^+ + bH^+ + cO_2 + dH_2O + 1CH_{1.8}O_{0.5}N_{0.2} + eHCO_3^- = 0$$

By definition the macro-chemical equation produces 1 C-mol of biomass. The yield on substrate is 0.086, indicating that for the production of 1 C-mol biomass there is needed a consumption of 1/0.086 C-mol oxalate or $1/0.086 \times {}^1/_2 = 5.815$ mol oxalate. The minus sign is due to the fact that oxalate is consumed. Hence there are 5 unknown stoichiometric coefficients (a, b, c, d, e), which can be calculated from the C, H, O, N and electric charge conservation as follows:

Conservation of C		$-5.815 * 2 + 1 + e = 0$
"	H	$4 + b + 2d + 1.8 + e = 0$
"	O	$-5.815 * 4 + 2c + d + 0.5 + 3e = 0$
"	N	$a + 0.2 = 0$
"	charge	$-5.815 * (-2) + a + b + e * (-1) = 0$

Solving these equations leads to a = -0.2; b = -0.8; c = -1.857; d = -5.42; e = $+10.63$. This gives the macro-chemical equation in Fig. 1 (*Ps. oxalaticus* on oxalate). Now it is easy, using tabulated values of $\triangle G_f^{01}$ (Table 2) of the various chemicals, to calculate $\triangle G_R^{01}$. This leads to:

$$\triangle G_R^{01} = 10.63 \ (-558) + 1 \ (-67) - 5.42 \ (-238) - 0.8 \ (-40) - 0.2 \ (-80) - 5.815 \ (-676) = -1048 \ kJ$$

Now clearly, this means that for the formation of 1 C-mol biomass 1048 kJ of Gibbs energy is dissipated. Hence $D_S^{01}/r_{Ax} = 1048$.

A second question is how to calculate an estimate of Y_{DX} if only the C-source, N-source, electron donor and -acceptor are known using eq. (3) (or Table 5). Suppose a microorganism grows anaerobically on methanol and produces acetate. The C-source is methanol and from eq. (3) one obtaines then (c = 1, $\gamma_D = 6$) for $D_S^{01}/r_{Ax} = 698$ kJ/C-mol.

The growth system contains biomass, NH_4^+, acetate, methanol, HCO_3^-, H^+ and H_2O as chemicals of interest. Hence one can write the macro-chemical equation as

$$fCH_3OH + \quad aNH_4^+ + \quad bH^+ + \quad cC_2H_3O_2^- + \quad dH_2O +$$
$$1CH_{1.8}O_{0.5}N_{0.2} + \quad eHCO_3^- = 0$$

We know from the estimated dissipation per C-mol biomass that the reaction Gibbs energy is -698 kJ. Now one can calculate straightforward the 6 coefficients a $\rightarrow$ f from the element and charge conservation relations and the Gibbs energy balance. The coefficients in the Gibbs energy balance follow from Tabulated Gibbs energy of formation (Table 2).

Conservation of C		$f + 2c + 1 + e = 0$
"	*H*	$4f + 4a + b + 3c + 2d + 1.8 + e = 0$
"	*O*	$f + 2c + d + 0.5 + 3e = 0$
"	*N*	$a + 0.2 = 0$
"	*charge*	$a + b - c - e = 0$
Balance of Gibbs energy		$-175f - 80a - 40b - 372c - 238d - 67 - 588e + 698 = 0$

Solving these equations leads to the following values: $a = -0.2$; $b = 2.866$; $c = 8.898$; $d = 12.964$; $e = -6.232$; $f = -12.564$. This leads then to $Y_{DX} = 1/12.564 = 0.080$ C-mol biomass/C-mol methanol.

Symbols

D_s^{01}	Gibbs energy dissipation in microbial growth reactors	kJ/m³h
r_{Ax}	Biomass production rate	C-mol/m³h
C_i	Concentration of compound	mol/ltr.
Y_{DX}	yield of biomass on electron donor	C-mol/(C)-mol
Y_{ATP}	yield of biomass on ATP	g/mol ATP
Y_{Ave}	yield of biomass on available electrons	g/mol electron
Y_C	carbon efficiency	(-)
η_o	oxygen efficiency	(-)
η_H	enthalpy efficiency	(-)
η^{BB}	Black box Gibbs energy efficiency	(-)
η^C	Conservation Gibbs energy efficiency	(-)
η^{EC}	Energy convertor Gibbs energy efficiency	(-)
γ_D	degree of reduction of electron donor (= organic substrate)	(-)
ΔG_R	Gibbs energy of reaction	kJ
ΔG_C	Gibbs energy conservative reaction	kJ
ΔG_{NC}	Gibbs energy non-conservative reaction	kJ
ΔG_f	Gibbs energy of formation of chemical compounds	kJ/mol
R	Gas constant	J/mol K
T	Absolute temperature	K

References

Asswell J & Ferry JG (1978). A new method of culturing methanogens with organic acids: characteristics of an isolate mass cultured with formate. Abstracts Ann. Meeting Am. Soc. Microbiol. I: 45

Battley EH (1960) A theoretical approach to the study of the thermodynamics of growth of Saccharomyces cerevisiae. Physiologia plantarum 13: 674–686

Battley EH (1987) Energetics of Microbial Growth. John Wiley and Sons

Blevins WT & Perry JJ (1971) Efficiency of a soil Mycobacterium during growth on hydrocarbons and related substrates. Zeitschrift für algemeine Mikrobiologie 11: 181–190

Bruinenberg PM, Van Dijken JP & Scheffers WA (1983) A theoretical analysis of NADPH production and consumption in yeasts. J. of Gen. Microbiol. 129: 953–964

Frankena J, van Verseveld HW & Stouthamer AH (1988) Substrate and energy costs of the production of exocellular enzymes by Bacillus licheniformis. Biotechnol. and Bioengin. 32: 803–812

Fuchs G, Thauer R, Ziegler H & Stichler W (1979) Carbon isotope fractionation by methanobacterium thermoautotrophicum. Arch. of Microbiol. 120: 135–139

Hazeu W, Bijleveld W, Grotenhuis JTC, Kakes E & Kuenen JG (1986) Kinetics and energetics of reduced sulfur oxidation by chemostat cultures of Thiobacillus ferrooxidans. A. van Leeuwenhoek 52: 507–518

Hazeu W, Schmedding DJ, Goddijn O, Bos P & Kuenen JG (1987) The importance of the sulfur oxidizing capacity of Thiobacillus ferrooxidans during leaching of pyrite. Proc. 4th Eur. Conf. Biotechnol. 3: 497. Elsevier Science Publishers, Amsterdam

Heijnen JJ & Roels JA (1981) A macroscopic model describing yield and maintenance relationships in aerobic fermentation process. Biotechnol. and Bioeng. 23: 739–763

Heijnen JJ & Van Dijken JP (1991) In search of a thermodynamic description of biomass yields for the chemotrophic growth of micro organisms. Biotechnol. and Bioeng. (in press)

Iehana M (1990) Kinetic analysis of the growth of Chlorella vulgaris. Biotechnol. and Bioeng. 36: 198–206

Kelly DP (1990) Energetics of chemolithotrophs. In: Krulwich TA (Ed) The Bacteria, Vol 12: 449–478. Bacterial Energetics, Academic Press, San Diego

Linton JD & Stephenson RJ (1978) A preliminary study on growth yields in relation to the carbon and energy content of various organic growth substrates. FEMS Microbiol. Let. 3: 95–98

Lynd L & Zeikus JG (1983) Metabolism of H₂-CO₂, methanol, and glucose by butyribacterium methylotrophicum. J. of Bacteriol. 153: 1415–1423

Mayberry WR. Prochazka GJ & Payne WJ (1967) Growth yields of bacteria on selected organic compounds. Appl. Microbiol. 15: 1332–1338

Meijer O & Schlegel HG (1983) Biology of aerobic carbon monoxideoxidizing bacteria. Ann. Rev. of Microbiol. 37: 277–310

Minkevich IG & Eroshin VK (1973) Productivity and heat generation of fermentation under oxygen limitation. Folia Microbiol. 18: 376–385

Morii H, Koga Y & Nagai S (1987) Energetic analysis of the growth of Methanobrevibacter arboriphilus A2 in Hydrogen-Limited continuous cultures. Biotechnol. and Bioeng. 29: 310–315

Van Niel E (1990) Nitrification by heterotrophic denitrifiers and its relationship to autotrophic nitrification. Ph Thesis, Delft University of Technology

Pronk JT, Meesters PJW, Van Dijken JP, Bos P & Kuenen JG (1990) Heterotrophic growth of Thiobacillus acidophilus in batch and chemostat cultures. Arch. of Microbiol. 153: 392–398

Robertson LA (1988) Aerobic denitrification and heterotrophic nitrification in Thiosphaera pantotropha and other bacteria. PhD. thesis, Delft University of Technology, The Netherlands

Roels JA (1983) Energetics and Kinetics in Biotechnology. Elsevier Biomedical Press

Rutgers M (1990) Control and thermodynamics of microbial growth. PhD thesis, University of Amsterdam

Schink B (1984b) Clostridium magnum sp.now., a non autotrophic homoacetogenic bacterium. Arch. of Microbiol. 137: 250–255

Schink B (1984a) Fementation of 2,3-butanediol by Pelobacter carbinolicus sp.nov. and Pelobacter propionicus sp.nov., and evidence for propionate formation from C_2 compounds. Arch. of Microbiol. 137: 33–41

Siegel RS & Ollis DF (1984) Kinetics of growth of the hydrogenoxidizing Bacterium Alcaligenes eutrophus (ATCC 17707) in chemostat culture. Biotechnol. and Bioeng. 26: 764–770

Von Stockar U & Birou B (1989) The heat generated by yeast cultures with a mixed metabolism in the transition between respiration and fermentation. Biotechnol. and Bioeng. 34: 86–101

Stouthamer AH (1988) Bioenergetics and yields with electronacceptors other than oxygen. In: Erickson LE & Yee-Chak Fung D (Eds) Handbook on Anaerobic Fermentation (345–427). Dekker Marcel Inc., New York

Stouthamer AH (1979) In search of a correlation between theoretical and experimental growth yields. In: Quayle JR (Ed) Microbial Biochemistry, Vol 21: 1–48. University Park Press, Baltimore

Stouthamer AH (1973) A theoretical study on the amount of ATP required for synthesis of microbial cell material. A. van Leeuwenhoek 39: 545–565

Sublette KL (1987) Aerobic oxidation of hydrogen sulfide by Thiobacullus denitrificans. Biotechnol. and Bioeng. 29: 690–695

Taylor GT & Pirt SG (1977) Nutrition and factors limiting the growth of a methanogenic bacterium. Arch. of Microbiol. 113: 17–22

Verduyn C (1991) Physiology of yeasts in relation to growth yields. A. van Leeuwenhoek 60 (this issue)

Van Verseveld HW (1979) Influence of environmental factors on the efficiency of energy conservation in Paracoccus denitrificans. PhD thesis. Free University of Amsterdam

Westerhoff HV & Van Dam K (1987) Mosaic Non-equilibrium Thermodynamics and the Control of Biological Free Energy Transduction. Elsevier, Amsterdam

Zehnder AJB (1989) Biology of Anaerobic Microorganisms. John Wiley and Sons

Antonie van Leeuwenhoek **60**: 257–273, 1991.

The use of stoichiometric relations for the description and analysis of microbial cultures

J.A. de Hollander
Gist-Brocades, PO Box 1, 2600 MA Delft, The Netherlands

Key words: elemental balance, metabolic balance, yield

Abstract

A general method is described, which enables the derivation of predictive fermentation equations for any microbiological process. The method combines the well-known achievements of the elemental balance approach with microscopic, metabolic balances and biochemical restrictions, using the key intermediates concept. Special attention is paid to the distinction between independent and dependent flow variables of a system. The method is fully illustrated for the very simple example of heterotrophic growth on a single substrate without product formation. Other examples include growth on mixed substrates and the description of catabolic and anabolic product formation.

Introduction

The quantitative description of microbial growth and product formation has matured considerably during the past few decades, not in the last place due to improvements in the available technical equipment. Petri dishes and shake flasks have been replaced by carefully controlled fermenters. Analytical techniques including automated enzymatical and chromatographical methods are available for accurate measurement of the consumption and accumulation of substrates and products. Further, modern fermenters are equipped with instruments for (on-line) measurement of gasses and volatile products in the exhaust air.

On the theoretical side, important progress has been made by application of elemental mass balancing methods. The mass balance of a fermentation becomes more and more recognized as a valuable tool for validation of the analytical data, for detection of measurement errors and/or unnoticed products (Wang & Stephanopoulos 1983), for estimation of variables for which no direct analytical methods are available (Humphrey 1974), and for improving the accuracy and reliability of fermentation parameter estimation (Solomon et al. 1982, 1984). The fundamentals of the mass balancing theory were developed in the late seventies by Minkevich, Erickson, Roels and others (Minkevich & Eroshin 1973; Erickson et al. 1978; Minkevich 1983; Roels 1980, 1983). The theory uses the formalism of linear algebra to express the relationships between measurable (macroscopic) flows. Practical applications are mostly in the field of numerical procedures to solve a system for unknown flows or to calculate maximum likelihood estimators in case of over-determined systems.

The macroscopic mass balance theory treats the microorganism as a black box: usually only flows to and from the system (the fermenter or the microorganism) are considered, and internal (metabolic) flows are neglected because they cannot be measured, and the microorganism is considered to be in a (pseudo) steady state, i.e. there is no internal accumulation of biochemical compounds.

Another branch of quantitative microbiology uses a different approach. Here the starting point is the cellular metabolism. Biochemical stoichiometries and ATP yields of various reactions are investigated, and growth and product formation rates are predicted based on internal metabolic rates (see Stouthamer 1979; Stouthamer & Van Verseveld 1985; Tempest & Neijssel 1984). Differences in macroscopic yields can very often be explained by differences in cellular metabolism, or those yield differences can be used to estimate certain metabolic parameters.

At present there is no established and general approach to evaluate and use the relations and laws of microbial metabolism for the development of model equations predicting growth and product formation. For simple cases there is no real need for such a method, because by algebraic manipulation of a few relations its is mostly possible to derive the required equations. However, sometimes quite a number of simplifications must be used, and when the systems which are studied are more complex, or when more accurate models are required, the development of the equations becomes laborious and is prone to errors. It is the purpose of this article to present a general method for developing equations for macroscopic flows in microbiological systems, using both the elemental mass balance principle and (microscopic) metabolic relations. Methods of linear algebra are used to derive equations rather than to analyze or reduce experimental data. For simple cases this may seem a complicated procedure for those who are not familiar with these methods, however it is a great advantage of this approach that it can easily be expanded to fairly complicated systems and descriptions.

It is not possible, and it would also not be very useful, to break down the complete microbial metabolism to the level of individual enzymatic reactions. In stead of this, a varying number of complete pathways or clustered pathways is considered. For aerobic heterotrophic growth the simplest subdivision of metabolism is: catabolism – anabolism – respiration. Further subdivisions can be applied when necessary. An explicit and central role in the approach is played by non-accumulating intermediates which are used in one metabolic compartment and consumed in another. These metabolites are designated as key intermediates, the most important of them being ATP and NADH, but other may arise. Using the pseudo steady state assumption for the key intermediates, the system of metabolic reactions can be solved, leading to the required equation predicting macroscopic flows.

The method presented in this article is fully illustrated with one simple example, followed by a limited treatment of a number of more complex cases. Practical application of the derived equations for analyzing actual fermentation data will be the subject of a following article.

Outline of the method, illustrated with a very simple example

The derivation of a stoichiometric description of microbial growth and product formation starts with a summing up of all relevant reactions of the metabolism. These include catabolic reactions involved in the break-down of carbon sources, oxidative reactions in case of the involvement of oxygen or other electron acceptors, and anabolic reactions leading to biomass and possibly product formation. The reactions of the intermediary metabolism can be specified with a varying degree of detail and complexity, in accordance with the available knowledge of biochemical stoichiometry and the degree of complexity which is wished for the resulting descriptive equation. Usually only overall equations for reaction sequences or complete segments of the metabolism are included. Once the microbial system is defined by writing the various partial reactions, a purely formal way of formula processing can be followed using the methods of matrix algebra, resulting in an expression for macroscopic (measurable) flows or reaction rates, with coefficients directly related to (microscopic) biochemical stoichiometric coefficients. Not every system of reactions can be solved. However the conditions for solvability are well defined and will be derived.

The principles will be illustrated with the simple case of aerobic heterotrophic growth on a single substrate which is completely oxidized to carbon dioxide in the catabolism, and with no other product than biomass. For this system a minimum of three reactions is required:

1. The anabolic reaction (biomass formation):

$$\sigma S + a_x ATP \rightarrow X + (\sigma - 1) CO_2 + \frac{1}{2}(\sigma\gamma_s - \gamma_x) NADH$$

2. The catabolic reaction:

$$S \rightarrow CO_2 + \frac{1}{2}\gamma_s NADH + a_s ATP$$

3. The respiratory reaction:

$$NADH + \frac{1}{2} O_2 \rightarrow H_2O + a_h ATP$$

(1)

By way of simplification and for clarity reasons, the counterparts of ATP and NADH (ADP and NAD) are omitted from the reaction equations throughout this paper. A biochemical reaction or any sequence of biochemical reactions must comply to the principle of the elemental balance. In this treatment a central role is played by the carbon balance and the 'reductance' or 'redoxon' balance (Minkevich & Eroshin 1973; Minkevich 1983). The elemental balance principle leads to balance equations for all the chemical compounds involved. However, because water production in biological systems normally cannot be measured, the use of separate hydrogen and oxygen balances is impracticable, and they are preferably replaced by a 'reductance' balance (Papoutsakis 1983; Tsai & Lee 1988). An extensive discussion of the reductance balance can be found in Roels, 1983. Care must be taken that every defined partial reaction has a closing carbon – and reductance balance. The degree of reductance (γ) of any compound can be calculated from its elemental composition.

For stoichiometric coefficients of intermediary reactions related to ATP formation or consumption a subscribed symbol a_i will be used, indicating that 'a' units of ATP are used or formed per unit of substance 'i'. The unit of carbon containing macroscopic elements such as substrates, biomass and products is the C-Mole (the amount containing 12 g carbon). Other species, like NADH and ATP and all non-carbon compounds are expressed in Molar units. The first reaction of the example is the biomass formation reaction. The coefficient a_x corresponds to the well-known parameter Y_{atp} (Bauchop & Elsden 1962; Stouthamer & Bettenhaussen 1973): $Y_{atp} = M_x/a_x$, with M_x the C-molecular weight of biomass. S symbolizes the carbon source and X is the biomass. The coefficient of S is not equal to 1 because the synthesis of biomass from most substrates is associated with net production of CO_2 (Oura 1972; Babel & Müller 1985; Gommers et al. 1988), not to be confused with the CO_2 production in catabolic reactions. For instance, for glucose a value for the coefficient σ of 1.1 is cited, whereas for ethanol the value is about 1.3. When net CO_2 fixation takes place, σ has a value lower than 1.

The biomass formation reaction can result in a deficit or a surplus of reducing equivalents, depending on σ and the degree of reduction of biomass and substrate. In the catabolic reactions of this example the carbon source is fully converted to carbon dioxide and reducing equivalents (NADH). ATP is formed via the process of substrate level phosphorylation (reaction 2). The third cluster of reactions is the oxidation of NADH by oxygen via the electron transport chain, with the concomitant formation of ATP. The coefficient a_h corresponds to the over-all P/O ratio, the stoichiometry of oxidative phosphorylation: Moles ATP formed per Mole-atom O_2 consumed.

The elemental mass balance

The macroscopic flows of a microbial fermentation system must comply to the constraints set by the elemental mass balance. These constraints can be expressed by the matrix equation:

$$\mathbf{B} \, \mathbf{R}_e = 0$$

(2)

with $\mathbf{R}_e$ the $n \times 1$ vector of macroscopic flows and $\mathbf{B}$ the corresponding $m \times n$ matrix with coefficients of the molecular compositions. Usually the system is over-defined, i.e. the number of measurable or known flows (n) is greater than m, the number of chemical elements in consideration. Using the mass balance restrictions we only need to know m-n flows for a complete description of the system. Therefore the external flows of a fermentation system can be subdivided into primary flows (vector R_{e1}) and secondary flows (vector R_{e2}), the latter following from the primary flows through the mass balance, according to the following procedure.

Equation 1 can be reformulated using matrix partitioning:

$$[B_1\ B_2] \begin{bmatrix} R_{e1} \\ R_{e2} \end{bmatrix} = 0$$

which is identical to:

$$B_1\ R_{e1} + B_2\ R_{e2} = 0$$

resulting in:

$$R_{e2} = -\ B_2^{-1}\ B_1\ R_{e1}$$

or

$$R_{e2} = B_e\ R_{e1} \tag{3}$$

under the condition that the partitioning is done such, that B_2 is a non-singular matrix of dimension m×m. For the simple example system two 'elemental' balances are considered: the carbon balance and the 'reductance' balance. The coefficients in an elemental balance equation are the corresponding coefficients in the molecular formula of the (bio)chemical substance. For the 'reductance' balance the respective values for the degree of reduction are filled in. For the system of growth on a single carbon substrate the vector of external flows is:

$$R_e = [r_x\ r_s\ r_o\ r_c]^T$$

and the matrix of coefficients:

$$B = \begin{bmatrix} 1 & 1 & 0 & 1 \\ \gamma_x & \gamma_s & -4 & 0 \end{bmatrix}$$

which results via formula 3 in a (matrix) expression for r_o and r_c as linear combinations of the other flows:

$$\begin{bmatrix} {}^1\!/_4\gamma_x & {}^1\!/_4\gamma_s \\ -1 & -1 \end{bmatrix} \begin{bmatrix} r_x \\ r_s \end{bmatrix} = \begin{bmatrix} r_o \\ r_c \end{bmatrix} \tag{4}$$

The number of balances can easily be increased when necessary. For instance a nitrogen balance must be included when the carbon source contains nitrogen, a sulfur balance can be included when a sulfur containing product is formed etc. After application of the elemental balance m-n = f degrees of freedom are left for the primary flow vector. In the following section it will be shown how metabolic stoichiometric relations provide a further restriction for the primary macroscopic flows. The partitioning of matrix B is not fully arbitrary. When one or more of the flows can be considered as independent variables of the system, such as the carbon source consumption rate in carbon-limited chemostats or fed-batch cultures, this flow should be included in the reduced flow vector R_{e1}. Further, unfavorable partitioning might lead in some cases to singular matrices in the calculation of the stoichiometric balance equation for R_{e1}.

Balancing the metabolic reactions

The general expression for a system of microbial reactions can be formulated as:

$$E = 0 \tag{5}$$

with E the vector of (bio)chemical species, e.g for equations 1:

$$E = [X\ S\ O_2\ CO_2\ NADH\ ATP]^T$$

and Z the stoichiometry matrix, which follows from the equations 1:

$$Z = \begin{bmatrix} 1 & -\sigma & 0 & \sigma-1 & \frac{1}{2}(\sigma\gamma_s-\gamma_x) & -a_x \\ 0 & -1 & 0 & 1 & \frac{1}{2}\gamma_s & a_s \\ 0 & 0 & -\frac{1}{2} & 0 & -1 & a_h \end{bmatrix}$$

Note that coefficients appearing on the left hand

side in the reaction equations get a negative sign in the formalism of the matrix notation. As a consequence of this convention, flows of substrates have negative values, and flows of products have positive values.

As the purpose of the following procedure is the derivation of a balance equation for the primary flows, a reduced form of equation 5 is used, where chemical species corresponding to secondary flows (O_2 and CO_2 in the example) are removed from the vector E and the corresponding columns in matrix Z. An expression for the primary external flows as linear combinations of 'internal' reaction rates is then obtained by transposing matrix Z' and multiplying with the vector of reaction rates:

$$Z'^T R_m = R'_{el} \qquad (6)$$

R'_{el} is the vector of primary flows, extended with zero elements corresponding to the key intermediates (ATP and NADH in this case). The matrix Z'^T has the dimension $(f + 1) \times k$; R_m is of dimension $k \times 1$ and R'_{el} of dimension $(f + 1) \times 1$, with l the number of conserved (internal) biochemical species (key intermediates), and k the total number of (clustered) metabolic reactions in consideration. For the example case the following matrices are found:

$$R_m = [\quad r_1 \quad r_2 \quad r_3 \;]^T$$

$$Z'^T = \begin{bmatrix} 1 & 0 & 0 \\ -\sigma & -1 & 0 \\ \hdashline \tfrac{1}{2}(\sigma\gamma_s - \gamma_x) & \tfrac{1}{2}\gamma_s & -1 \\ -a_x & a_s & a_h \end{bmatrix}$$

$$R'_{el} = \begin{bmatrix} r_x \\ r_s \\ \hdashline 0 \\ 0 \end{bmatrix} \qquad E' = \begin{bmatrix} X \\ S \\ NADH \\ ATP \end{bmatrix}$$

A solution for the matrix equation is achieved by partitioning of the matrices in correspondence with the distinction between external flows and directly associated metabolic reactions on the one hand, and balanced internal flows on the other hand. The partitioning is already indicated in the matrix expressions. The following system of matrix equations results:

$$Z_1\, R_m = R_{el}$$
$$Z_2\, R_m = 0$$

which can have the solution:

$$(Z_2\, Z_1^{-1})\, R_{el} = 0,$$

or

$$Z_3\, R_{el} = 0 \qquad (7)$$

which is a balance equation for the primary external flows in terms of coefficients (in Z_3) which are algebraic expressions of stoichiometric or physiological parameters.

A number of dimensional constraints must be fulfilled to make the indicated solution possible. Matrix Z_1 must be square to make inversion possible (and, of course, the matrix must be non-singular). As the dimension of Z_1 is $f \times k$, this would imply that $f = k$, or: the number of metabolic reactions must be equal to the number of primary external flows. In general this will not be the case. However, very often one or more of the metabolic reactions will be purely internal, i.e. the reaction has no direct relation with one of the primary external flows. In the example this is the case for reaction 3: oxidation of NADH. A column of zero's in sub-matrix Z_1 makes the elimination of one of the reactions possible, leading to the following reduced matrix expressions:

$$Z_1'\, R_m' = R_{el}$$
$$Z_2'\, R_m' = 0$$

with

$$\mathbf{z}_1' = \begin{bmatrix} 1 & 0 \\ -\sigma & -1 \end{bmatrix}$$

$$\mathbf{z}_2' = [-a_x + \tfrac{1}{2}(\sigma\gamma_s - \gamma_x)a_h \quad a_s + \tfrac{1}{2}\gamma_s a_h]$$

The solution for $\mathbf{Z}_3$ is then:

$$\mathbf{Z}_3 = [a_x + \sigma a_s + \tfrac{1}{2}\gamma_x a_h \quad a_s + \tfrac{1}{2}\gamma_s a_h]$$

A sufficient condition for solvability of a microbial stoichiometric reaction system is therefore: The system of reaction equations must be formulated such, that there are just as many (lumped) internal reactions related to external flows as there are 'free' flows according to the mass balance. Or, when this condition is not fulfilled, the number of conserved (internal) biochemical species must be at least one more than the difference between the number of metabolic reactions and the degree of freedom of the system: $l \geq k - f + 1$. The column dimension of $\mathbf{Z}_2$ after the eventual elimination of one or more of the internal reactions is $g = f - k + l$. The parameter g will be called the exuberance of the metabolic system. Although these conditions may seem rather restrictive, all realistic cases studied so far appeared in a form that made a solution possible, or they could be modified to such a form. Sometimes this modification requires introduction of additional relations between the internal or external flows, not directly following from metabolic stoichiometries. In one of the more complicated examples such a case is treated.

Development of a predictive fermentation model

The final expression for the metabolic balance, equation 7, represents g linear balance equations between the primary external flows. The form of the equation is similar to equation 2, the elemental balance equation, and accordingly its further processing is equivalent. $\mathbf{Z}_3$ may be partitioned into a $g \times g$ square matrix $\mathbf{Z}_d$ and a $g \times (f - g)$ matrix $\mathbf{Z}_i$,

implying a partitioning of the primary flows vector $\mathbf{R}_{e1}$ into $\mathbf{R}_{d1}$, a vector with dependent flows, and $\mathbf{R}_i$, a vector containing the independent flows of the system. The dimension of $\mathbf{R}_i$ is $f' = f - g$ and it represents the remaining degree of freedom for the metabolic system, after application of the elemental *and* metabolic balance. $\mathbf{R}_{d1}$ is predicted through the metabolic balance when R_i is known: equation 7 can be processed to (cf. equation 2 and 3):

$$\mathbf{R}_{d1} = -\mathbf{Z}_d^{-1}\mathbf{Z}_i\mathbf{R}_i \tag{8}$$

The complete vector of dependent flows ($\mathbf{R}_d$) is the combination of $\mathbf{R}_{d1}$ and $\mathbf{R}_{e2}$. The inversion of $\mathbf{Z}_d$, and further multiplication with $\mathbf{Z}_i$, as indicated in equation 8, can be done easily numerically when all stoichiometric coefficients have known values. However, for developing an explicit equation relating dependent and independent flows, containing unknown coefficients as parameters, it appeared useful to express equation 8 in a slightly different way:

$$\mathbf{R}_{d1} = -\mathbf{D}_m^{-1}\mathbf{Z}_c\mathbf{Z}_i\mathbf{R}_i = -\mathbf{D}_m^{-1}\mathbf{C}_1\mathbf{R}_i \tag{8'}$$

Here inversion of $\mathbf{Z}_d$ was done using the matrix with the so-called cofactors of $\mathbf{Z}_d$ (Bronson 1989), transposing this matrix (symbolically: $\mathbf{Z}_c = (\mathbf{Z}_d^c)^T$), and dividing by the determinant of $\mathbf{Z}_d$. This determinant is a important descriptor of the metabolic system: its reciprocal appears as a factor in all expressions for relations between external flows; I propose the name Metabolic Determinant ($\mathbf{D}_m$). The multiple of the matrices $\mathbf{Z}_c$ and $\mathbf{Z}_i$ is the matrix with coefficients of the linear relations between dependent and independent primary flows, hence the symbol $\mathbf{C}_1$. The metabolic determinant for the example case is simply the first element of $\mathbf{Z}_3$, because here $g = 1$:

$$\mathbf{D}_m = a_x + \sigma a_s + \tfrac{1}{2}\gamma_x a_h$$

It is interesting to give an interpretation of this metabolic determinant: it represents the amount of microbial energy contained in biomass: $a_x = $ the amount of ATP needed for biosynthesis of one C-Mole biomass; $\sigma a_s = $ the amount of ATP which would have been formed when the substrate for biomass synthesis would have been used in catabolism; $\frac{1}{2}\gamma_x a_h$ = the energy gain (ATP/C-Mole) of 'biological combustion' of the biomass. Such an interpretation as given here appears to be possible in cases where the carbon (or energy) substrate is included in the independent flow vector. In other cases interpretation of D_m is less obvious.

The full expression for the dependent flows of a microbial system follows from the equations 8 in combination with the elemental balance equations, and can be expressed in a form analogous to equation 8':

$$R_d = - D_m^{-1} \, C \, R_i \tag{9}$$

For practical situations the primary dependent flow (s) can be evaluated by use of equation 8', followed by application of equation 3 for the secondary flows. However for theoretical reasons it is illustrative to develop an expression for the coefficients matrix C in formula 9. For this purpose equation 2 is reformulated using a three-fold partitioning:

$$\begin{bmatrix} B_1 & B_2 & B_3 \end{bmatrix} \begin{bmatrix} R_{d1} \\ R_{d2} \\ R_i \end{bmatrix} = 0 \tag{10}$$

or

$$B_1 \, R_{d1} + B_2 \, R_{d2} + B_3 \, R_i = 0$$

By combining with equation 8' and after some algebraic manipulation the following expression results:

$$R_{d2} = - D_m^{-1} \, C_2 \, R_i$$

with

$$C_2 = B_2^{-1} \, (D_m \, B_3 - B_1 \, C_1)$$

By adding equation 8' to this set of linear equations the complete description for the system is obtained

(expression 9) with C being the vertical concatenation of C_1 and C_2. Using this strictly formal procedure the following complete result was obtained for the case of aerobic heterotrophic growth on a single substrate:

$$\begin{bmatrix} r_x \\ r_o \\ r_c \end{bmatrix} = - \frac{1}{a_x + \sigma a_s + \frac{1}{2}\gamma_x a_h} \begin{bmatrix} a_s + \frac{1}{2}\gamma_s a_h \\ -\frac{1}{4}\gamma_s a_x - \frac{1}{4}(\sigma\gamma_s - \gamma_x) a_s \\ a_x + (\sigma - 1) a_s + \frac{1}{2}(\gamma_x - \gamma_s) a_h \end{bmatrix} r_s \tag{11}$$

Equation 9 gives a description of a microbial system in terms of a relationship between independent and dependent flows or reaction rates. Expressions like equation 11, or the general form, equation 9, can be used to evaluate the effect of variation in physiological parameters on mascroscopic variables. For instance, an expression for the RQ (respiratory coefficient), can be derived easily from equation 11. With such an equation it can be illustrated why and under which circumstances the RQ of a fermentation deviates from the 'theoretical' value of 1.0 for glucose and analogous carbon substrates.

Usually a relation between flows is expressed as a (macroscopic) yield coefficient. Starting from equation 9 it is easy to develop a general expression for such yield coefficients. A yield coefficient Y_{ij} is defined as the amount of i formed per amount of j consumed. If we have p dependent flows and q independent flows, then we can define $p \times q$ yield coefficients. A matrix expression for the yield coefficients is:

$$Y = D_m^{-1} \, C \, \Phi \tag{12}$$

where Y is the $p \times q$ dimensional matrix of yield coefficients, C is the same matrix with expressions of physiological coefficients as in equation 9 and Φ is a $q \times q$ dimensional square matrix with ratios of independent flows (with 'one's' at the main diagonal positions). It follows that one particular macroscopic yield coefficient can always be written as the sum of a constant and a series of linear combinations of q-1 (independent-) flow ratios.

A computer program for solving the matrix equations

The matrix approach is powerful, because it gives formal rules for the derivation of metabolic balance equations which also can be applied for very complicated systems. Most computer programs working with matrices permit manipulations with numerical matrices only. Such programs can be used for consistency checking, error reduction or calculations of unknown flows. For the elaboration of the metabolic balance equations in this paper, a programming tool called CML (character matrix language) was used which enables the processing of formula's in character matrices. A part of the source listing of the program used for the example treated above is given as an illustration. The procedures discussed in the example can easily be recognized. The program runs under DOS on a PC; further information can be obtained from the author.

The treatment of maintenance coefficients

In the equations derived so far, no attention was paid to the occurrence of maintenance coefficients, which normally are introduced to describe non-constancy of yield factors (macroscopic or microscopic) in dependency of some specific rate, mostly the microbial growth rate. The classical equation for the relation between growth yield and growth rate is: $1/Y'_{xs} = 1/Y_{xs} + m_s/\mu$ (Pirt 1965). Here we are interested in the way how microscopic yield coefficients determine the macroscopic yield; the classical equation for growth rate dependency of microscopic yield is: $1/Y'_{atp} = 1/Y_{atp} + m_{e}/\mu$ (Stouthamer & Bettenhausen 1973). As the relation between Y_{atp} and the coefficient a_x is already given, the latter equation can easily be transformed to: $a'_x = a_x + m_x/\mu$ with m_x the maintenance coefficient in Mole ATP per C-Mole biomass per time unit. In the case of chemostat studies, where usually a series of steady states is evaluated over a range of growth rates, the coefficient a_x in the formulas such as in equations 11 should be replaced by the expanded form including the maintenance coefficient, as given here. This substitution will make the matrix equations non-linear, but this problem can be tackled by using iterative numerical methods for parameter estimation, as will be illustrated in a following paper. Also for (fed-) batch cultures, where often a trajectory of growth rates is followed during a single fermentation, this approach can be followed.

The concept of maintenance energy can also be

```
" the stoichiometry matrix:"

Z = ( 1, -sig, .5*sig*gamx-.5*gamx, -a(x);   "biomass formation"
      0, -1 ,      .5*gams       , a(s); "catabolism      "
      0,  0 ,        -1          , a(h) }; "respiration      "

Z' = TRANSP(Z);          " transpose the matrix            "
Z1 = Z'[1:2,1:2];        " partition into Z1 (rows 1 and 2) "
Z2 = Z'[3:4,1:3];        "            and Z2  (rows 3 and 4) "
Z2 = REDUCE(Z2);         " reduce Z2 to one row            "

Z3 = Z2*INV(Z1);         " solve the metabolic balance equation "

PRINT(Z3);               " and print the result            "
```

applied for the description of non-constant yield coefficients of product formation in a similar way as for biomass formation. For instance when extracellular protein (enzyme) production is studied, an energetic coefficient a_p can be defined. One can imagine that the specific energy costs of protein synthesis and excretion are not fully proportional to the rate of protein production, because the maintenance of the cellular machinery for this process will require energy, also when the conditions are such that no product is formed. By analogy with the ATP-per-biomass coefficient a_x one can substitute a_p by a form like $a_p + m_p/q_p$ with q_p the specific rate of product formation and m_p a 'maintenance coefficient' for the product forming cellular apparatus.

A few more complicated examples

Energetically distinguishable forms
of reducing equivalents

In the example of heterotrophic aerobic growth treated in the previous section, the only form of reducing equivalents taken into consideration was NADH. This is a simplification which is not always justified, as reducing equivalents may enter the respiratory chain at various levels, and accordingly have different energy yields in oxidative phosphorylation. For instance, NADH formed in mitochondria of eucaryotic organisms is passed to the respiratory chain via a 'Site I' coupled NADH-dehydrogenase, whereas cytoplasmic NADH is oxidized by a 'Site I' uncoupled enzyme (De Vries & Grivell 1988). Some catabolic reactions use FAD as a reducing equivalent acceptor, which links to the respiratory chain at a position below 'Site I', and leading to a lower energy gain during the oxidation process. Another example is the methanoldehydrogenase in *Paracoccus denitrificans*, which delivers reducing equivalents at the level of 'Site III' (Van Verseveld & Stouthamer 1980). In such situations equation 11 will not be correct in predicting biomass yields, or parameters estimated using this model equation will give a more or less wrong value

(especially for a_h, the P/O ratio). This problem can be solved by taking into account the various forms of reducing equivalents; here we differentiate between NADH and FADH. A factor α is defined, which gives the fraction of reducing equivalents formed in the catabolism of the substrate which is in the form of FADH. This fraction is known in many cases where the metabolic routes are known. The energy gain in oxidative phosphorylation is defined as a_{h1} for the oxidation of NADH and as a_{h2} for the oxidation of FAD. The matrix of stoichiometric coefficients then becomes (cf. equation 5):

$$\mathbf{E} = \begin{bmatrix} X & S & NADH & FADH & ATP \end{bmatrix}^T$$

$$\mathbf{Z} = \begin{bmatrix} 1 & -\sigma & \frac{1}{2}(\sigma\gamma_s-\gamma_x) & 0 & -a_x \\ 0 & -1 & \frac{1}{2}(1-\alpha)\gamma_s & \frac{1}{2}\alpha\gamma_s & a_s \\ 0 & 0 & -1 & 0 & a_{h1} \\ 0 & 0 & 0 & -1 & a_{h2} \end{bmatrix} \begin{matrix} \text{(growth)} \\ \text{(catabolism)} \\ \text{(NADH oxidation)} \\ \text{(FADH oxidation)} \end{matrix}$$

In comparison with equation 5 the number of reactions is increased with one; the columns for oxygen and carbon dioxide are left out for simplicity reasons (i.e. $\mathbf{Z}$ is actually the reduced stoichiometry matrix). Using the same procedure as for the more simple case with only one form of reducing equivalents, the following equation for the macroscopic biomass yield Y_{xs} was obtained:

$$Y_{xs} = \frac{a_s + \frac{1}{2}(1-\alpha)\gamma_s a_{h1} + \frac{1}{2}\alpha\gamma_s a_{h2}}{a_x + \sigma a_s + \frac{1}{2}\gamma_x a_{h1} - \frac{1}{2}\alpha\sigma\gamma_s(a_{h1} - a_{h2})}$$

This yield expression can be compared with the first row of equation 11, expressing r_x as linear function of r_s. Both the numerator and denominator of the yield coefficient expression are changed. In the numerator the energy gain in oxidative phosphorylation is a weighted average of the two P/O ratio's, as expected. In the denominator (= Metabolic Determinant) a correction term appears, related to the difference in P/O ratio. This is a result of the assumption that reducing equivalents produced in biomass formation are in the form of NADH (cf. the first row in the stoichiometry matrix).

Aerobic heterotrophic growth on two substrates

The general description of two-carbon-sources growth requires the extension of equations 1 with at least two more equations: anabolic and catabolic reactions of the second substrate. Here we assume that the reduction equivalents in various reactions are indistinguishable, so a total of 5 reactions is sufficient to describe the metabolic system in the form of equation 5:

$$\mathbf{Z}\,\mathbf{E} = 0$$

$$\mathbf{E} = [X \quad S_1 \quad S_2 \quad NADH \quad ATP]$$

$$\mathbf{Z} = \begin{bmatrix} 1 & -\sigma_1 & 0 & \frac{1}{2}(\sigma_1\gamma_{s1}-\gamma_x) & -a_{x1} \\ 1 & 0 & -\sigma_2 & \frac{1}{2}(\sigma_2\gamma_{s2}-\gamma_x) & -a_{x2} \\ 0 & -1 & 0 & \frac{1}{2}\gamma_{s1} & a_{s1} \\ 0 & 0 & -1 & \frac{1}{2}\gamma_{s2} & a_{s2} \\ 0 & 0 & 0 & -1 & a_h \end{bmatrix} \begin{matrix} \text{(growth on S1)} \\ \text{(growth on S2)} \\ \text{(catabolism } S_1) \\ \text{(catabolism } S_2) \\ \text{(respiration)} \end{matrix}$$

In these equations two different biomass yield coefficients a_x are introduced, one for each substrate. This is necessary, because energy costs for biosynthesis of macromolecules are not the same for substrates entering at different levels in the metabolism. In general growth on low molecular weight substrates requires more energy (lower Y_{atp}, higher a_x) than growth on substrates like glucose (Stouthamer 1973). It is not possible in this case to partition the matrix Z in such a way that a square submatrix Z_1 of dimension f × f is obtained, related to external flows only. In other words: the system as formulated now cannot be solved. A solution can be found however, when an extra relation is added, for instance regarding the flows of the two substrates towards the anabolic pathways. When microorganisms are grown in batch culture, usually one substrate is consumed first, and growth proceeds on the second substrate when the first substrate is depleted from the medium (diauxic growth). Under carbon-limited conditions however, very often two substrates may be consumed simultaneously, but there is a preference of anabolism for one substrate (e.g. glucose) above the alternative (e.g. ethanol) which is predominantly used for energy generating purposes. When the flux of the second substrate is relatively high, also this alternative substrate may be used in anabolic reactions, thereby changing the energetics of growth of the organism. So at least two metabolic states can be distinguished, as illustrated in Fig. 1. These different states can be defined in mathematical form by the following equations:

– State 1, substrate 2 goes exclusively to catabolism, anabolic reactions start exclusively from substrate 1: $r_2 = 0$
– State 2, substrate 1 goes exclusively to anabolism : $r_3 = 0$
– An intermediate state (state 3) may be defined, where anabolic routes are equally provided by substrates 1 and 2 (for instance such a situation can occur when the substrates are very similar): $r_1 = r_2$

These constraints can be put into matrix form:

$$\begin{bmatrix} 0 & 1 & 0 & 0 & 0 \\ 0 & 0 & 1 & 0 & 0 \\ -1 & 1 & 0 & 0 & 0 \end{bmatrix} \begin{bmatrix} r_1 \\ r_2 \\ r_3 \\ r_4 \\ r_5 \end{bmatrix} = 0$$

Of these 3 equations, only one at the time is valid. For each metabolic state the appropriate row of equation 14 can be selected and added to the equations analogous to equations 6. By adding such a constraint, one of the reactions related to external flows can be eliminated (r_2 or r_3 in this case), resulting in a solvable metabolic system, according to the described procedures. The first row of equation 8 for this system reads:

$$r_x = -D_m^{-1} [c_1\ c_2] \begin{bmatrix} r_{s1} \\ r_{s2} \end{bmatrix}$$

from which the following equation for the macroscopic yield of biomass on carbon Y_{xc} can be derived:

$$Y_{xc} = -r_x/(r_{s1}+r_{s2}) = -D_m^{-1}[c_1\ c_2] \begin{bmatrix} 1-\o \\ \o \end{bmatrix}$$

or

$$Y_{xc} = D_m^{-1}\,\mathbf{C}\,\boldsymbol{\Phi}$$

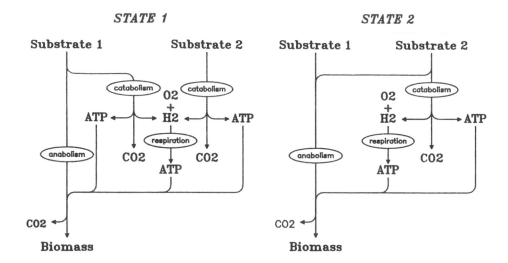

Fig. 1. Simplified metabolic schemes for heterotrophic microbial growth on two substrates. *(Left)* state 1 of metabolism, the second substrate is used for energy generation, not for anabolism. *(Right)* state 2 of metabolism, substrate 1 is used only as a biosynthesis precursor; the second substrate provides the energy and is also used as an additional biosynthesis precursor.

with $\emptyset = r_{s2}/(r_{s1} + r_{s2})$, the flow of substrate 2 relative to the total flow of carbon sources. This equation is analogous to the general equation 12 given earlier. The definition of Φ is slightly different here, because a yield on total carbon source is expressed, rather than a yield on one particular carbon source. The following expressions were found for the metabolic determinant and the coefficient vector $\mathbf{C}$ in this linear equation:

state 1:

$$\mathbf{C} = \begin{bmatrix} a_{s1}+\tfrac{1}{2}\gamma_{s1}a_h & a_{s2}+\tfrac{1}{2}\gamma_{s2}a_h \end{bmatrix}$$

$$D_m = a_{x1} + \sigma_1 a_{s1} + \tfrac{1}{2}\gamma_x a_h$$

state 2:

$$\mathbf{C} = \begin{bmatrix} \sigma_2/\sigma_1 a_{s2}+\delta_e/\sigma_1+\tfrac{1}{2}\gamma_{s1}a_h & a_{s2}+\tfrac{1}{2}\gamma_{s2}a_h \end{bmatrix}$$

$$D_m = a_{x1} + \delta_e + \sigma_2 a_{s2} + \tfrac{1}{2}\gamma_x a_h$$

where δ_e is $a_{x2} - a_{x1}$, the difference between energetic coefficients for growth on S_1 and S_2.

A similar equation was derived by Roels and coworkers (Roels 1983; Bonnet et al. 1980, 1984), using less general algebraic methods. For experimental data for growth of *Saccharomyces cerevisiae*

on glucose plus ethanol, parameters similar to a_x and a_h were estimated by linear regression. Using the reported parameter values (and associated assumptions about stoichiometric coefficients) the equations for 'state 1' and 'state 2' were plotted (Fig. 2). The two lines cross in the physically possible interval for $\emptyset$, and it can be calculated that the crossing point is at a value for $\emptyset$ where precisely all of substrate 1 (glucose) goes to anabolism, and all catabolic energy generation starts from substrate 2 (ethanol). The intercept of the line 1 for state 1 at the left vertical axis is the yield of biomass on pure glucose. Going to the right along the $\emptyset$-axis, auxiliary substrate is provided which is used for energy generation. This may have a positive or negative effect on the over-all yield of biomass on carbon, depending on the energy content of the cosubstrate and the metabolic pathway used for its catabolism. In this particular case the line is nearly horizontal, despite of the high degree of reduction of ethanol. This can be explained by the relatively low P/O ratio of *Saccharomyces* and the energy needed for activation of ethanol. In the right part of the figure an increasing amount of ethanol is used for anabolic purposes, resulting in a lower over-all yield. The switch-over between the metabolic states when the ratio of substrates is changed was not clearly no-

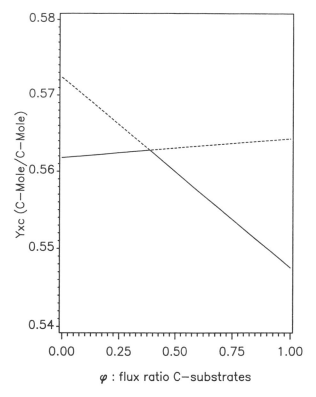

_φ_ : flux ratio C—substrates

Fig. 2. Predicted biomass yields on total consumed carbon for _Saccharomyces cerevisiae_ growing on glucose plus ethanol, as a function of the flux ratio ø. The lines for state 1 and state 2 of the metabolism (cf. Fig. 1 and the text) were calculated using parameter values as provided by Bonnet et al. 1980. The transition between the states occurs at the intersection of the two straight lines (_left_: state 1, _right_: state 2). The solid line represent the model equation under the restrictions as discussed in the text.

ticed by the original authors. In this case the changes in yields are rather small, however for other organisms or substrates this may be different. The advantage of a secondary energy generating substrate has been put forward by Babel and others (Babel & Müller 1985b; Babel 1986; Gommers et al. 1986; Van Verseveld & Stouthamer 1980). The stoichiometric balance equations as derived here, may predict the point where further increase of the relative flux of the secondary substrate becomes disadvantageous.

Anabolic and catabolic product formation

Microbial product formation may be directly relat-ed with carbon source breakdown (catabolic product formation) or not (anabolic product formation). Examples of the first category are products resulting from fermentative reactions (e.g. ethanol), partial oxidations (e.g. gluconic acid, citric acid) or over-flow metabolites (e.g. mannitol). The second group of product formation reactions includes secondary metabolite formation and excretion of macromolecules: polysaccharides, proteins (enzymes) etc. All these different forms of microbial product formation can be described with essentially the same metabolic balance equation. For the simple case of one heterotrophic substrate one can start with the system of equations 1 (equations for growth, catabolism and respiration) and add an equation for the product forming reaction, which is written in analogy to the biomass formation reaction:

$$\sigma_p\ S + a_p\ ATP$$
$$\rightarrow P + (\sigma_p - 1)\ CO_2 + {}^1\!/_2(\sigma_p\gamma_s - \gamma_p)\ NADH$$

The coefficient a_p is the amount of ATP used per amount of product formed. In the case of catabolic product formation usually this will be a negative figure (i.e. ATP is formed). The coefficient σ_p is the amount of carbon source used for product formation, which is different from 1 when net CO_2 production or CO_2 formation occurs in the pathway from the substrate to product. As an example: glycolytic ethanol formation by yeast reads:

$$1{}^1\!/_2 CH_2O - {}^1\!/_2 ATP \rightarrow CH_3O_{1/2} + {}^1\!/_2 CO_2$$

with $\sigma_p = 1{}^1\!/_2$, $a_p = -{}^1\!/_2$ and ${}^1\!/_2(\sigma_p\gamma_s - \gamma_p) = 0$ (the reaction is redox-neutral). The matrix expression for the stoichiometric system follows automatically from the 4 reaction equations:

$$\mathbf{Z}\ \mathbf{E} = 0$$

$$\mathbf{E} = [X \quad P \quad S \quad NADH \quad ATP]^T$$

$$\mathbf{Z} = \begin{bmatrix} 1 & 0 & -\sigma_x & {}^1\!/_2(\sigma_x\gamma_s-\gamma_x) & -a_x \\ 0 & 1 & -\sigma_p & {}^1\!/_2(\sigma_p\gamma_s-\gamma_p) & -a_p \\ 0 & 0 & -1 & {}^1\!/_2\gamma_s & a_s \\ 0 & 0 & 0 & -1 & a_h \end{bmatrix}$$

Using the standard procedure, the following equation for the flow of biomass related to the flows of product and substrate is obtained:

$$r_x = -D_m^{-1} \, C \begin{bmatrix} r_p \\ r_s \end{bmatrix}$$

with

$$C = [a_p + \sigma_p a_s + {}^1\!/_2 \gamma_p a_h \quad a_s + {}^1\!/_2 \gamma_s a_h]$$

and

$$D_m = a_x + \sigma_x a_s + {}^1\!/_2 \gamma_x a_h$$

The element in C corresponding to r_s is the same as in the example without product formation (equation 11); also the 'metabolic determinant' is identical. Anabolic product formation always results in a decreased biomass yield Y_{xs}, because all terms in the first element of C (corresponding to r_p) normally have positive values: the sign of r_p is positive (because it is a flow rate of a product), therefore the partial derivative $\partial r_x / \partial r_p$ is negative. For catabolic products however, it is (at least theoretically) possible that partial oxidation of the carbon source leads to a higher yield of biomass on the carbon source. If $a_p < 0$ (i.e. ATP is formed in the product forming reaction), *and* $\sigma_p a_s < - a_p$ *and* $\gamma_p a_h$ is low (the product is rather oxidized and/or the P/O ratio (a_h) is low), then such a situation might occur. Clearly such a condition will be rather rare, and is certainly not fulfilled in the case of ethanol formation by yeast, given as an example of catabolic product formation.

In equation 15 the biomass formation rate is defined as a linear function of the flows of substrate and product. According to the definition of equation 8, r_p is then an independent variable of the system. Very often this will not be correct: product formation (especially anabolic product formation) normally can not be controlled by the experimenter and should be considered as a dependent flow from

an experimental point of view. Only in very special cases the rate of product formation can be controlled independently from the rate of carbon-source utilization. One example is a situation where product formation depends on the supply of a specific precursor substance. For the general case however, one would like to have an equation which predicts biomass formation and product formation from the rate of carbon-source utilization. The metabolic balance method *as such* does not provide a solution for this. For the development of such an equation, further relationships between the various reaction rates must be included. These relationships can follow from physiological or biochemical knowledge, for instance about the occurrence of a maximal activity of a certain enzyme or pathway. Alternatively, there may be experimental evidence that there exists a relation between the specific rate of product formation (q_p) and the specific growth rate (μ) of the organism. Such a relation can be used to eliminate r_p from the vector of independent flows and transfer it to the dependent flows vector. The procedure is illustrated for the case where a phenomenological relation is known or supposed between q_p and μ:

$$q_p = f(\mu)$$

The functional relationship can have any suitable form, for instance a straight line function, or a Monod-type saturation formula. For the purpose of developing a predictive fermentation equation a proportional relation between r_p and r_x must be defined:

$$r_p = \beta \, r_x \tag{16}$$

The coefficient β in this equation will generally not be a constant, but a function of μ. Equation 16 can now be added to the metabolic balance equation 7, for the example of this paragraph:

$$\begin{bmatrix} a_x + \sigma_x a_s + \tfrac{1}{2}\gamma_x a_h & a_p + \sigma_p a_s + \tfrac{1}{2}\gamma_p a_h & \vdots & a_s + \tfrac{1}{2}\gamma_s a_h \\ \beta & -1 & \vdots & 0 \end{bmatrix} \begin{bmatrix} r_x \\ r_p \\ r_s \end{bmatrix} = 0$$

As a consequence of the introduction of a phenomenological equation into the metabolic balance equation, the 'exuberance' of the system is increased from 1 to 2, and therefore 2 dependent flows (r_x and r_p) are predicted from one independent flow (r_s). Then the fermentation equation can be derived according to equation 8', after partitioning of the balance matrix and flow vector into dependent and independent parts, as indicated by the dotted lines. The following equation results:

$$\begin{bmatrix} r_x \\ r_p \end{bmatrix} = -D_m^{-1} \; C \; r_s$$

with

$$D_m = a_x + \beta a_p + (\sigma_x + \beta \sigma_p) a_s + \tfrac{1}{2}(\gamma_x + \beta \gamma_p) a_h$$

and

$$C = \begin{bmatrix} a_s + \tfrac{1}{2}\gamma_s a_h \\ \beta a_s + \tfrac{1}{2}\beta \gamma_s a_h \end{bmatrix}$$

Further development of a full description of the system, also including r_o and r_c in the dependent flows vector, is possible according to the procedure of equation 10 and further. The resulting expressions become rather complicated however, and direct numerical evaluation of these flows through the mass balance (equation 3) will be preferred in general. The transfer of the product formation flow to the dependent flows vector has the consequence that one extra parameter (β) is introduced in the metabolic system. In general β will not be a constant; for instance β can be dependent of μ. This does not present a problem for the numerical evaluation of the fermentation equation, as was already noticed in the paragraph about maintenance coefficients, where a_x was made dependent of μ.

Discussion

The theme of this study is the elaboration of a general method for the development of a fermenta

tion equation, using elemental balances and metabolic balances in a combined way. Most of the concepts crucial in this approach (degree of reduction balance, biological energy (ATP-) balance, non-accumulating key intermediates) have been presented and used more or less extensively in various forms by others. The power of the present method is its comprehensiveness, its general applicability, and its link with experimental design by emphasizing the distinction between independent and dependent flow variables of a microbial system. The concept of breaking down the cellular metabolism into a limited number of partial reactions, consisting of clustered individual enzymatic reaction, is implicit in many publications in the field of quantitative microbiology. An explicit formulation in relation to balancing method can be found in Minkevich, 1983. The usefulness of considering metabolic balances was demonstrated by Verhoff & Spradlin (1975) for the citric acid fermentation by *Aspergillus niger*. This approach was further refined and extended by Papoutsakis (Papoutsakis 1983; Papoutsakis & Meyer 1984a, 1984b), who showed that considering the metabolic topology results in a reduced degree of freedom for a systems description, as compared with the macroscopic mass balance method. The procedures of Verhoff & Spradlin, Papoutsakis (and Aiba & Matsuoka 1979, as another example) require a careful and rather complicated elaboration of algebraic equations. Tsai & Lee (1988) proposed an alternative, the application of Gibb's rule, which ends up with the same results, but which is simpler. These latter authors used Papoutsakis' method to perform a statistical analysis and data adjustment similar to a treatment based on elemental balances, as reviewed by Wang & Stephanopoulos (1983). All these papers are mainly interested in checking the internal consistency of fermentation data or discrimination between different possible pathways. Here emphasis is on the elaboration of predictive fermentation equations, which enables the estimation of physiological parameters of the microorganism. This theme is more extensively illustrated in a second article with examples from the literature.

By using macroscopic mass balancing techniques a relation can be set-up between 2 subsets of the external flows of a microbial fermentation system (equation 3). The secondary flow vector has the dimension m (the number of chemical elements in consideration), whereas the primary flows vector has the dimension of the degree of freedom of the system f. When m-n = f flows are known, the other flows follow through equation 3 as a linear combination of the known flows. By correlating the macroscopic flows with microscopic (metabolic) reaction rates, another partitioning of the macroscopic flows was obtained: a vector of 'dependent' flows predicted by a vector of 'independent' flows (equation 8). A parameter g was defined, the exuberance of the metabolic system, following from the number of (clustered) metabolic reactions and the number of key intermediates involved with these reactions. For solvable systems g must be at least one, which is also the most common value for g. The dimension of the vector of independent flows is f' = f-g; in other words: application of metabolic balancing reduces the degree of freedom of the system with at least one. The elements in the coefficients matrix defining the relationship between the two flow vectors consist of more or less complicated formulas build-up with physiological, microscopic parameters. When these parameters are known, the 'dependent' flows can be predicted for values of the 'independent' flows. When the parameters are not known, they can be estimated by measuring one or more of the dependent flows. The metabolic exuberance parameter g is in principle identical to the parameter 'd' defined by Tsai & Lee (1988b). These authors state that application of the metabolic balance in addition to the elemental mass balance is advantageous (reduces the degree of freedom) when 'd' is greater than 1. This is equivalent to the statement in the paragraph 'outline of the method' that for solvable metabolic systems g must be at least 1. However, the concept of metabolic exuberance is used with more freedom than the parameter 'd' of Tsai & Lee, which follows strictly from the pathway stoichiometry: allowance can be made for metabolic restrictions such as maximal activities of pathway enzymes. Such restrictions increase g and accordingly decrease the degree of freedom of the system.

Most fermentation systems are in principle over-determined systems, which means: parameters may be estimated independently using different types of dependent flow or yield variables. Rather than the method of correcting measured flows before using them for parameter estimation (de Kok & Roels 1980; Wang & Stephanopoulos 1983), I support the procedure of multivariate parameter estimation using scaled and weighted (non-linear) multivariate least squares or maximum likelihood procedures with the raw experimental data (Oner et al. 1983 Solomon et al. 1984), because confidence intervals obtained by the former method are likely to be an underestimate.

Basically two different experimental systems can be used for studying the quantitative relations of microbial growth and product formation: the chemostat and the fed-batch culture. In the chemostat the data which are obtained primarily are yields for compounds in the culture broth (biomass, products) and flow rates for gaseous compounds (mostly CO_2 and O_2). These data can be used directly with equations of the form 8 and/or 13 for the estimation of parameters in **C**. For (fed-)batch cultures the situation is somewhat more complicated. The primary type of observation is masses of accumulated biomass and product. Yields in the situation of the fed-batch culture are usually calculated as cumulative yields, which are not ratios of flow rates, so equation 13 is not applicable. Calculating flow rates from accumulated masses requires numerical differentiation, which is a very dangerous technique when the data are scarce and noisy, which is usually the case. A common situation for the fed-batch culture is the occurrence of one or more limiting substrates for which flows can be measured accurately and which can be considered as independent variables. By integration of equation 8 one arrives at a prediction of accumulated masses as a function of time: $W(t) = \int R_d dt$. Usually the integration must be done numerically; only in very special situations an analytical solution is possible (for instance when R_i is a constant, see Van Verseveld et al. 1986). The integrated (dependent) flow vector can be used for parameter estimation via non-linear regression techniques.

The designation 'dependent' and 'independent' with respect to the macroscopic flows should not be taken too literal. In the cases of single- and two-substrates growth in continuous culture the terms are fully justified: dilution rates and substrate concentrations in the feed medium can be controlled in principle by the experimenter, so substrate flows are indeed independent variables fully determining the dependent flows of biomass, oxygen and carbon dioxide. For the case of a product forming fermentation the situation is more complicated. In the example treated in this paper the flows of substrate and product were in first instance both considered as independent flows. In fact this is only justified for those situations where the rate of product formation can be controlled independently from the substrate flow. Such a situation is rare; an example could be a product forming reaction which can be controlled by a limiting flow of a precursor molecule. Generally the product formation rate should be considered as a dependent variable, and using it as a neo-independent variable should be confined (when possible) to situations where only a minor fraction of the total carbon flux goes to the product and/or when the product formation can be measured very accurately. An alternative is given in the treatment of the example: the product flux can be transferred to the vector of dependent flows when a kinetic relation between product formation rate and for instance biomass growth rate is defined. The system equations are extended with at least one more parameter, and the degrees of freedom of the system is decreased with one.

References

Aiba S & Matsuoka M (1979) Identification of metabolic model: citrate production from glucose by *Candida lipolytica*. Biotechnol. Bioeng. 21: 1373–1386

Babel W & Müller RH (1985a) Correlation between cell composition and carbon conversion efficiency. Appl. Microbiol. Biotechnol. 22: 201–207

Babel W & Müller RH (1985b) Mixed substrate utilization in microorganisms: biochemical aspects and energetics. J. Gen. Microbiol. 131: 39–45

Babel W (1986) Increase and limits of growth yields for heterotrophic microorganisms. Acta Biotechnol. 6: 305–309

Bauchop T & Elsden SR (1960) J. Gen. Microbiol. 23: 457–479

Bonnet JABAF, De Kok HE & Roels JA (1980) The growth of *Saccharomyces cerevisiae* CBS 426 on mixtures of glucose and ethanol: a model. A. van Leeuwenhoek 46: 565–576

Bonnet JABAF, Koellman CJW, Dekkers-De Kok HE & Roels JA (1984) The growth of *Saccharomyces cerevisiae* CBS 426 on mixtures of glucose and succinic acid: a model. Biotechnol. Bioeng. 26: 269–272

Bronson R (1989) Theory and problems of matrix operations. Schaum's Outline Series. McGraw-Hill Book Company, New York

De Vries S & Grivell LA (1988) Purification and characterization of a rotenone-insensitive NADH: Q6 oxidoreductase from mitochondria of *Saccharomyces cerevisiae*. Eur. J. Biochem. 176: 377–384

Erickson LE, Minkevich IG, Eroshin VK (1978) Application of mass and energy balance regularities in fermentation. Biotechnol. Bioeng. 20: 1595–1621

Erickson LE (1979) Energetic efficiency of biomass and product formation. Biotechnol. Bioeng. 21: 725–743

Geurts TGE, De Kok HE & Roels JA (1980) A quantitative description of the growth of *Saccharomyces cerevisiae* CBS 436 on a mixed substrate of glucose and ethanol. Biotechnol. Bioeng. 22: 2031–2043

Gommers PJF, Van Schie BJ, Van Dijken JP & Kuenen JG (1988) Biochemical limits to microbial growth yields: an analysis of mixed substrate utilization. Biotechnol. Bioeng. 32: 86–94

Humphrey AE (1974) Current developments in fermentation. Chem. Eng. 81(26): 98–112

Minkevich IG & Eroshin VK (1973) Productivity and heat generation of fermentation under oxygen limitation. Folia. Microbiol. 18: 376–386

Minkevich IG (1983) Mass-energy balance for microbial product synthesis – biochemical and cultural aspects. Biotechnol. Bioeng. 25: 1267–1293

— (1985) Estimation of available efficiency of microbial growth on methanol and ethanol. Biotechnol. Bioeng. 27: 792–799

Müller RH, Sysoev OV & Babel W (1986) Use of formate gradients for improving biomass yield of *Pichia pinus* growing continuously on methanol. Appl. Microbiol. Biotechnol. 25: 238–244

Niranjan SC & San K-Y (1989) Analysis of a framework using material balances in metabolic pathways to elucidate cellular metabolism. Biotechnol. Bioeng. 34: 496–501

Oner MD, Erickson LE & Yang SS (1983) Estimation of true growth and product yields in aerobic cultures. Biotechnol. Bioeng. 25: 631–646

Oura E (1972) The effects of aeration on the energetics and biochemical composition of baker's yeast. PhD thesis, University of Helsinki

Papoutsakis ET (1984) Equations and calculations for fermentations of butyric acid bacteria. Biotechnol. Bioeng. 26: 174–187

Papoutsakis ET & Meyer CL (1985a) Equations and calculations of product yields and preferred pathways for butanediol and mixed-acid fermentations. Biotechnol. Bioeng. 27: 50–66

— (1985b) Fermentation equations for propionic acid bacteria and production of assorted oxychemicals from various sugars. Biotechnol. Bioeng. 25: 76–80

Pirt SJ (1965) The maintenance energy of bacteria in growing cultures. Proc. Roy. Soc. London, Ser. B 163: 224–231

Roels JA (1980) Application of macroscopic principles to microbial metabolism. Biotechnol. Bioeng. 22: 2457–2514

— (1983) Energetics and kinetics in Biotechnology (p 117). Elsevier Biomedical Press, Amsterdam

Solomon BO, Erickson LE, Hess JE & Yang SS (1982) Maximum likelihood estimation of growth yields. Biotechnol. Bioeng. 24: 633–649

Solomon BO, Oner MD, Erickson LE & Yang SS (1984) Estimation of parameters where dependent observations are related by equality constraints. AIChE Journal 30: 747–757

Stouthamer AH (1973) A theoretical study on the amount of ATP required for synthesis of microbial cell material. A. van Leeuwenhoek 39: 545–565

Stouthamer AH & Bettenhaussen CW (1973) Utilization of energy for growth and maintenance in continuous and batch culture of micro-organisms. Biochim. Biophys. Acta 301: 53–70

Stouthamer AH (1979) The search for correlation between theoretical and experimental growth yields. In: Quayle JR (Ed) Intern. Rev. Biochem., Vol 21, University Park Press, Baltimore

Stouthamer AH & Van Verseveld HW (1985) Stoichiometry of microbial growth. In: Bull AT & Dalton H (Eds) Comprehensive Biotechnology, Vol 1 (pp 215–238). Pergamon Press, Oxford

Tempest DW & Neijssel OM (1984) The status of Y_{ATP} and maintenance energy as biologically interpretable phenomena. Ann. Rev. Microbiol. 38: 459–486

Tsai SP & Lee YH (1988a) Application of metabolic pathway stoichiometry to statistical analysis of bioreactor measurement data. Biotechnol. Bioeng. 32: 713–715

— (1988b) Application of Gibbs' rule and a simple pathway method to microbial stoichiometry. Biotechnol. Progr. 4: 82–88

— (1989) A criterion for selecting fermentation stoichiometry methods. Biotechnol. Bioeng. 33: 1347–1349

Van Verseveld HW & Stouthamer AH (1980) Two-(carbon) substrate-limited growth of Paracoccus denitrificans. A direct method to determine the P/O ratio in growing cells. FEMS Microbiol. Lett. 7: 207–211

Van Verseveld HW, De Hollander JA, Frankena J, Braster M, Leeuwerik FJ & Stouthamer AH (1986) Modelling of microbial substrate conversion, growth and product formation in a recycling fermentor. A. van Leeuwenhoek 52: 325–342

Verhoff FH & Spradlin JE (1976) Mass and energy balance analysis of metabolic pathways applied to citric acid production by Aspergillus niger. Biotechnol. Bioeng. 18: 425–432

Wang NS & Stephanopoulos G (1983) Application of macroscopic balances to the identification of gross measurement errors. Biotechnol. Bioeng. 25: 2177–2208

Symbols

scalars

a_h	stoichiometry of oxidative phosphorylation (P/O ratio)
a_i	energetic coefficient, relating 'a' Mole ATP to one unit 'i'
a_p	reciprocal yield of product of ATP, Mole ATP/C-Mole
a_x	reciprocal yield of biomass on ATP, Mole ATP/C-Mole
D_m	metabolic determinant
f	degree of freedom for the macroscopic flows after application of the mass balance
f'	degree of freedom when also the metabolic balance is applied; $f'=f-g$
g	exuberance of the metabolic system
k	number of (clustered) metabolic reactions
l	number of internal (conserved) biochemical species in consideration
m	number of chemical "elements" in consideration (carbon, nitrogen, redoxon (degree of reduction), etc.)
n	number of (measurable) macroscopic flows
M	C-molecular weight
m_e	maintenance coefficient Mole ATP/g/h
m_x	maintenance coefficient Mole ATP/C-Mole/h
P	product
r	flows of a component in a fermentation system (reaction rate), (C-)Mole/h
S	substrate
X	biomass
Y_{ij}	Macroscopic yield coefficient: ratio between a dependent and an independent flow rate
Y_{atp}	yield of biomass on ATP, g/Mole
γ	degree of reduction
σ	stoichiometric coefficient of the carbon source in anabolic or product formation reactions.

matrices and vectors

B	matrix with elemental compositions
B_e	matrix with coefficients relating primary and secondary macroscopic flows according to the mass balance
C	matrix with expressions of stoichiometric coefficients, defining the linear relation between dependent and independent flows
E	vector of biochemical species (internal and external)
R	vector with flows (reaction rates)
R_e	vector with external flows
R_{e1}	vector with primary external flows
R_{e2}	vector with secondary external flows
R_m	vector with internal (microscopic) flows
R_d	vector with dependent (external) flows
R_i	vector with independent (external) flows
z	matrix with stoichiometric coefficients
z_d	submatrix related to the dependent flows
z_i	submatrix related to the independent flows

Antonie van Leeuwenhoek **60**: 275–292, 1991.
© 1991 *Kluwer Academic Publishers. Printed in the Netherlands.*

Application of a metabolic balancing technique to the analysis of microbial fermentation data

J.A. de Hollander
Gist-Brocades, P.O. Box 1, 2600 MA Delft, The Netherlands

Key words: elemental balance, metabolic balance, yield, *Rhizobium trifolii*, *Clostridium butyricum*, *Candida lipolytica*, citric acid production

Abstract

A general method for the development of fermentation models, based on elemental and metabolic balances, is illustrated with three examples from the literature. Physiological parameters such as the (maximal) yield on ATP, the energetic maintenance coefficient, the P/O ratio and others are estimated by fitting model equations to experimental data. Further, phenomenological relations concerning kinetics of product formation and limiting enzyme activities are assessed. The results are compared with the conclusions of the original articles, and differences due to the application of improved models are discussed.

Introduction

In a preceding article a general method was developed, which uses both elemental mass balance as well as metabolic balance principles for the elaboration of model equations for microbial growth and product formation (De Hollander 1991). Here the method will be illustrated with three cases taken from the literature: polysaccharide forming aerobic cultures of *Rhizobium trifolii*, anaerobic fermentation by *Clostridium butyricum*, and citric acid production from n-alkanes by *Candida lipolytica*.

Fermentations are analyzed in terms of mass flows to and from the system, which is defined as the total mass of (active) microbial cells in the fermenter. Flows are partitioned in two ways into different categories. The first division is between primary and secondary flows. The number of primary flows is equal to the degree of freedom determined by the elemental mass balance, i.e. the total number of flows minus the number of relevant chemical elements present in the compounds in consideration. Each of the secondary flows can be calculated as a linear combination of primary flows. The second division is between independent and dependent flows. The independent flows are a subset of the primary flows. Each of the dependent flows follows from the independent flows via a (pseudo-)linear equation, with coefficients composed of relevant metabolic parameters of the system. The way in which the flows are divided into the mentioned categories is partly a matter of choice, and partly imposed by the experimental conditions. For example, in the case of a carbon source-limited continuous culture, the experimentalist can choose the flow rate of carbon source to the system (via the dilution rate and the concentration of the carbon source in the fresh medium feed). Then, the independent flow is r_s, the flow rate of carbon source, and all other flows (biomass formation rate, nitrogen source consumption rate, oxygen consumption rate, carbon dioxide evolution rate, and others) are dependent flows. In the development of the linear equations predicting the dependent flows from the independent flows, a central role is played by the concept of the metabolic key intermediate. The complexity of the complete cellular metabolism is not suited for use in an analysis of macroscopic phenomena. Therefore the metabolism must be divided into major pathways or clustered reactions of special relevance for the problem. Key intermediates are those metabolites that define the relation between the rates of the various partial reactions of the metabolism. For example, a key intermediate can be formed in one reaction in a particular metabolic compartment, and be used as a substrate in another metabolic compartment. Also metabolites at branching or converging points of pathways can be defined as a key intermediate. The quintessence of the key intermediate concept is their non-accumulating nature, i.e. at steady state their concentration (related to biomass) is constant. For the simple system of two reactions joined by one specific key intermediate, this implies that both reactions must have the same rate in (pseudo-) steady state conditions. For more complicated situations the relations between the rates of internal metabolic reactions can best be assessed using linear algebra or matrix methods, as discussed in detail in the preceding article. Examples of key intermediates are ATP, the link between energy-generating and energy-consuming pathways, and NADH, the link between reductive and oxidative reactions. In practically every case these compounds are necessary key intermediates. Other examples will appear in the treatment of cases from the literature.

The development of fermentation models proceeds in a straightforward manner, once the partial reactions of the metabolic system are formulated. Care must be taken that every partial reaction is defined such that balance principles are obeyed. Putting the stoichiometric coefficients of these reactions into a matrix is followed by a number of standard operations, resulting in an expression for the dependent flows as a function of the independent flows. In a number of cases extra relationships must be introduced in order to make a satisfactory solution of the system possible. This can take the form of a metabolic restriction not directly following from known biochemical data, or a phenomenological relation between two or more macroscopic flows.

Taking the metabolic balance into consideration, results in a reduction in the degree of freedom of the system, as determined in first instance by the elemental mass balance. In other words: when certain biochemical stoichiometries or relationships are known and taken into account, less measurements are necessary for a complete description of a system. Or, when sufficient measurement data are available, these data can be used to estimate parameters of the metabolic system. It is this latter application that will be illustrated in this article.

Methods and symbolism

Matrix equations were processed using the CML program shortly described before (De Hollander 1991). The program runs under DOS on a PC.

Parameters were estimated by fitting data to multi-response fermentation models, using a program called CURV. This multi-purpose numerical program uses FORTRAN routines from the NAG library (Numerical Algorithms Group, Oxford, UK), with a user-interface developed under SAS® software (SAS Institute Inc., Cary, NC, USA). The program was run on a IBM-RS6000 system.

For the expression of metabolic reactions a coherent system of symbols is used, which is described shortly. For stoichiometric coefficients related to ATP production or consumption (energetic coefficients) the symbol a_i is used, indicating that 'a' Moles of ATP are formed/consumed per C-mole 'i' converted in the reaction. The symbol σ is used as coefficient of the carbon source in anabolic reactions; σ differs from 1 when carbon dioxide is used or produced in these reactions. The symbol may acquire an appropriate subscript. Degree of reduction is expressed by γ_i, with the subscript indicating the compound. In general the following characters are used to indicate substances: S, main carbon substrate (often glucose); P, product; X, biomass. Reaction rates (flows) are indicated with a lower case 'r', whereas specific rates per biomass have the subscripted symbol q. Flows to the system have negative values; flows from the system (products) have positive values. For vectors of flows a bold type uppercase **R** is used. All vectors and matrices are in bold uppercase type. Other symbols are explained in the text when they appear. The reader may also refer to the preceding article (De Hollander 1991) for a more comprehensive list of symbols.

Growth and polysaccharide production by *Rhizobium trifolii* in mannitol/asparagine – limited chemostat cultures

Rhizobium trifolii strain S1537 (De Hollander et al. 1979) is not able to grow on a medium without an organic carbon source. Growth yields and extracellular polysaccharide formation were studied in a series of carbon-limited chemostat cultures with mannitol as carbon source and asparagine as nitrogen source. Under the employed conditions, asparagine was also used as a supplementary carbon source and excess nitrogen beyond the need for biosynthesis was excreted as ammonia. The case is used as a typical example of a product forming aerobic fermentation system with an organic nitrogen source.

The metabolic scheme, as used for the development of the metabolic balance equation, is depicted in Fig. 1. The following set of equations can be formulated as a description of the metabolic system. All reactants are expressed in C-Mole units.

1. Biomass formation

$$\sigma_x \text{ Glucose} + a_x \text{ ATP} + 1/c_{nx} \text{ NH}_3 \rightarrow$$
$$\text{biomass} + \tfrac{1}{2}(\sigma_x\gamma_s - \gamma_x) \text{ NADH} + (\sigma_x - 1) \text{ CO}_2$$

Although the actual carbon source in the study with *Rhizobium* was mannitol, glucose is defined here as the substrate for biomass formation. Energy costs of biomass formation (a_x) depend on the carbon substrate which is used. It is therefore advantageous to define a yield on ATP ($Y_{atp} \approx 1/a_x$) for a standardized substrate. For this purpose pyruvate was used in an early quantitative study by Von Meyenburg (1969) or a hypothetical (set of) precursor(s) (Roels 1983). I prefer glucose, because it is the most widely used substrate in fermentation studies. As a consequence, glucose is a key inter-

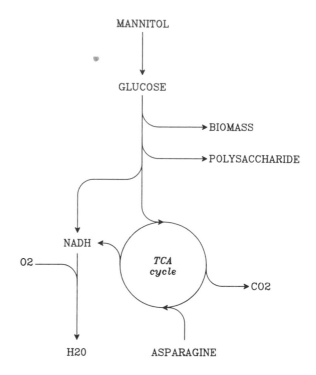

Fig. 1. Simplified metabolic scheme for growth of *Rhizobium trifolii* on a mixed carbon source: mannitol and asparagine.

mediate in the metabolic scheme, according to the definition given in the introduction. The coefficient c_{nx} is the nitrogen content of the biomass (Mole/C-Mole). The net production or consumption of NADH in this reaction is of course dependent on the value of the coefficients σ_x and the degree of reduction of biomass.

2. Extracellular polysaccharide (EPS) formation

$$\sigma_p \text{ Glucose} + a_p \text{ ATP} \rightarrow$$
$$\text{EPS} + \tfrac{1}{2}(\sigma_p\gamma_s - \gamma_p) \text{ NADH} + (\sigma_p - 1) \text{ CO}_2$$

This reaction is fully analogous to the biomass formation reaction.

3. Conversion of mannitol to glucose

$$\text{Mannitol} + a_m \text{ ATP} \rightarrow$$
$$\text{Glucose} + \tfrac{1}{2}(\gamma_m - \gamma_s) \text{ NADH}$$

Because glucose is defined as a key intermediate for biosynthesis and carbon source catabolism, the actual carbon source is oxidized to the glucose level in this hypothetical reaction. The coefficient a_m involves the energy costs of mannitol uptake.

4. Asparagine uptake and catabolism

$$Asparagine \rightarrow$$
$$\tfrac{1}{2}NH_3 + CO_2 + \tfrac{1}{2}\gamma_a \ NADH + a_a \ ATP$$

Although part of the amino acids may receive their amino-group via transamination reactions, the formalism is used that all nitrogen in asparagine enters the biosynthetic routes via intracellular ammonia (cf. reaction 1); excess nitrogen is excreted as ammonia. The energy gain of the reaction (a_a) is the amount of ATP formed in the breakdown of asparagine via the tricarboxylic acid cycle, corrected for the energy costs of asparagine uptake.

5. Catabolic breakdown of glucose

$$Glucose \rightarrow CO_2 + \tfrac{1}{2}\gamma_s \ NADH + a_s \ ATP$$

6. Respiration/oxidative phosphorylation

$$NADH + \tfrac{1}{2}O_2 \rightarrow (H_2O) + a_h \ ATP$$

This is the major energy generating pathway. No distinction is made between different forms of reducing equivalents. This leads to slightly biased estimations of the parameter values. In the preceding article it was discussed how a differentiation between for instance NADH and FADH can be incorporated in the equations.

A dimensional analysis reveals that the metabolic system has four degrees of freedom: Seven external flows are defined (flows of mannitol, asparagine, biomass, polysaccharide, ammonia, carbon dioxide and oxygen, $n = 7$). There are three elements in consideration (carbon, 'reductance', and nitrogen, $m = 3$). After application of the elemental balance, 4 degrees of freedom are left ($f = n - m$). The exuberance of the metabolic system was defined as $g = f - k + 1$, with k the number of metabolic reactions ($k = 6$), and l the number of key intermediates ($l = 3$: glucose, NADH and ATP). Via the formula $f' = f - g$ it follows that

application of the metabolic balance reduces the degrees of freedom to $f' = 3$.

This result is not in accordance with the experimental setup. The following flows are defined as primary flows: r_m (mannitol), r_a (asparagine), r_x (biomass) and r_p (polysaccharide). Of these four, three can be chosen freely according to the metabolic balance. However, experimentally only r_m and r_a can be varied independently. This problem is solved by defining r_p to be a function of r_x. This is not an uncommon procedure in microbiology, where it often appears that product formation varies with the physiological state of the cell, reflecting the specific growth rate. For convenience reasons a proportional relation is used:

$$r_p = \beta \ r_x$$

with β a pseudo constant. Adding this phenomenological relation to the system of reaction equations, the degree of freedom is reduced to the appropriate level of 2, with r_m and r_a as independent variables. The secondary flows r_n (ammonia), r_o (oxygen) and r_c (carbon dioxide) follow from the primary flows according to the elemental balance relations:

$$\left. \begin{aligned} r_n &= -c_{nx}r_x - \tfrac{1}{2}r_a \\ r_o &= \tfrac{1}{4}(\gamma_m r_m + \gamma_a r_a + \gamma_x r_x + \gamma_p r_p) \\ r_c &= -(r_m + r_a + r_x + r_p) \end{aligned} \right\} \quad (1)$$

For the development of a predictive equation for r_x and r_p as a function of the independent variables r_m and r_a the system of 6 metabolic reaction, extended with the phenomenological relation between r_p and r_x, is written in matrix form:

$$
\begin{bmatrix}
1 & 0 & 0 & 0 & 0 & 0 \\
0 & 1 & 0 & 0 & 0 & 0 \\
0 & 0 & -1 & 0 & 0 & 0 \\
0 & 0 & 0 & -1 & 0 & 0 \\
-\sigma_x & -\sigma_p & 1 & 0 & -1 & 0 \\
\frac{\sigma_x\gamma_s-\gamma_x}{2} & \frac{\sigma_p\gamma_s-\gamma_p}{2} & \frac{\gamma_m-\gamma_s}{2} & \frac{\gamma_a}{2} & \frac{\gamma_s}{2} & -1 \\
-a_x & -a_p & -a_m & a_a & a_s & a_h \\
\beta & -1 & 0 & 0 & 0 & 0
\end{bmatrix}
\begin{bmatrix}
r_1 \\
r_2 \\
r_3 \\
r_4 \\
r_5 \\
r_6
\end{bmatrix}
=
\begin{bmatrix}
r_x \\
r_p \\
r_m \\
r_a \\
0 \\
0 \\
0 \\
0
\end{bmatrix}
$$

or

$$\mathbf{Z} \, \mathbf{R}_m = \mathbf{R}'_{e1}, \text{ partitioned: } \begin{bmatrix} \mathbf{Z}_1 \\ \mathbf{Z}_2 \end{bmatrix} \mathbf{R}_m = \begin{bmatrix} \mathbf{R}_{e1} \\ \mathbf{0} \end{bmatrix}$$

The development of a predictive fermentation equation proceeds as follows: The matrix $\mathbf{Z}$ and vector $\mathbf{R}'_{e1}$ are partitioned as indicated by the dotted lines. The balance equation for the primary external flows can be derived as:

$$\mathbf{Z}_3 \, \mathbf{R}_{e1} = 0, \text{ with } \mathbf{Z}_3 = \mathbf{Z}_2 \, \mathbf{Z}_1^{-1}$$

where $\mathbf{Z}_1$ and $\mathbf{Z}_2$ are the sub-matrices of $\mathbf{Z}$, after elimination of two of the reactions r_4 to r_6. A further partitioning of $\mathbf{Z}_3$ into $\mathbf{Z}_d$ and $\mathbf{Z}_i$, according to the distinction between dependent and independent flows leads to:

$$\mathbf{R}_{d1} = - \, (\mathbf{Z}_d^{-1} \, \mathbf{Z}_i) \, \mathbf{R}_i$$

$$\begin{bmatrix} r_x \\ r_p \end{bmatrix} = - \, \frac{1}{D_m} \mathbf{C} \begin{bmatrix} r_m \\ r_a \end{bmatrix} \qquad (2)$$

where D_m is the metabolic determinant:

$$D_m = a_x + \beta a_p + (\sigma_x + \beta \sigma_p)a_s + \tfrac{1}{2}(\gamma_x + \beta \gamma_p)a_h$$

and $\mathbf{C}$ the linear coefficients matrix:

$$\mathbf{C} = \begin{bmatrix} a_s - a_m + \tfrac{1}{2}\gamma_m a_h & a_a + \tfrac{1}{2}\gamma_a a_h \\ \beta(a_s - a_m + \gamma_m a_h) & \beta(a_a + \tfrac{1}{2}\gamma_a a_h) \end{bmatrix}$$

The expressions for D_m and $\mathbf{C}$ follow straightforward from the metabolic stoichiometry equations, applying matrix operations according to the procedures discussed more extensively in the preceding article. It can be observed that the coefficients in $\mathbf{C}$ are closely associated with the energy gain in the breakdown and oxidation of the two catabolic substrates, and that the metabolic determinant is a weighted sum of the biological energy contents of the products biomass and extracellular polysaccharide. This is a rather general result for these kind of metabolic systems.

The resulting predictive model (a combination of equations 1 and 2) permits the calculation of five dependent flows (r_x, r_p, r_n, r_o and r_c) from two independent flows (r_m and r_a) when the parameters in the model have known values. Not all parameters are known however, and some of them may be estimated using experimental values in a curve-fitting procedure. Unfortunately, only data for three of the five dependent flows were reported, so less than the maximum possible information can be used for the estimation of parameter values. The purpose of the original study was to gather some information about the energetic efficiency of *Rhizobium*. From linear regression coefficients of specific rates as function of μ (the specific growth rate, Fig. 2), point estimates and 95% confidence intervals were calculated for Y_{atp}^{max}, the maximal bioenergetic yield, and m_e the energetic maintenance coefficient, assuming various values for the P/O ratio. Here the same parameters are estimated, using the improved model and a simultaneous multivariate non-linear regression procedure with the directly observed variables X (biomass, g/l), P (extracellular polysaccharide, g/l) and r_o (oxygen uptake rate, Mole/l/h) as a function of r_m, r_a and μ. For this purpose the following substitutions were done in equation 2:

$$a_x = m_e/\mu + 1/Y_{atp}^{max}$$

introducing the maintenance coefficient, and

$$\beta = b_0/\mu + b_1 \text{ (or: } q_p = b_0 + b_1 \, \mu\text{)}$$

replacing the proportional relation between q_p and μ by a linear relation, as in the original work. Appropriate factors with molecular weights M_x and M_p for biomass and polysaccharide were introduced for the conversion of C-Mole units into gram units. The energetic coefficients (a_m, a_s, a_p, a_a) were kept identical to corresponding values in the original study. Several new parameters were introduced in the improved model. Values for the degree of reduction follow from the elemental composition of the compounds, as reported for biomass and polysaccharide ($\gamma_x = 4.19$, $\gamma_p = 3.81$), and as known for mannitol and asparagine ($\gamma_m = 4.33$, $\gamma_a = 3$). The σ_x coefficient was given a reasonable value of 1.1 for bacterial growth, according to liter-

280

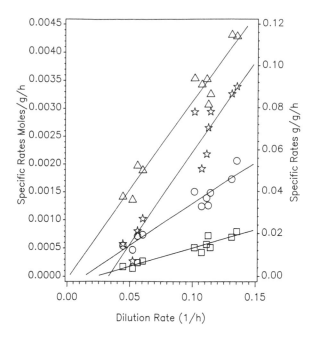

Fig. 2. Reconstruction of the originally published figure, giving experimental data for mannitol/asparagine limited chemostat cultures of *Rhizobium trifolii* (De Hollander et al. 1979). Symbols: Δ q_o, $\star$ q_p, $\bigcirc$ q_m, $\square$ q_a. The best fit linear regression lines are drawn.

ature; σ_p could be calculated from the biosynthetic pathway and molecular composition of the polysaccharide ($\sigma_p = 1.02$).

The curve-fit procedure revealed that the experimental data for the *Rhizobium trifolii* chemostat cultures do not comply with the law of the elemental balance. A good correspondence between model and experimental data could be obtained for separate fits of X and r_o, but by no way in a simultaneous fitting procedure. In general such a discrepancy indicates that there are systematic errors in one or more of the measurements, or there is an unnoticed product or substrate. In this case the most probable explanation is the occurrence of a bias in the measurements of the oxygen uptake rate. This bias was estimated to be 15%, which is not an impossible value, taking into consideration the type of apparatus used at the time, and the low level of oxygen depletion of the fermentation air. Alternatively, a good fit could be obtained by assuming a bias in the carbon source concentration in the feed of 7%. Experimental data and fitted lines

Table 1. Point estimates and 95% confidence intervals for Y_{atp}^{max} and m_e for *Rhizobium trifolii*.

	Y_{atp}^{max} (g/Mole)	m_e (mMole/g/h)
Original data	14.8 (11.1–18.6)	0.5 (− 1.2–2.1)
New estimates	11.8 (9.3–14.2)	0.2 (− 1.0–1.4)

Original data compared with values obtained with an improved model. A P/O ratio of 1.5 was used in the calculations. Experimental data from De Hollander et al. 1979.

are shown in Fig. 3. Estimated values for Y_{atp}^{max} and m_e are compared with the original values in Table 1.

Although the Figs 2 and 3 have a dissimilar appearance, the data have the same origin and contain the same amount of information. The common procedure of plotting specific metabolic rates against the specific growth rate (or dilution rate) involves transformation of measured variables. In such procedures a distortion of variance distribution is inevitable, and this may result in biased estimators for the parameters and unnoticed outliers. In this case the biomass concentration at $\mu =$

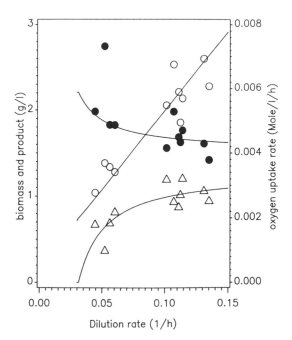

Fig. 3. Data from Fig. 2, replotted in the form of original experimental observations. Symbols: $\bigcirc$ r_o (oxygen uptake rate), $\bullet$ X (biomass concentration), $\triangle$ P (polysaccharide concentration). The lines are the best-fit curves predicted by the model.

0.05 in Fig. 3 falls clearly outside the expected region. Consequently the observation was skipped from the dataset before performing the final parameter estimation. Plotting specific rates has the advantage that the data reflect more closely the metabolic state of the microorganism, and thus this technique should preferably be used for the mental process of model development. Also, linear regression can be done even without a computer, an aspect which was of relevance not so long ago. However, from a statistical point of view non-linear regression with data as close as possible to the original observations is to be preferred.

The difference in regression procedure is one reason for the difference in estimated parameter values and variances. In the improved model advantage is taken from the fact that the metabolic system is over-determined. The same parameters could have been estimated by using only 2 of the 3 available response variables. By using a simultaneous optimization procedure, with as many variables as measured, the reliability of the resulting estimators is increased.

Growth of *Clostridium butyricum* on glucose and mannitol

The balancing of flows to and from key intermediates of the metabolism is the most important tool for the development of comprehensive predictive/descriptive fermentation models. For relatively simple and linear metabolic systems, consideration of the intermediates NADH and ATP is usually sufficient. They provide the link between catabolic and oxidative reactions (NADH) and energy generating and energy consuming reactions (ATP). However, when one or more of the highways of the metabolism includes a converging or a diverging pathway, it is necessary to include other key intermediates as well in the description of the system. The fermentation of glucose/mannitol mixtures to acetate and butyrate by *Clostridium butyricum* (Crabbendam et al. 1985) is treated as an example for such systems. Figure 4 shows a simplified scheme of the metabolic pathways; it is an adaptation of Fig. 1 in the original paper. The major

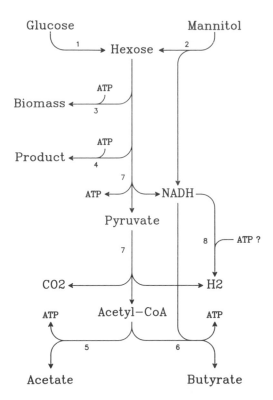

Fig. 4. Simplified metabolic scheme for growth of *Clostridium butyricum* on glucose plus mannitol, used as a base for the development of a predictive fermentation equation based on the metabolic balance principle.

products of fermentation are acetate and butyrate (and CO_2 of course). It was shown that the ratio of these products in steady state carbon-limited continuous cultures is dependent on the ratio of the two carbon sources in the medium (glucose and mannitol). Other products are biomass, hydrogen and extracellular protein. Further potential fermentation products (lactate, ethanol, butanol) were formed in negligible amounts. Hydrogen can be formed via two mechanisms: (1) as a result of the clastic decarboxylation of pyruvate to acetyl-CoA, and (2) generated from NADH, a reaction catalyzed by ferredoxin. Crabbendam et al. raised the question whether this latter reaction, which is thermodynamically unfavorable, requires the input of metabolic energy. It will be shown how a comprehensive predictive model can be used to answer this question, using the maximum of available experimental information. All data were taken from Table 4 in the original paper, which collects the results

282

of a series of chemostat experiments with varying ratios of mannitol and glucose.

The metabolic scheme of Fig. 4 represents a system of 8 metabolic reactions (pathways). The metabolic key intermediates are NADH, ATP, 'hexose' and acetyl-CoA. The intermediate 'hexose' is introduced for a convenient description of the effect of changing the relative fluxes of glucose and mannitol; it can be thought to be identical to fructose-6-phosphate, the converging point of the catabolic pathways of glucose and mannitol. Acetyl-CoA is a necessary key intermediate because it is a branching point in the pathways to acetate and butyrate. Each reaction is described with the reactants expressed in C-Molar units:

1. Glucose uptake
    glucose → 'hexose'

It is assumed that the uptake of the carbon sources does not require energy because of the presence of PEP-transferase systems in Clostridium. Although the actual converging point is fructose-6-phosphate, 'hexose' is formally described as an unphosphorylated intermediate; the ATP consumption in the activation of the carbon sources is accounted for in reaction 7. This is done so, in order to get a correct estimate for Y_{atp}, the energy costs of biomass formation from glucose.

2. Mannitol uptake
    mannitol → 'hexose' + $\frac{1}{6}$ NADH

All reducing equivalent are expressed as NADH, justified by the existence of equilibrium between the NADH and NADPH redox systems in Clostridium.

3. The anabolic reaction (biomass formation)
    σ_x 'hexose' + a_x ATP →
    biomass + $\frac{1}{2}(4\sigma_x - \gamma_x)$ NADH + $(\sigma_x - 1)$ CO$_2$

As usual it is assumed that there is a net formation of carbon dioxide in the anabolic reactions.

4. The product formation reaction
    σ_p 'hexose' + a_p ATP →
    product + $\frac{1}{2}(4\sigma_p - \gamma_p)$ NADH + $(\sigma_p - 1)$ CO$_2$

This reaction is fully analogous to the biomass formation reaction. Rates of product formation were not given in the paper of Crabbendam et al., however they could be calculated from the reported carbon recovery percentages. The precise nature of the product(s) is unknown, however it was assumed that it consists largely of extracellular protein. Hence a significant energy expenditure in this reaction can be expected, symbolized by the coefficient a_p.

5. Acetate formation
    acetyl-CoA → acetate + $\frac{1}{2}$ATP

6. Butyrate formation
    acetyl-CoA + $\frac{1}{2}$NADH → butyrate + $\frac{1}{4}$ATP

The coefficients in the last two equations may seem unusual on first sight, but they follow from the convention to express all carbon-containing compounds in C-Molar units.

7. Acetyl-CoA formation
    $1\frac{1}{2}$ 'hexose' → acetyl-CoA + $\frac{1}{2}$ATP + $\frac{1}{2}$NADH + $\frac{1}{2}$H$_2$ + $\frac{1}{2}$CO$_2$

This equation represents a condensation of a number of reactions, including the Embden-Meyerhoff pathway and the decarboxylation of pyruvate to acetyl-CoA under formation of hydrogen.

8. Ferredoxin catalyzed hydrogen formation from NADH
    NADH + a_h ATP → H$_2$
The coefficient a_h represents here the possible need for metabolic energy to drive this reaction.

In total there are 8 macroscopic flows which can be measured in principle: flows of glucose, mannitol, biomass, product, acetate, butyrate, hydrogen and carbon dioxide. By application of the carbon balance and the 'reductance' balance 6 degrees of freedom remain and 2 external flows can be skipped for the evaluation of the metabolic balance equation. Hydrogen and carbon dioxide were chosen as secondary (dependent) flows. The system of metabolic reactions in matrix expression becomes then:

$$\begin{bmatrix} -1 & 0 & 0 & 0 & 0 & 0 & 0 & 0 \\ 0 & -1 & 0 & 0 & 0 & 0 & 0 & 0 \\ 0 & 0 & 1 & 0 & 0 & 0 & 0 & 0 \\ 0 & 0 & 0 & 1 & 0 & 0 & 0 & 0 \\ 0 & 0 & 0 & 0 & 1 & 0 & 0 & 0 \\ 0 & 0 & 0 & 0 & 0 & 1 & 0 & 0 \\ 1 & 1 & -\sigma_x & -\sigma_p & 0 & 0 & -1\tfrac{1}{2} & 0 \\ 0 & 0 & 0 & -1 & -1 & 1 & 0 \\ 0 & \tfrac{1}{6} & \tfrac{4\sigma_x-\gamma_x}{2} & \tfrac{4\sigma_p-\gamma_p}{2} & 0 & -\tfrac{1}{3} & \tfrac{1}{3} & -1 \\ 0 & 0 & -a_x & -a_p & \tfrac{1}{3} & \tfrac{1}{3} & \tfrac{1}{3} & -a_h \end{bmatrix} \begin{bmatrix} r_1 \\ r_2 \\ r_3 \\ r_4 \\ r_5 \\ r_6 \\ r_7 \\ r_8 \end{bmatrix} = \begin{bmatrix} r_g \\ r_m \\ r_x \\ r_p \\ r_a \\ r_b \\ 0 \\ 0 \\ 0 \\ 0 \end{bmatrix}$$

Partitioning into external and internal parts leads to a 6 × 6 matrix associated with the external flows (Z_1) and a 4 × 8 matrix Z_2 describing the intermediates balance. This latter system of equations can easily be reduced to the dimension 2 × 6 by elimination of two of the flows, hence the resulting balance equation for external flows:

$$\mathbf{z}_3 = \begin{bmatrix} -1 & -1 & -\sigma_x & -\sigma_p & -1\tfrac{1}{2} & -1\tfrac{1}{2} \\ 0 & \tfrac{1}{6}a_h & -\tfrac{1}{3}(4\sigma_x-\gamma_x)a_h-a_x & -\tfrac{1}{3}(4\sigma_p-\gamma_p)-a_p & 1-\tfrac{1}{2}a_h & \tfrac{1}{4} \end{bmatrix}$$

could be processed further to an equation predicting 2 dependent flows from 4 independent flows.

From an experimental point of view the Clostridium fermentation system has only two independent variables which can be varied at will: glucose and mannitol flows. Mass balance and metabolic relations determine the other macroscopic flows. In order to cope with this situation, the metabolic balance equations should be extended with two more relations between the external flows. In this case that is easily done by assuming 'laws' for the rate of extracellular product formation and the ratio of acetate and butyrate formation.

The calculated data for product formation rates clearly indicated that the amount of products formed was not constant, but varied with the mannitol/glucose ratio in the feed medium. The most obvious form for a linear relation between the rate of product formation and the other macroscopic flows in this case is:

$$r_p = b_1 r_g + b_2 r_m,$$

which can be transformed to:

$$q_p/q_g = b_1 + b_2\ q_m/q_g$$

The experimental data were plotted in the form of the latter equation (Fig. 5) and it appears that a linear relation is indeed a reasonable first approximation.

The ratio of acetate and butyrate formation in carbon-limited cultures was found to be constant for cultures growing on glucose, but varied strongly with the amount of mannitol provided to the culture. Mannitol, being more reduced than glucose, influences the redox balance of the metabolic system, and it is therefore not surprising to find that the more mannitol is present, the more butyrate is formed, the latter compound being more reduced than acetate. However, the correlation is not fixed, because the organism has an alternative outlet for excess NADH: ferredoxin catalyzed hydrogen formation. The relation can be described phenomenologically by plotting the mean degree of reduction of the fermentation products acetate and butyrate against the mean degree of reduction of the substrates glucose and mannitol (Fig. 6). The strikingly linear relation provides a kind of explanation for the fact that Clostridium cannot grow on mannitol as a sole carbon source. Redox stress forces the organism to increase the degree of reduction of the fermentation products. The maximum degree of reduction ($\gamma = 5$) is reached at a mannitol/glucose ratio of about 3. Further increase of the mannitol/glucose ratio would require the additional formation of a more reduced product (e.g. ethanol or butanol, $\gamma = 6$), however the metabolic potential for such an escape is apparently not present. The linear relation between the degrees of reduction results in a non-linear relation between the flows of acetate, butyrate, glucose and mannitol. The further development of a fermentation model requires a pseudo linear relation between the flows of acetate and butyrate. The mean degree of reduction of the products was calculated as

$$\gamma_p = \beta\gamma_a + (1 - \beta)\,\gamma_b,$$

with β defined as the ratio of the acetate flow and the total flow of fermentation products:

$$\beta = q_a / (q_a + q_b)$$

The following pseudo linear relation follows then for the acetate and butyrate flows:

$$(1 - \beta)\,q_a = \beta\,q_b$$

The value of β follows from the linear relation between the degree of reduction of products and substrates (Fig. 6) and can be expressed as a function of α, the fraction mannitol in the mixed carbon source. β is evaluated during the numerical processing of the fermentation equation. We now can add two phenomenologically defined balances to the two balance equations following from metabolic stoichiometries:

$$\begin{bmatrix} b_1 & b_2 & 0 & 1 & \cdot & 0 & 0 \\ 0 & 0 & 0 & 0 & \cdot & 1-\beta & -\beta \end{bmatrix} \mathbf{R}_{e1} = \mathbf{0}$$

Taking together all 4 balance equations, followed by partitioning of the matrix according to the division in dependent and independent flows, gives the following linear expression:

$$\mathbf{R}_d = -\,\mathbf{Z}_d^{-1}\,\mathbf{Z}_i\,\mathbf{R}_i,$$

with

$$\mathbf{R}_d = \begin{bmatrix} r_x \\ r_p \\ r_a \\ r_b \end{bmatrix} \qquad \mathbf{R}_i = \begin{bmatrix} r_g \\ r_m \end{bmatrix} \qquad \mathbf{Z}_i = \begin{bmatrix} -1 & -1 \\ 0 & \tfrac{1}{6}a_h \\ 0 & 0 \\ b_1 & b_2 \end{bmatrix}$$

$$\mathbf{Z}_d = \begin{bmatrix} -\sigma_x & -\sigma_p & -1\tfrac{1}{2} & -1\tfrac{1}{2} \\ -a_x - \tfrac{1}{2}(4\sigma_x - \gamma_x)a_h & -a_p - \tfrac{1}{2}(4\sigma_p - \gamma_p)a_h & 1-\tfrac{1}{2}a_h & \tfrac{3}{4} \\ 0 & 0 & 1-\beta & -\beta \\ 0 & 1 & 0 & 0 \end{bmatrix}$$

Technically it is possible to process this equation further by performing the indicated matrix oper

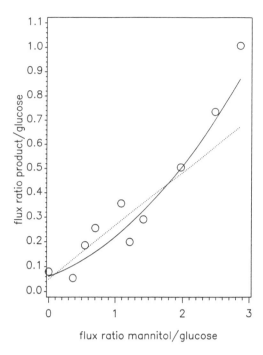

Fig. 5. Product formation in *Clostridium butyricum* fermentations described according to the equation: $q_p/q_g = b_1 + b_2 q_m/q_p$ (dashed line), or the equation: $q_p/q_g = b_1 + b_2 q_m/q_g + b_3(q_m/q_g)^2$ (solid line). Experiments of Crabbendam et al. 1985.

ations, however the resulting equations are very complicated in this case and do not provide useful information about the properties of the system. Therefore the matrix processing was done in numerical form by the computer program for parameter estimation.

The important bioenergetic parameters of the system are: a_x (or Y_{atp}), a_p, the energy costs of product formation, and a_h, the possible energy cost of ferredoxin catalyzed hydrogen formation. These parameters were estimated, along with the coefficients of the phenomenological relations, by fitting the model equation to observed concentrations of biomass, acetate, butyrate and products, using a maximum likelihood procedure. Although the initial assumption of a linear relation between q_p, q_g and q_m was reasonable, a significant improvement of the fit between model and experimental data was obtained by introducing a quadratic term in the relation, as illustrated in Figs 5 and 7. The analysis

Table 2. Estimated values for the bioenergetic parameters of *Clostridium butyricum*, testing the hypothesis: ferredoxin-catalyzed hydrogen formation is not energy dependent.

		$a_h = 0$	a_h is estimated
Y_{atp}	(g/Mole)	10.4 ± 0.4	10.2 ± 1.1
a_p	(Mole/C-Mole)	0.19 ± 0.08	0.22 ± 0.20
a_h	(Mole/Mole)	–	-0.1 ± 0.5

Experimental data from Crabbendam et al. 1985

was focussed on the question whether ferredoxin-catalyzed hydrogen formation is a energy-dependent process. This was tested by comparing results of an optimization with a_h fixed at zero with those of an optimization where a_h was a free parameter. There was no difference found in the sum of squared residuals, so the conclusion must be that hydrogen formation is not energy dependent. The data for the relevant parameters in Table 2 show that when a_h is estimated, its value in not significantly different from zero. A reasonable value for the energy costs of hydrogen formation would be $a_h = 1$, a value which has a very low likelihood. The same conclusion was reached by the authors of the original article, using a more qualitative reasoning. The estimated value of Y_{atp} is in the range of what can be expected. The energy costs of product formation could not be estimated very accurately from the available data. Its point estimate 0.2 translates to approximately 100 g protein per Mole ATP. This is a rather high figure, considering the theoretical energy costs of protein biosynthesis. Probably other (less expensive) products were formed as well.

Formation of citric and isocitric acid from n-alkanes and glucose by *Candida lipolytica*

Citric acid production by the yeast *Candida lipolytica* is treated as an example of aerobic catabolism-related microbial product formation. Data were taken from a paper by Aiba & Matsuoka (1978), studying citric acid production in steady-state nitrogen-limited chemostat cultures of this yeast, using n-alkanes as a carbon source. Citric acid pro-

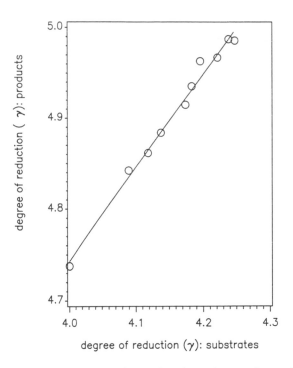

Fig. 6. Ratios of fermentation products in continuous cultures of *Clostridium butyricum* as function of the mannitol/glucose ratio in the feed medium. The mean degree of reduction of the fermentation products acetate and butyrate is plotted against the mean degree of reduction of the carbon source substrates.

duction from alkanes is especially interesting, because here we have a situation of a highly compartmentalized major pathway (Fig. 8). Citric acid formation occurs in the mitochondria, whereas the reformation of C4-carboxylic acids (the immediate precursors of citric acid) takes place via the glyoxylate pathway located in microbodies. The formation of the key intermediate acetyl-CoA via β-oxidation of alkane-derived carboxylic acids occurs in the cytosol. A major issue in citric acid fermentation research is the question how citric acid formation is regulated: in principle the organism could choose for complete oxidation of the carbon source via the TCA cycle. In the paper by Aiba an Matsuoka the question is approached alternatively: it was found that citric acid formation increased with the dilution (and growth-) rate up to a certain value, beyond which citric acid production fell to a considerably lower level and the TCA cycle apparently became predominantly active. It will be illustrated how this behavior can be described using

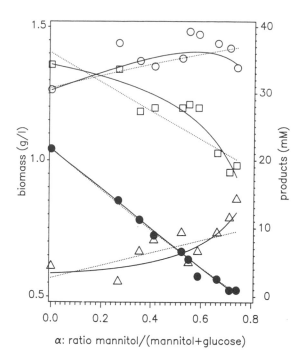

Fig. 7. Biomass (□) and products concentrations (○, butyrate; ●, acetate; △, unknown products) in continuous cultures of *Clostridium butyricum* as function of the α, the fraction mannitol in the mixed carbon source. Lines are best-fit model predictions. Unknown products formation rate is described by a first order polynomial (dashed lines) or second order polynomial (solid lines), cf. Fig. 5. Experiments of Crabbendam et al. 1985.

the metabolic balance method and one simple assumption about a maximal activity of the glyoxylate pathway. The glyoxylate pathway is important for citric acid formation as well as for the formation of biomass precursors. At higher growth rates there may be a competition between both processes for oxaloacetate, which is apparently solved by increased operation of the TCA cycle, leading to a diminishing of citric acid production.

The measurable flows of the system

In principle all relevant flows were measured in the work by Aiba & Matsuoka: nitrogen assimilation (ammonia + yeast extract), alkane consumption, citric and isocitric acid formation, biomass formation, oxygen consumption and carbon dioxide evolution. However the latter two flows were mea-

sured with an indirect method, and they will not taken into consideration in the calculations in this article. Three elemental balances must be considered: carbon, nitrogen and 'reductance' which gives the following expression for the elemental balance:

$$\begin{bmatrix} 0 & 1 & 1 & 1 & 0 & 1 \\ 1 & 0 & 0 & c_{nx} & 0 & 0 \\ 0 & 6+\frac{2}{n} & 3 & \gamma_x & -4 & 0 \end{bmatrix} \begin{bmatrix} r_n \\ r_a \\ r_{cit} \\ r_x \\ r_o \\ r_c \end{bmatrix} = 0$$

where three simultaneous linear equations express respectively the carbon, nitrogen and reductance (γ) balances.

The degree of reduction of n-alkanes is dependent on the chain length n, and is given by $6 + \frac{2}{n}$. The coefficient c_{nx} is the nitrogen content of biomass. Having six flows and three balances, the system is over-determined, and three secondary flows may be selected, which can be written as a linear function of the primary flows:

$$\begin{bmatrix} r_x \\ r_o \\ r_c \end{bmatrix} = \begin{bmatrix} \frac{-1}{c_{nx}} & 0 & 0 \\ \frac{-\frac{1}{4}\gamma_x}{c_{nx}} & \frac{1}{4}(6+\frac{2}{n}) & \frac{1}{4} \\ \frac{1}{c_{nx}} & -1 & -1 \end{bmatrix} \begin{bmatrix} r_n \\ r_a \\ r_{cit} \end{bmatrix}$$

From an experimental point of view the system has one independent flow which can be set by the experimenter: r_n, the nitrogen assimilation flow. By considering the metabolic balance it will be shown how the other primary flows r_a and r_{cit} can be predicted from r_n. In principle one equation could then be developed, predicting all dependent flows from one independent flow.

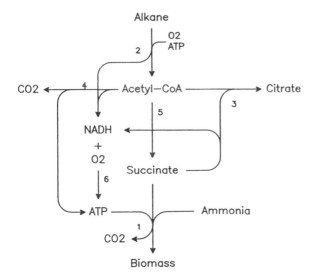

Fig. 8. Simplified metabolic scheme of *Candida lipolytica* growing on and producing citric acids from n-alkanes. The scheme provides the base for the fermentation equation derived from the metabolic balance expression. The central role of acetyl-Coa is stressed in this scheme. No account is made for the compartmentalization of the reactions into various organelles, such as the mitochondrion and the glyoxysome.

The metabolic balance equation, derived from Fig. 8

For the evaluation of the metabolic balance 4 key intermediates are defined: ATP, NADH, acetyl-CoA and succinate, participating in 6 lumped reactions (pathways). Acetyl-CoA was chosen, because it forms the central link between the initial breakdown of alkanes and all other pathways; succinate is considered as a branching point between anabolism an product formation. Other C4-dicarboxylic acids could have been chosen, e.g. oxaloacetate or malate, giving identical results.

1. The anabolic reaction (biomass formation)
Succinate is the starting intermediate for biosynthesis as well as for citric acid formation. The route to gluconeogenesis proceeds via oxidation to oxaloacetate, which is decarboxylated to phosphoenolpyruvate. The following equation represents the over-all reaction:

$$\sigma \text{ succinate} + c_{nx} \text{ NH}_3 + a_x \text{ ATP} \rightarrow$$
$$\text{biomass} + \tfrac{1}{2}(3\tfrac{1}{2}\sigma - \gamma_x) \text{ NADH} + (\sigma - 1) \text{ CO}_2$$

The coefficient σ must have a value which deviates significantly from 1, more than in the case of growth on glucose, because of the initial decarboxylation step in the route to gluconeogenesis from oxaloacetate. The σ coefficient does not have the exact stoichiometric value, because part of the biosynthesis pathways start directly from TCA intermediates, and there may be other decarboxylating or carboxylating reactions involved in biosynthesis.

2. Oxidation of n-alkane to Acetyl-CoA
After initial oxidation of the alkane to the corresponding carboxylic acid, breakdown to acetyl-CoA follows via the β-oxidation pathway. This reaction requires a net investment of ATP in the activation of the carboxylic acid. As the reactants are expressed in C-Mole units, the ATP stoichiometry is dependent on the chain length of the alkane (n). The following over-all reaction equation is used:

$$\text{alkane} + \tfrac{2}{n} \text{ ATP} + \tfrac{1}{n} \text{ O}_2 \rightarrow$$
$$\text{acetyl-CoA} + (1 - \tfrac{1}{n}) \text{ NADH}$$

Although the reaction for alkanes with an uneven number of carbon atoms proceeds slightly different (the final product of the β-oxidation being propionyl-CoA), the over-all equation is still approximately valid.

3. Citric and isocitric acid formation

$$\text{succinate} + \tfrac{1}{2} \text{ acetyl-CoA} \rightarrow$$
$$1\tfrac{1}{2} \text{ citrate} + \tfrac{1}{2} \text{ NADH}$$

is the appropriate expression for product formation in C-Mole units. Citric acid producing yeasts always show considerable by-product formation in the form of isocitric acid. For the purpose of assessing the over-all metabolic balance, these two products can be taken together, and the above equation is therefore equally valid for the by-product formation.

4. The tricarboxylic acid cycle
Complete oxidation of acetyl-CoA produced from alkanes follows the equation:

$$acetyl\text{-}CoA \rightarrow CO_2 + 2\ NADH + \tfrac{1}{2}\ ATP$$

5. The glyoxylate pathway

The immediate products of the glyoxylate pathway are malate and succinate, however the overall equation can be expressed as the formation of one Mole of succinate from 2 Moles of acetyl-CoA (cf. Lehninger, p 466), in C-Moles:

$$acetyl\text{-}CoA \rightarrow succinate + \tfrac{1}{4}\ NADH$$

6. Respiration

Citric acid formation is a strictly aerobic process: the reducing equivalents in the form of NADH must be reoxidized via the respiratory chain. Associated with this reaction is the formation of ATP, which may be used in the biosynthesis (reaction 2) and for alkane activation (reaction 1). It is important to note that for citric acid producing Aspergillus it is observed that an alternative respiratory chain is operative (Kubicek & Röhr 1986), leading to a considerable uncoupling between respiration and ATP formation. It will be seen that this may be an important aspect of citric acid formation.

$$NADH + \tfrac{1}{2}\ O_2 \rightarrow H_2O + a_h\ ATP$$

The metabolic balance equation for the 6 reactions as given becomes then:

$$
\begin{bmatrix}
-c_{nx} & 0 & 0 & 0 & 0 & 0 \\
0 & -1 & 0 & 0 & 0 & 0 \\
0 & 0 & 1\tfrac{1}{2} & 0 & 0 & 0 \\
\hdashline
0 & 1 & -\tfrac{1}{2} & -1 & -1 & 0 \\
-\sigma & 0 & -1 & 0 & 1 & 0 \\
1\tfrac{3}{4}\sigma - \tfrac{1}{3}\gamma_x & 1 - \tfrac{1}{n} & \tfrac{1}{2} & 2 & \tfrac{1}{4} & -1 \\
-a_x & -\tfrac{2}{n} & 0 & \tfrac{1}{2} & 0 & a_h
\end{bmatrix}
\begin{bmatrix}
r_1 \\ r_2 \\ r_3 \\ r_4 \\ r_5 \\ r_6
\end{bmatrix}
=
\begin{bmatrix}
r_n \\ r_a \\ r_{cit} \\ 0 \\ 0 \\ 0 \\ 0
\end{bmatrix}
$$

or:

$$
\begin{bmatrix}
\mathbf{Z}_1 \\
\mathbf{Z}_2
\end{bmatrix}
\mathbf{R}_m =
\begin{bmatrix}
\mathbf{R}_{el} \\
0
\end{bmatrix}
$$

where the matrices are partitioned according to the discrimination between internally balanced flows and external flows. The metabolic balance is solved according to the formula:

$$(\mathbf{Z}_2\ \mathbf{Z}_1^{-1})\mathbf{R}_{el} = 0,\ \text{or}\ \mathbf{Z}_3\ \mathbf{R}_{el} = 0$$

which can be done when $\mathbf{Z}_1$ is square and singular (this condition is fulfilled when the zero columns are omitted) and the dimensions of the two submatrices $\mathbf{Z}_2$ and $\mathbf{Z}_1$ and the vector $\mathbf{R}_{el}$ are compatible with the indicated matrix multiplications. These dimension are respectively 4×6, 3×3 and 3×1, so in its original form the system is insolvable, as $\mathbf{Z}_2$ should have a horizontal dimension of 3. By elimination of 3 equations of the system $\mathbf{Z}_2$ $\mathbf{R}_m = 0$ the matrix $\mathbf{Z}_2$ can be reduced to a dimension of 1×3, which would make the system solvable. The exuberance of the metabolic system then is $g = 1$, and with 3 primary external flows there would be two independent flows (e.g. r_n and r_a) predicting one primary dependent flow (r_{cit}), and through the mass balance the secondary flows. However, the nitrogen-limited chemostat system has only one independent variable: the alkane consumption flow cannot be chosen independently from the nitrogen assimilation flow, so the solution obtained is not suited for a predictive description of the experimental observations.

A solution for this problem can be found by introducing the following hypothesis: there exist two states of the metabolism of Candida lipolytica in nitrogen-limited chemostat cultures. State 1 is characterized by a low basal level of TCA cycle operation and a relatively high rate of citrate production. In state 2 citrate production is diminished in favor of CO_2 production via the TCA cycle. At low specific growth rates the organism is in state 1. The basic rate of the TCA cycle is adjusted to the use of TCA intermediates for biosynthetic purposes; it is therefore natural to assume a proportional relation between this rate and the growth rate:

$$r_4 = \alpha\ r_1$$

In state 1 the glyoxylate pathway is operational for supplying precursors for growth as well as for regenerating oxaloacetate for the biosynthesis of citric and isocitric acid. It is assumed that there is a limiting capacity of the glyoxylate pathway, which may be dependent on the specific growth rate:

$$q_{gly} = \beta\,\mu \text{ or } r_5 = \beta\,r_1$$

The limiting activity may be determined for instance by the enzyme isocitrate lyase (ICL) as suggested by Aiba & Matsuoka, or by another step, for instance one of the involved intra-organellar transport processes. As a consequence of the limiting capacity of the glyoxylate pathway, citrate production will be inhibited and the TCA cycle will operate at a rate above the basic level: state 2. Going from low growth rates to higher growth rates, there will be a critical growth rate at which there is a transition from state 1 to state 2. The occurrence of such a transition can be deduced qualitatively from the experimental data (Fig. 9).

The two metabolic restrictions: basic level operation of the TCA cycle at low growth rates and a limiting capacity of the glyoxylate pathway at higher growth rates, can be expressed by the matrix equation:

$$\begin{bmatrix} \alpha & 0 & 0 & 0 & -1 & 0 & 0 \\ \beta & 0 & 0 & 0 & 0 & -1 & 0 \end{bmatrix} \mathbf{R_m} = \mathbf{0}$$

For each of the metabolic states the appropriate row of the above matrix is selected and included in the matrix $\mathbf{Z}_2$, which will then have the dimension 5×6. The metabolic exuberance then becomes $g = 2$, leaving one degree of freedom for the metabolic system. Two separate fermentation equations are the result of the processing of the modified metabolic balance equation, both predicting the dependent flows r_a and r_{cit} from the independent flow r_n:

$$\begin{bmatrix} r_a \\ r_{cit} \end{bmatrix} = -\mathbf{Z}_d^{-1}\,\mathbf{Z}_i\,r_n,$$

For state 1 (basic level of TCA cycle):

$$\mathbf{z}_d = c_{nx}\begin{bmatrix} -1 & -1 \\ \tfrac{2}{n}+(\tfrac{1}{n}-1\tfrac{1}{4})\,a_h & \tfrac{1}{4}a_h \end{bmatrix}, \quad \mathbf{z}_i = \begin{bmatrix} \sigma+\alpha \\ a_x-\tfrac{1}{3}\alpha+\tfrac{1}{2}\gamma_x a_h-1\tfrac{1}{4}(\sigma+\alpha)\,a_h \end{bmatrix}$$

$$\mathbf{z}_d = c_{nx}\begin{bmatrix} 0 & -\tfrac{2}{3} \\ \tfrac{2}{n}-\tfrac{1}{2}+(\tfrac{1}{n}-3)\,a_h & -\tfrac{1}{6}-\tfrac{1}{2}a_h \end{bmatrix}, \quad \mathbf{z}_i = \begin{bmatrix} \sigma-\beta \\ a_x+\tfrac{1}{3}\beta+\tfrac{1}{2}\gamma_x a_h-1\tfrac{1}{4}(\sigma-\beta)\,a_h \end{bmatrix}$$

The matrix expressions for the dependent flow rates were included in a curve-fitting program and

evaluated further numerically. The following additional relationships were assumed:

- The ratio of citrate formation and isocitrate formation is a constant, independent of the growth rate.
- The relation between q_{gly} and μ is linear for the region of growth rates where the glyoxylate cycle is limiting.
- The nitrogen content of the cells (c_{nx}) varies linearly with the specific growth rate.

The model fits reasonably well to the experimental data, as shown in Figs 9 and 10. The dashed line in Fig. 10 for the specific activity of the glyoxylate pathway at lower growth rates indicates that for this region the line merely represents an extrapolation of the line at higher growth rates; the actual specific activity might be different. It must be said that the original hypothesis of a TCA cycle running at a rate which is proportional to the growth rate during state 1 of metabolism, appeared not compatible with the experimental data. The final fit, as shown in the figures, was obtained by assuming a constant specific turnover rate of the TCA cycle, independent of the specific growth rate. The greatest deviations between the model and experimental data are found for the alkane consumption rate at lower dilution rates. Improvements might be obtained by a further refinement of the hypothesis for TCA cycle operation in state 1 at low specific growth rates. Figure 9 seems to indicate that the assumption of a linear relation between the nitrogen content of the cells and the growth rate, resulting in the biomass concentration being a hyperbolic function of the growth rate, can be refined. It is clear however, that a very simple assumption for a metabolic restriction can explain the observed pattern of citrate and isocitrate production by *Candida lipolytica* in nitrogen-limited chemostat cultures. A number of estimated parameter values are included in Table 3. Due to the shift in metabolism (state 1 → state 2) it appeared possible to get estimates for both Y_{atp} and the P/O ratio. Normally this can be done only when special experimental set-ups are used (Stouthamer & Van Verseveld 1985). The rather large uncertainty in the estimate for Y_{atp} is

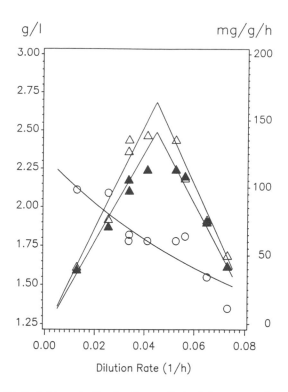

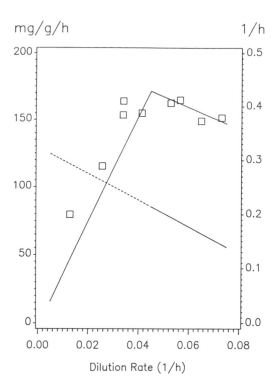

Fig. 9. Acid production in nitrogen-limited chemostat cultures of *Candida lipolytica* with n-alkanes as the main carbon source. Symbols: $\bigcirc$, biomass (g/l); $\triangle$, q_{cit}; $\blacktriangle$, q_{isocit} (mg/g/h). Data from Aiba & Matsuoka 1978. Lines are best-fit model predictions.

Fig. 10. The same set of experiments as Fig. 9: specific rates of alkane consumption in nitrogen-limited chemostat cultures of *Candida lipolytica* ($\square$, mg/g/h) with the best-fit model predictions. The specific activity of the glyoxylate pathway (dimension 1/h, or C-Mole glyoxylate turnover per C-Mole biomass per hour) was calculated for the region where this activity is supposed to be limiting for citrate production.

probably caused by the misfit between model and alkane consumption rate at lower growth rates. Nevertheless it can be concluded that the energetic efficiency of biomass formation in *Candida* in these conditions has a rather normal value, in comparison with other data for yeast, taking into account that the gluconeogenesis route must be followed for a considerable part of the cellular constituents. The ability to produce citric acids seems to be related to a very low efficiency of oxidative phosphorylation, permitting the oxidation of the surplus of reduction equivalents without the concomitant ex-

cessive formation of ATP. An explanation for the low P/O ratio can possibly be found in the presence of an alternative, SHAM (salicylhydroxamic acid) -sensitive respiration, such as found in citric acid producing *Aspergillus niger* (Kubicek & Röhr 1986; Kubicek et al. 1980; Kirmura 1987) although the importance of the presence of such a system was not confirmed for *Candida lipolytica* (Akimenko et al. 1979).

Discussion

The case of growth and polysaccharide production by *Rhizobium trifolii* was used to illustrate the importance of checking data consistency and avoiding data transformation procedures. It is of relevance to mention here that this particular dataset was

Table 3. Parameters estimated for nitrogen-limited citrate producing cultures of *Candida lipolytica* growing on n-alkanes.

Y_{atp}	9.4 ± 3.2
P/O ratio	0.4 ± 0.1
Ratio citrate/isocitrate	1.16 ± 0.04

Experimental data from Aiba & Matsuoka 1978.

re-investigated already before by Erickson & Hess (1981). These authors used a model which is superior to the model in the original publication, because mass balance requirements were incorporated. Although the model derived in this paper is still more accurate, the differences are probably insignificant in view of the variances of the experimental data. The most important improvements of the present analysis are: firstly, all experimental data are used in a single simultaneous optimization procedure, and secondly, experimental data are used in their pure form, i.e. without transformation to some linear function. It is illustrative to note that Erickson and Hess identified an outlier in the dataset, as was done in the present analysis. However, they identified the experimental run at the highest growth rate as the outlier, whereas Fig. 2 shows that these data are within the expected range, and that the biomass concentration measurement at $\mu = 0.05$ clearly is the outlier. They also noted a slight distortion in the 'available electron balance' (comparable with the 'reductance balance' defined in this and the preceding article), however the data were called reasonably consistent. With the present model it is really impossible to obtain a good fit without the introduction of a bias factor for the oxygen flow deviating significantly from one. Apparently this analysis is far more critical for consistency of experimental data than the procedures used by Erickson & Hess.

The system of acetate and butyrate formation by *Clostridium butyricum* could be solved by incorporating two so-called phenomenological equations in the balance expression. These are relations that do not follow from theoretical considerations or biochemical stoichiometries. By plotting experimental observations or functional combinations of observations, one can try to find a descriptive formulation of the observed phenomena. This can be an arbitrary, but preferably simple function. Sometimes (not necessarily) such a relation can give a hint about the metabolic background of the observed behavior. In the case of Clostridium a very convincing linear relation between the degree of reduction of products and substrates was found, with a slope of practically 1.0. In other words: the organism is forced to channel extra reduction equivalents from mannitol towards butyrate, and can not use the alternative outlet of hydrogen production for this purpose. This also explains the inability of Clostridium to grow on mixtures of mannitol and glucose when the ratio between these substrates exceeds the value of 3. The second phenomenological equation introduced in this case, for the rate of unknown product formation, does not suggest any metabolic explanation. By fitting the model to the experimental data, the conclusion could be drawn that hydrogen formation in this organism is not energy-driven, as was also concluded by Crabbendam et al. Further it appeared that the unknown product (estimated from the reported carbon recoveries) is probably not of a proteinaceous nature, as suggested, because the energy costs associated with this product formation are extremely low.

For *Candida lipolytica* growing on alkanes another type of relation was incorporated in the metabolic stoichiometry model. Two regions of different metabolic behavior were identified, one at low growth rates, where citrate production is the predominant metabolic pathway, and another at higher growth rates, where carbon dioxide production increases at the cost of a diminishing citrate production. This behavior can be explained perfectly by assuming a limiting capacity of the glyoxylate cycle at higher growth rates and a basal level of the tricarboxylic acid cycle at lower growth rates. These two alternate metabolic restrictions provided the extra equation required to solve the metabolic system. This aspect is absent in the original paper, which is one of the early examples of the use of metabolic balances in the interpretation of macroscopic phenomena. The fermentation model permitted the determination of two bioenergetic parameters, Y_{atp} and the P/O ratio, of which the latter has an very low value. It can be argued that a low efficiency of energy conservation is a prerequisite for a high yield of citrate in citric acid fermentations. Apparently the yeast is able to manipulate its efficiency of oxidative phosphorylation, whereas the efficiency of anabolic processes remains at a normal value.

With the three examples, reanalyzing literature data, it is illustrated that the method of (matrix-)

processing of metabolic equations is comprehensive and general applicable. In all three cases original conclusions could be reinforced, more accurate estimates of parameters could be obtained or new conclusions could be drawn.

References

Akimenko VK, Finogenova TV, Ermakova IT & Shishkanova NV (1979) The cyanide resistance of respiration and the overproduction of citric acids in *Candida lipolitica*. Microbiology USSR 48: 503–509

Aiba S & Matsuoka M (1978) Citrate production from n-alkane by *Candida lipolytica* in reference to carbon fluxes in vivo. Eur. J. Appl. Microbiol. Biotechnol. 5: 247–261

Crabbendam PM, Neijssel OM & Tempest DW (1985) Metabolic and energetic aspects of the growth of *Clostridium butyricum* on glucose in chemostat culture. Arch. Microbiol. 142: 375–382

De Hollander JA (1991) The use of stoichiometric relations for the description and analysis of microbial cultures. A. van Leeuwenhoek 60 (3/4): this issue

De Hollander JA, Bettenhaussen CW & Stouthamer AH (1979) Growth yields, polysaccharide production and energy conservation in chemostat cultures of *Rhizobium trifolii*. A. van Leeuwenhoek 45: 401–415

Erickson LE & Hess JL (1981) Analysis of growth and polysaccharide yields in chemostat cultures of *Rhizobium trifolii*. Ann. NY Acad. Sci. 369: 81–89

Kirimura K, Hirowatari Y & Usami S (1987) Alterations of respiratory systems in *Aspergillus niger* under the conditions of citric acid fermentation. Agric. Biol. Chem. 51: 1299–1303

Kubicek CP & Röhr M (1986) Citric acid fermentation. Crit. Rev. Biotechnol. 3: 331–373

Kubicek CP, Zehentgruber O, Housam El-Kalak, & Röhr M (1980) Regulation of citric acid production by oxygen: effect of dissolved oxygen tension on adenylate levels and respiration in *Aspergillus niger*. Eur. J. Appl. Microbiol. Biotechnol. 9: 101–115

Lehninger AL (1975) Biochemistry, 2nd edition. Worth Publishers Inc, New York

Stouthamer AH & Van Verseveld HW (1985) Stoichiometry of microbial growth. In: Bull AT & Dalton H (Eds) Comprehensive Biotechnology, Vol 1 (pp 215–238). Pergamon Press, Oxford

Roels JA (1983) Energetics and Kinetics in Biotechnology. Elsevier Biomedical Press, Amsterdam

Von Meyenburg HK (1969) Energetics of the budding cycle of *Saccharomyces cerevisiae* during glucose limited aerobic growth. Arch. Mikrobiol. 66: 289–303

Antonie van Leeuwenhoek **60**: 293–311, 1991.

Metabolite production and growth efficiency

J.D. Linton
Shell Research Limited, Sittingbourne, Kent ME9 8AG, UK

Key words: antibiotics, exopolysaccharides, growth efficiency, metabolite production rate, organic acids, yields

Abstract

The capacity to sustain the large fluxes of carbon and energy required for rapid metabolite production appears to be inversely related to the growth efficiency of micro-organisms. From an overall energetic point of view three main classes of metabolite may be distinguished. These are not discrete categories, as the energetics of biosynthesis will depend on the precise biochemical pathways used and the nature of the starting feed stock(s). (1) For metabolites like exopolysaccharides both the oxidation state and the specific rate of production appear to be inversely related to the growth efficiency of the producing organism. Maximum rates of production are favored when carbon and energy flux are integrated, and alteration of this balance may negatively effect production rates. (2) The production of metabolites like organic acids and some secondary metabolites results in the net production of reducing equivalents and/or ATP. It is thought that the capacity of the organism to dissipate this product-associated energy limits its capacity for rapid production. (3) For metabolites like biosurfactants and certain secondary metabolites that are composed of moieties of significantly different oxidation states production from a single carbon source is unfavorable and considerable improvements in specific production rate and final broth concentration may be achieved if mixed carbon sources are used. By careful selection of production organism and starting feedstock(s) it may be possible to tailor the production, such that the adverse physiological consequences of metabolite overproduction on the production organism are minimized.

Introduction

The identification of the major flux-control points in the biosynthesis of a given metabolite is a prerequisite to a rational approach to increasing specific rates of production and has been extensively discussed (Kacser & Burns 1973; Kell 1989). However, overcoming the major flux-control points within a given pathway (eg. by overexpression of enzymes) may not lead to metabolite overproduction on its own, because the physiological consequences of overproduction may be unfavorable to the organism (eg. instability, growth disadvantage). Additional steps will have to be taken to anticipate and alleviate the physiological conse-

quences of metabolite over-production, be it via energy dissipation or the amplification of pathway enzymes, to ensure that deregulation of important flux-control points does not prove toxic to the cell. This communication will be confined to discussing the effect of growth efficiency and feedstock choice on the energetic consequences of metabolite over-production on the producing organism.

Economic considerations

In general the price of a given chemical is inversely related to the volume of its production (Dunnill 1983; King 1982; Tonge & Cannell 1983) and bio-

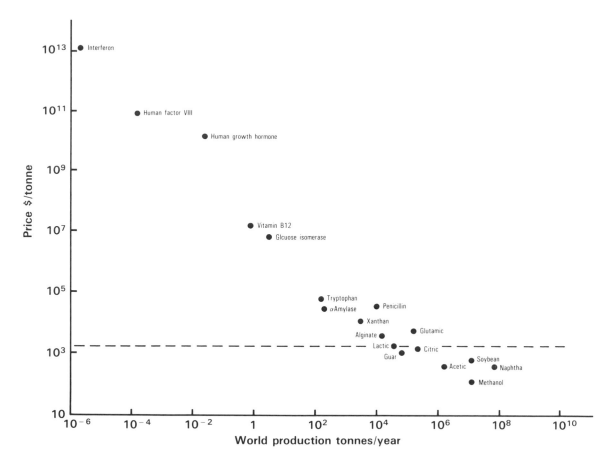

Fig. 1. The relationship between the price of various chemicals and the volume in which they are produced. (Data taken from Dunill 1982; Tong & Cannell 1983).

technological processes have to conform to this relationship if they are to be competitive (Fig. 1). For the production of bulk chemicals like ethanol and acetic acid, bio-routes have to compete with conventional chemical synthesis, whereas for the production of natural oils, proteins and poly saccharides that are also produced in large tonnages, competition is from agriculture. High yields, high product concentration and high productivity are characteristics of bulk production processes be they chemical, agricultural or bio-based (Fig. 2).

With respect to bio-routes, these criteria can be met when fermentative metabolism is employed to produce chemicals like ethanol. Under these conditions, energy is generated by substrate-level phosphorylation, and organic compounds function as electron acceptors for the reoxidation of reduced

cofactors produced during metabolism. Thus, in a fermentative organism, high growth rate is obligately coupled to a high rate of product formation, although the converse does not necessarily hold. Fermentative metabolism, however, leads to a fairly limited range of chemicals and therefore aerobic processes are usually employed in bio-processes.

The production of chemicals via aerobic routes has to address problems inherent to this type of metabolism. In most aerobic organisms grown under conditions optimal for product formation in continuous culture (ie. when growth is limited by some nutrient other than the carbon source), the yield of the product is inversely related to growth rate. Although the biomass yield is low under these conditions of carbon excess, it increases with increasing growth rate at the expense of product yield, (Linton & Musgrove 1983) Fig. 3b. This is likely to be a general phenomenon with aerobic processes, because at low growth rates the difference between the potential substrate uptake rate and the *in situ* rate needed to satisfy the demand for cell material is large. Therefore, relatively more substrate can be directed towards product formation without affecting cell production, Fig. 3a. In the absence of a change in growth efficiency, an increase in growth rate will require an increase in the specific substrate uptake rate. Therefore with increasing growth rate the difference between the *in situ* and potential substrate uptake rates gets smaller and at u_{max} they are equal. Consequently, the relative proportion of substrate that can be directed to products, (i.e., the Yp) decreases with increasing growth rate for aerobic cultures. This inverse relationship between yield and specific production rate (Fig. 3b) will hold for most aerobic processes with the exception of those where the substrate molecule undergoes very little change, e.g., the production of gluconate or possibly in the polymerisation of the substrate into certain types of polysaccharide.

Therefore, although high yields of products can be achieved at low growth rates, the specific rates of production are low and it is therefore difficult to achieve the high productivities (kg product m^{-3} h^{-1}) required for the production of bulk chemicals. These considerations are also relevant to batch culture systems, where metabolite production occurs after growth has ceased and substrate uptake is similarly regulated. Aerobic organisms appear to control the rate of substrate utilization and product formation when growth is slow or absent. Clearly, if process organisms could be made to express their potential substrate uptake rates in the absence of growth and yet maintain a tight control of respiratory activity (so that substrate is not combusted to CO_2 but channelled to the desired product), much higher rates of product formation might be obtained. Yields of product near to the theoretical would be obtained and yet the rate of product formation would be high. To achieve this, effort will have to be directed towards understanding how substrate uptake can be manipulated and

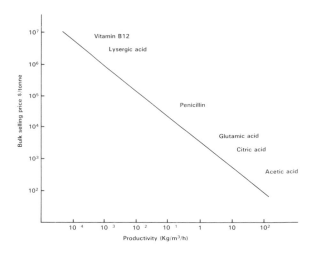

Fig. 2. The relationship between the price of various chemicals and the productivity at which they are operated. (Based on Andrew 1983).

maintained at elevated levels in non-growing cultures. The regulation and control of substrate uptake and metabolism (eg. protein turnover, maintenance energy requirements, genetic regulation) in the absence of growth is the key to improving the productivity and operational life-time of bioprocesses and this area requires more attention if bioprocesses are to become competitive with chemical synthesis or agriculture for the production of bulk chemicals. The importance of substrate uptake on metabolite production rate will be discussed later in this communication.

For high value chemicals that are produced on a smaller scale by multi-step synthesis from relatively cheap carbon feedstocks (such as sugars or plant oils) the yield value is not as important as that for the production of bulk chemicals. For these processes, product concentration and recovery, specificity of the conversion and productivity are more important than yield. However, where the starting material represents a major cost in the overall process, as in the case of most biotransformations, a high yield is essential.

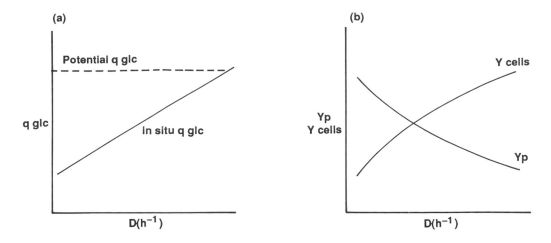

Fig. 3. The theoretical relationship between the *in situ* and potential q glucose (a) and biomass and product yield (b) as a function of dilution rate in a nitrogen-limited culture of a typical obligately aerobic microorganism.

Practical approach to improving yields and rates of metabolite production

Fermentation processes generally use cheap sources of carbon and energy such as molasses or plant oils. Complex nutrient sources such as soybean meal and yeast extract may be added as cheap sources of nitrogen and vitamins respectively. In most cases the optimal combination of nutrients for a given process has been derived by trial and error over a number of years. In the recent past the complexity of production media discouraged attempts to quantify the metabolic fate of the nutrients supplied because suitable analytical techniques were not available. However, modern analytical techniques allow accurate and rapid estimation of the major carbon compounds present in the starting medium as well as the majority of metabolites produced from these compounds. Unfortunately, there are very few examples in the open literature where mass balances have been constructed for the production of an industrially important metabolite. Whether this is because detailed analysis of medium components and metabolic products has not been routinely carried out or because the information is considered industrially sensitive and therefore not published is not clear.

In the context of this discussion growth under conditions of carbon excess is defined as the supply of carbon source in excess of that required for growth. It does not necessarily equate with the presence of excess carbon source in the culture medium. For example, in continuous or fed- batch production the carbon source can be supplied in excess of that required for cell production but at a rate that is equivalent to or less than that required for product formation. Under these conditions the concentration of carbon source in the medium is effectively zero, but growth is nevertheless under conditions of carbon excess.

Mass – balances

In order to improve the yield of a given metabolite in a rational way it is necessary to make some estimate of the fate of the nutrients that are required for its synthesis. Carbon and nitrogen balances allow estimates to be made of the proportion of the carbon and nitrogen source that is converted into microbial biomass, CO_2 and both desired and unwanted products. This information is crucial to the development of a rational approach to improving yields and rates of metabolite production. For further details regarding the construction of mass balances readers are referred to Stouthamer & van Verseveld (1987); Linton et al. (1987a, b).

In its simplest form a mass-balance may be represented as follows:

Input (g carbon or nitrogen)	=	Output (g carbon or nitrogen)	Recovery (%)
Substrate Carbon	= bacterial + carbon	CO_2 + product carbon	100
		or	
Nitrogen	= bacterial nitrogen	+ product nitrogen	100

Although virtually all industrial fermentations are operated as batch or fed-batch cultures a great deal of useful information for process optimization can be obtained from continuous culture systems.

Maximum attainable yields

Where the biosynthetic pathway is known for a given metabolite, it is possible to estimate the theoretical yield from a given carbon source or a mixture of carbon sources. If the biosynthesis is energy- requiring, then the theoretical yield will be influenced by the ATP/O quotient of the organism used to produce the given metabolite. Where the production of a given metabolite leads to the net production of ATP and/or reducing equivalents the ATP/O quotient will influence the extent to which energy dissipation will be an integral part of rapid metabolite production (Stouthamer & van Verseveld 1987; Linton 1990). Consequently, to estimate the potential for yield improvement for a given primary or secondary metabolite it is necessary to determine the ATP/O quotient of the producing organism.

Determination of ATP/O quotients

The ATP/O quotient is defined as the amount of ATP produced per 0.5 moles O_2 consumed. Unfortunately, it is not possible to measure the ATP/O quotient precisely and in practice it is usually ob-tained from growth yields of microorganisms measured under conditions of carbon-limitation in che-mostat culture. Growth yields are usually determined under these conditions because they favor the most efficient conversion of the carbon source into cellular material and the highest efficiency of energy conservation.

In order to estimate the ATP/O quotient of a given organism the steady state respiration rate (q_{O_2}) is measured as a function of dilution rate in continuous culture. The maximum growth yield from oxygen ($Y_{O_2}^{max}$) can be obtained from the reciprocal of the slope of the q_{O_2} against dilution rate plot (Stouthamer & Bettenhaussen; 1973). The ATP/O quotient is derived from the $Y_{O_2}^{max}$ using the equation:

$$Y_O^{max} = Y_{ATP}^{max} \cdot ATP/O$$

It is stressed that a number of assumptions have been made in calculating the ATP/O quotient from the above equation. The theoretical ATP requirement for cell biosynthesis from glucose (in mineral salts medium) has been estimated to be 28.8 g dry wt. mol^{-1} ATP, which is considerably higher than that usually estimated from growth yields and known routes of ATP synthesis (Stouthamer 1979). This discrepancy between the theoretical Y_{ATP}^{max} and that derived from yield determination (12–14 g. mol^{-1} ATP) is an important source of error in determining the 'real' ATP/O quotient. The other assumption implied in the equation is that ATP synthesis via substrate-level phosphorylation is not important. The influence of the latter is inversely proportional to the ATP/O quotient and could result in an over estimation by approximately 7 and 30% at ATP/O quotients of 3 and 1, respectively. However, as long as organisms growing under carbon-limited conditions on the same carbon source are compared, the difference in the Y_{ATP}^{max} is minimized and the ATP/O quotient will be valid as a comparative measure of growth efficiency. The ATP/O quotient obtained in this manner is only an approximation, but it nevertheless allows worthwhile analysis of the energetics of metabolite production which would otherwise not be possible.

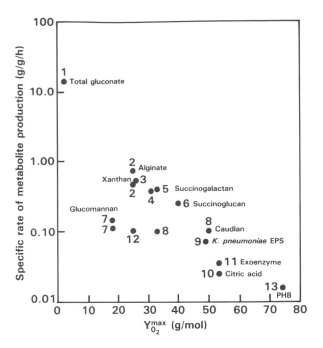

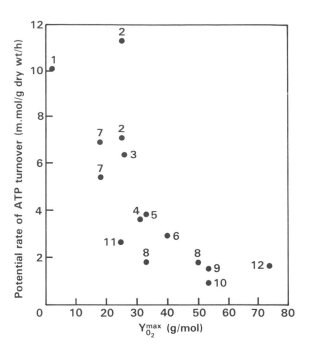

Fig. 4. The highest specific rates of metabolite production reported as a function of $Y_{O_2}^{max}$ of the producing organism. 1, total gluconate; 2, alginate; 3&5, xanthan; 4, succinogalactan; 6, succinoglycan; 7, glucomannan; 8, curdlan; 9, *K pneumoniae* exopolysaccharide; 10, citric acid; 11, exoenzyme; 12, rhamnolipid; 13, PHB (from Linton 1990).

Fig. 5. The potential rate of ATP turnover associated with metabolite production as a function of the $Y_{O_2}^{max}$ of the producing organism. 1, total gluconate; 2, alginate; 3&5, xanthan; 4, succinogalactan; 6, succinoglucan; 7, glucomannam; 8, curdlan; 9, exoenzyme; 10, citric acid; 11, alginate; 12, PHB (from Linton 1990).

Scope for yield improvement

As discussed above the theoretical yield for a given metabolite can be estimated if the biosynthetic pathway for its synthesis and the ATP/O quotient of the producing organism is known. The scope for yield improvement is essentially the difference between the theoretical and observed yield (the latter must be corrected for substrate and oxygen requirement for growth and maintenance). The use of this procedure will be discussed in further detail later in this communication.

Metabolite overproduction

Rapid metabolite production necessitates large fluxes of carbon through the metabolic system of an organism and it has been suggested (Linton 1990) that the capacity to sustain these high fluxes is inversely related to the growth efficiency ($Y_{O_2}^{max}$)

of the producing organism. Thus the rate of metabolite production by a wide range of organisms appears to be inversely related to their $Y_{O_2}^{max}$, Fig. 4. Similarly, if the stoichiometry of synthesis is estimated, it can be seen that the rate of ATP turnover associated with metabolite production (either directly or via the complete oxidation of reducing equivalents produced as a consequence of metabolite production) is also inversely related to growth efficiency Fig. 5. Organisms possessing low growth efficiencies appear to have a considerably higher capacity to turnover ATP (dissipate energy) than those possessing higher growth efficiencies.

In general three main classes of metabolite may be distinguished. (1) Metabolites such as exopolysaccharides, for which part of the substrate has to be oxidized to provide for ATP needed for biosynthesis. For polysaccharides, both the oxidation state and the rate of production appears to be inversely related to the growth efficiency of the producing organism. (2) The production of metabolites such as organic acids and certain secondary metabolites is accompanied by the net generation of ATP and /or reducing equivalents. It is believed that the rate of production of this class of metabolite is limited by the capacity of the producing organism to dissipate the energy produced concomitantly. (3) Metabolites like biosurfactants and certain secondary metabolites that contain moieties of significantly different oxidation state such as sugars and fatty acids. Production of these molecules from a single substrate requires high rates of ATP turnover and in some cases rates of production and final broth concentrations may be improved significantly if ATP turnover is reduced by the provision of ready made precursors. The production of each of these classes of metabolite will be discussed separately but it should be noted that all metabolites will not fall neatly into each of these categories. In some cases one or more categories may be involved in the production of a single end product, for example the class into which the biosynthesis of certain antibiotics falls is dependent on the substrate(s) used for their production.

Exopolysaccharide production

For an exopolysaccharide of known composition the theoretical yield from the carbon source and oxygen can be estimated from the ATP/O quotient. Although the yield from the carbon source is virtually independent of the ATP/O quotient the yield

from oxygen is proportional to it (Jarman & Pace 1984). For example the yields of Xanthan gum from oxygen have been calculated to be 5.3, 10.6 and 15.97 g $g^{-1}O_2$ at ATP/O quotients of 1, 2 and 3, respectively. The productivity of processes for the production of viscous exopolysaccharides is oxygen mass- transfer limited. It is therefore surprising that, although xanthan production has been optimised for over 30 years, the ATP/O quotient of X campestris was not established until fairly recently (Rye et al. 1988). Clearly, prior to process optimization, it would be helpful to have some idea regarding the scope for improvement. In order to do this it is necessary to correct the observed yield for the amount of glucose and oxygen needed for cell production and to compare these corrected yields with the theoretical calculated from the ATP/O quotient estimated for the given organism.

When the experimental yields of EPS produced by *Agrobactrium radiobacter* (Linton et al. 1987a), *Erwinia herbicola* (Linton et al. 1988), *Xanthomonas campestris* (Rye et al. 1988) and *Pseudomonas mendocina* (Rye, Jones & Conant, unpublished.) were corrected for cell production they were found in all cases to be > 70% of the theoretical. Bearing in mind that small changes in the ATP/O quotient have a relatively large effect on the EPS yield from O_2 these yields are remarkable close to the theoretical, which suggests that EPS production is a very efficient process and there is little scope for major yield improvements, Table 1. It should be noted that all the above EPS producers have been shown

Table 1. A comparison of theoretical and observed yields for exopolysaccharide from oxygen and the carbon source.

Organism	Exopolysaccharide	Yields of EPS (g g^{-1})			
		Corrected for cell production		Theoretical*	
		Yglc	Y_{O_2}	Yglc	Y_{O_2}
A. radiobacter	Succinoglycan	0.74	5.7	0.83	8.1
E. herbicola	Succinogalactan	0.69	8.15	0.83	7.8
X. campestris	Xanthan	0.95	4.0	0.80	5.32
P. mendocina	Alginate	0.52	0.85	0.45	0.91

* Calculated from ATP/O quotients derived from carbon-limited $Y_{O_2}^{max}$ values. The observed yields have been corrected for cellular requirements for oxygen and carbon source. The theoretical yields were calculated from ATP/O quotients derived from the $Y^{max} O_2$ values of carbon-limited cultures.

to possess ATP/O quotients considerably lower than those expected from the composition and properties of their respiratory systems (Rye et al. 1988; Cornish et al. 1987; Jones 1988).

ATP requirement for EPS synthesis

When EPS producing organisms such as *Xanthomonas campestris P. aeruginosa* (Jarman & Pace 1984), *Methylophilus sp.* (Linton et al. 1986) and *A. radiobacter* (Linton et al. 1987a) are grown in continuous culture under conditions optimal for EPS production the ATP requirement for EPS biosynthesis may amount to more than twice that required for cell biosynthesis Table 2. EPS production is therefore a major event in physiological terms and unless it offers the producing organism a selective advantage, it would be expected to be rapidly selected against in continuous culture. High rates of EPS production therefore require high rates of ATP synthesis and use. In the organisms studied to date a significant proportion (30–70%) of the ATP requirement is generated during the production of the oxidized substituents on these polysaccharides, (eg. acetate, succinate, uronic acids) Table 2. Synthesis of the EPS sugar backbone is an energy- requiring process, whereas the production of the oxidized moieties is an energy generating process. Carbon and energy flux and EPS production appear to be integrated in these

organisms, the extent of this integration will depend on the oxidation state of the particular EPS, the ATP/O quotient of the producing strain and the degree of reduction of the carbon source used for production.

Carbon and energy fluxes to EPS

It has been found that optimal production of EPS occurs when carbon and energy flux is integrated, and alteration of this integration (eg. by changing the oxidation state of the carbon source) has a marked effect on specific production rate. Under conditions optimal for EPS production from substrates such as glucose, xylose and gluconate that are at the same oxidation state as EPS, virtually all the respiratory activity that occurs over and above that required for growth appears to satisfy the energy demands of EPS production. The proportion of respiratory activity dissociated from EPS synthesis increases sharply with substrates either more oxidized or reduced than glucose. This probably occurs because carbon and energy flux is not favorable, for example, with substrates like ethanol, oxidation to the level of a hexose generates considerably more energy than can be turned over in EPS synthesis. Succinoglycan synthesis from ethanol by *A. radiobacter* would lead to an overproduction of 91 ATP/ repeat unit and therefore a high rate of EPS production from this reduced feedstock would

Table 2. The rate of ATP utilisation for growth and EPS production for variousorganisms growing in chemostat culture (D = 0.05 h^{-1}).

Organism	Exopolysaccharide	ATP/0 quotient	Rate of ATP utilization for: (mmol ATP/g dry wt.h^{-1})		The proportion of ATP required for EPS synthesis provided by the oxidised moieties (%)*
			Cell production*	EPS production**	
A. radiobacter	Succinoglycan	1.5	3.6	4.8	30–50
E. herbicola	Succinogalactan	1.25	3.6	5.2	50
X. campestris	Xanthan gum	1.0	3.6	9.0	30
P. mendocina	Alginate	1.0	3.6	9.94	40
Methylophilus sp.	Glucomannan	1.0	4.7***	5.4	60–75

* Assuming Y$_{ATP}$ = 14 g mol^{-1}.
** 1 ATP/hexose transported.
*** on methanol Y$_{ATP}$ = 10.6 g mol^{-1}.
The proportion of the ATP requirement for EPS synthesis provided by the production of the oxidised constituents is also shown.

require some form of energy dissipation. In practice, the specific rate of EPS production decreases when the substrate is more reduced than glucose and in the case of succinoglycan the specific production rate from ethanol is two orders of magnitude lower than that with glucose (Linton et al. 1987b).

Physiological role of EPS production

In order to determine the physiological role of EPS production, mutants of *A. radiobacter* and *E. herbicola* unable to produce EPS were selected, either via Tn5 mutagenesis or via chemostat selection. These mutants were then subjected to nitrogen-limitation in the presence of excess glucose at a dilution rate optimal for EPS production. Under these conditions no EPS was produced by these mutants, which appeared to control the rate of substrate uptake to approximately 25% of that of the EPS producing wild type. However, both mutants exhibited respiration rates that were similar to those of the EPS-producing strains Table 3. Thus ATP turn over rather than EPS production *per se* appears to be the physiologically important event in these organisms. EPS production is only one means of achieving this and is a consequence of a rapid flux of glucose into the cell coupled to partial oxidation of the glucose to EPS. In the EPS minus mutants, glucose flux is tightly regulated but

a small proportion is completely oxidized to CO_2 to generate a similar ATP turnover rate.

This probably explains why EPS- minus mutants are readily selected when polysaccharide producing strains are cultured for extended periods under conditions of carbon excess.

Even substrates of identical oxidation state, such as glucose and fructose, can markedly effect both the rate and yield of EPS production if they are metabolized via different metabolic pathways. For example, in *P. mendocina*, gluconate (glucose) and fructose enter central metabolism via different routes. Gluconate (glucose) is cleaved into two C-3 intermediates, glyceraldehyde-3 phosphate and pyruvate, but only the former appears to be incorporated into alginate (Anderson et al. 1987). The fate of the pyruvate will determine the extent to which carbon and energy flux is integrated. Complete oxidation of pyruvate to CO_2 will result in the production of potentially more energy than that required for alginate synthesis:

2 gluconate + 3 ATP + GDP + 2NADP + 8 NAD = (alginate) + 2 NADPH2 + 8 NADH2 + 6 CO_2 + 3 ADP + 3Pi + GTP

Mass balances for alginate production from glucose and gluconate confirm that the pyruvate is indeed combusted to CO_2 (Linton 1990). Fructose on the other hand is incorporated into alginate directly (Anderson et al. 1987) and the overall carbon and energy flux in the synthesis of alginate is considerable more favorable, as 66% of the energy requirement for total synthesis is produced during the production of alginate monomers (*P. mendocina*, ATP/O = 1).

Fructose + 3ATP + GTP + 2 NAD = (Alginate) + 2NADH2 + 3ADP + GDP + 4Pi

It has been found that the experimental yields of alginate from oxygen during growth on fructose are considerably higher than those observed when glucose or gluconate are used as feedstocks (Linton 1990). Moreover, the specific rate of alginate production from fructose, glucose and gluconate are, respectively 0.35, 0.2 and 0.15 g alginate g^{-1} h^{-1}

Table 3. A comparison of the rate of glucose uptake and respiratory activity of exopolysaccharide plus (EPS +) and minus (EPS −) strains of *A. radiobacter* and *E. herbicola* grown under conditions of carbon excess in nitrogen-limited chemostat culture, D = 0.05 h^{-1} (Linton 1990).

Organism		Rate of substrate consumption (mmol g^{-1} h^{-1})	
		q_{glc}	q_{O_2}
A. radiobacter	EPS +	2.72	2.66
A. radiobacter (Tn.5 mutant)	EPS −	0.67	2.4
E. herbicola	EPS +	3.06	4.97
E. herbicola	EPS −	1.78	4.97

and correlate with the extent to which carbon and energy flux is integrated (Linton 1990).

The need for integration of carbon, energy and reductant flux is clearly seen in the case of EPS production from methanol. An examination of the three major pathways of C-1 assimilation shows that one, the ribulose monophosphate pathway (RuMP), is energetically considerably more favorable for EPS synthesis in terms of carbon, energy and reductant balance than the others (Linton et al. 1986). The dihydroxyacetone and serine pathways are respectively ATP and NADH- limited for EPS synthesis. It has been found that organisms possessing the RuMP pathway exhibit the highest rates and yields of EPS production (Linton et al. 1986; Grinberg et al. 1984).

Influence of substrate uptake on EPS production rate

The identification of the major flux-control points in biosynthesis of a given metabolite is a prequisite to a rational approach to increase specific rates of production (Kell 1989). In the case of succinoglycan synthesis by *A. radiodacter* it has been shown that glucose uptake is a major kinetic control point in EPS synthesis (Cornish et al. 1988 a, b). *Agrobacterium radiobacter* produces two periplasmic proteins in response to glucose limited growth. These proteins were shown to be glucose binding proteins and were called GBP-1 and GBP-2 respectively (Cornish et al. 1988 a). Extended growth of *A. radiobacter* in glucose-limited continuous culture led to the selection of mutants AR-18 and AR-9 in which GBP-1 or GBP-2 but not both were hyper-produced. The increase in the binding protein level correlated with a significant increase in the glucose uptake capacity of these mutants. However, when these mutants were cultured under conditions of glucose excess the uptake was repressed to a similar level in the wild type strain and AR-9. In AR-18 repression was incomplete and the glucose uptake capacity was approximately a factor 2 higher than that of the wild type and AR-9. The higher capacity for glucose uptake of AR-18 was accompanied by a significant increase in the specific rate of succinoglycan production from 0.21 g^{-1} h^{-1} to $0.31 \text{ g g}^{-1} \text{h}^{-1}$. The increase in the rate of EPS production was specifically attributed to the higher capacity for glucose uptake in AR-18 as the rate of EPS production on sucrose was similar to that of the wild type (Cornish et al. 1988b). Flux control analysis indicated that glucose uptake was a major kinetic control point (control coefficient > 0.8) in EPS production in this organism (Cornish et al. 1988b).

Impact of molecular genetics on increasing EPS production rate

The genes involved in the synthesis of xanthan have been identified and cloned (Betlac et al. 1987). In addition, a range of mutants altered in their capacity to produce xanthan have been produced and their biochemical lesions characterized (Betlac et al. 1987). This elegant work has shown in principle that various forms of xanthan gum can be produced. One particular modification, the poly-trimer, has been shown to have improved rheological properties. Similarly, the genetic loci involved in succinoglycan synthesis have been identified and various mutants generated (Glazebrook & Walkeret 1989; Doherty et al. 1988). Nevertheless, there have been no well substantiated reports that genetically engineered strains exhibiting either improved rates of production or altered chemical composition have been used commercially.

The cloning and amplification of EPS- associated loci without establishing the flux-control points in the biosynthesis is the main reason for the apparent lack of success in achieving improved production rates. As discussed above the use of molecular genetics to deregulate substrate uptake will undoubtedly lead to further improvements in specific production rates. However, repeated flux-control analysis will be necessary to identify further kinetic control points in order to maximize the impact of genetic manipulation.

With respect to altering the composition of exopolysaccharides, the need to maintain carbon, energy and reductant flux integrated for optimal production must also be taken into account in strain construction. Mutants that produce modified forms of xanthan (Betlac et al. 1988), have been reported. Theoretically, nine variations of the xanthan repeat unit are possible. A truncated form of xanthan, the poly-trimer, was reported to possess interesting rheological properties. However, it cannot be produced economically because its specific production rate is two orders of magnitude lower than that of native xanthan. On closer examination it is evident that the energetics of biosynthesis of the poly-rimer are considerably less favorable than for xanthan gum (Linton 1990). Whether this alteration in production is a consequence of the altered energetics of synthesis or due to the specificity of the polymerase or a combination of both remains to be established.

In *Rhizobium meliloti*, 16 loci have been implicated in succinoglycan synthesis and mutations in some of these loci cause alterations in chemical composition or specific production rate. Some of these mutants were kindly provided to the author by Professor Walker and their capacity to produce EPS examined in batch fermentations. The results confirmed that mutation in the regulatory region *exo*S resulted in a significant increase in the specific rate of production (q_p = 0.09 g EPS g^{-1} h^{-1}) compared to the wild type (q_p = 0.034 g g^{-1} h^{-1}, Eagle R, Jones DS, Cornish A & Linton JD, unpublished). However, the specific production rate of succinoglycan by the deregulated mutant is considerably slower than that observed with *A. radiobacter* producing a similar EPS. Moreover, as discussed above, the energetic consequences of alterations in chemical composition can be calculated and the production of the desuccinylated succinoglycan shown to be the most unfavorable single change in composition. The mutant producing such an EPS (*exo* H) exhibited an exceedingly low q_p of < 0.002 g g^{-1} h^{-1}. For the impact of molecular genetics on EPS composition and rate to be maximized emphasis will have to be given to quantifying specific production rates and greater considerations given to the energetic consequences of the proposed changes in composition. In addition, the specificity

of the biosynthetic systems involved will have to be established in order to determine whether genes from different organisms can be mixed to produce polymers by design.

Metabolites that result in the net production of ATP and/or reducing equivalents

As discussed earlier, the production of organic acids such as citric acid or glutamic acid results in the net generation of ATP and NADH. The organisms used for the commercial production of these and related metabolites are generally obligate aerobes, and regeneration of NAD occurs via respiration. For metabolite production to proceed rapidly for any length of time, some means of energy dissipation will have to occur. It has been suggested that the capacity to dissipate this metabolite associated energy limits the rate of production of this class of metabolite (Linton & Rye 1989; Kola et al. 1987; Stouthamer & van Verseveld 1985). Organisms possessing low ATP/O quotients appear to have a disproportionally higher capacity for energy dissipation than those with high ATP/O quotients. Thus, the organism that produces gluconic acid at the fastest rate reported has a $Y_{O_2}^{max}$ of only 2.1 (Weenk et al. 1984). Organisms possessing high ATP/O quotients also produce gluconic acid rapidly (Linton et al. 1984). Indeed, *Aspergillus niger* is used for the commercial production of gluconic acid. This apparent anomaly can be explained by the operation of the glucose oxidase catalase complex that effectively dissipates the product associated energy allowing high rates of production in these organisms. Energy dissipation occurs because reducing equivalents are not fed into the electron transport chain but instead result in the generation of H_2O_2 which is cleaved to produce H_2O + O_2 by the action of catalase.

Moreover, in some cases, the glucose oxidase-catalase complex is secreted into the culture medium when glucose is present in excess (nitrogen-limitation) and the level of secretion appears to be related to the external glucose concentration (Linton et al. 1984). In this case the rapid conversion of glucose into gluconate is thought to act as a means of controlling glucose uptake as no over metabolism of gluconate or fructose occurs under similar conditions (Linton et al. 1984).

Similarly, the use of *A niger* to produce citric acid would appear to be far from ideal, as it has a relatively high growth efficiency and therefore a limited capacity for energy dissipation. However, industrial process has evolved empirically over more than 50 years and close examination indicates that conditions to aid energy dissipation have been encouraged. This is achieved by separating the growth and production phases. The former is controlled at approximately 5 whereas the latter is held between 2.5 and 3.5. It is well known that organic acids in the undissociated form diffuse into the cell causing uncoupling between respiration and energy conservation (Hueting & Tempest 1977; Linton et al. 1981). In the case of citric acid, reduction of the pH value from 3.5 to 2.0 caused a 5-fold increase in specific production rate (Kristiensen & Charley 1981). The stimulation of citric acid production rate caused by the addition of organic acids (Miall 1972) is probably related to this uncoupling effect.

Prospects for improvement in industrial production rates

After over 30 years of mutation and selection, the maximum specific production rate of glutamic acid by *Corynebacterium glutamicum* remains comparatively low (between 0.04 and 0.06 g g^{-1} h^{-1}, Tanaka et al. 1960, 1969). From the relationship between metabolite production rate and growth efficiency, Fig. 4, these comparatively low specific rates of production suggest that *C. glutamicum* possesses a relatively high ATP/O quotient. This has indeed been shown to be the case, a $Y_{O_2}^{max}$ of 78 g mol^{-1} O$_2$ was estimated for *C glutamicum* grown in carbon-limited chemostat culture (Linton 1990). The growth efficiency (Y_{O_2}) of *C glutamicum* appears to be unaffected by the nature of the nutrient-limitation in chemostat culture. That is, the Y_{O_2} of cultures grown under carbon-limited and carbon-excess conditions were indistinguishable. Moreover, none

of these cultures excreted significant levels of metabolites into the culture broth. These observations suggest that substrate uptake and metabolism is tightly regulated in this organism even though growth may be limited by a nutrient other than the carbon source which is present in excess (Holroyd, Jones & Linton unpublished).

In contrast, biotin-limited cultures were found to have considerably lower growth efficiencies and a significant amount of the carbon source supplied was converted into metabolic products under these conditions (Jones & Linton unpubl., Linton 1990). Biotin-limitation is employed in industrial processes for the production of glutamic acid and it is generally believed that these conditions facilitate glutamate production by aiding excretion (Nakayama 1976; Nakaayama 1986) or blocking fatty acid synthesis (Hoischen & Kramer 1990). However, biotin-limitation probably effects a number of membrane mediated processes due to its effect on fatty acid synthesis, including energy conservation and, as discussed above it is this increased capacity for energy dissipation under biotin-limitation that facilitates higher specific production rates (Linton 1990). The importance of energy dissipation on production rate has been demonstrated for keto-gluconate- producing cultures of *Klebsiella aerogenes*, where the addition of the uncoupler 2, 4-dinitrophenol caused a marked stimulation in production rate, as expected, because it aids energy dissipation (Neijssel 1977).

To obtain significantly higher specific rates of organic and amino acid production it may be worthwhile examining the possibility of using organisms possessing lower growth efficiencies. For example *Bacillus acidocaldarius* has been shown to have a very low growth efficiency and to be capable of growing at pH 3 (Farrand et al. 1983) This organism appears to have physiological characteristics suited to amino acid production, ie. low growth efficiency coupled to a capacity to grow at low pH values. Gluconobacter, the organism used to produce gluconic acid industrially has precisely these characteristics (Weenk et al. 1984). This does not mean that all organisms possessing low growth efficiencies will necessarily be capable of rapid metabolite production. What is suggested is that these organisms can dissipate disproportionally greater amounts of energy than more efficient organisms and ought to be capable of producing metabolites at high rates if they are genetically manipulated to take up the carbon source rapidly, and blocked in the appropriate part of the pathway. Clearly, additional basic characteristics such an ability to withstand the high osmotic pressure caused by concentrated solutions of organic acids / salts must be kept in mind during the initial selection of a suitable organism.

Overcoming the major flux control points within a given metabolic pathway may, by itself, not lead to rapid metabolite overproduction if the energetic consequences of the alterations are unfavorable. In addition, the physiological consequences of metabolite overproduction will have to be anticipated and appropriate steps taken to allow the organism to cope with these problems. For example, via energy dissipation or by increasing the biosynthetic pathways concerned so that if substrate uptake is deregulated it does not enter the cell at a rate that exceeds flux to the desired product and proves lethal.

Metabolites that contain moieties of differing oxidation state

EPS production has a high demand for ATP and in many cases part of this can be provided for by the synthesis of the oxidized moieties that comprise these molecules. For metabolites such as biosurfactants and antibiotics that contain moieties of differing oxidation level such as carbohydrate and lipid moieties synthesis from a single carbon source has a high ATP demand. Thus, rapid production of these molecules leads to rapid rates of ATP turnover per mole metabolite. The capacity to handle the rapid rates of ATP turnover concomitant with rapid production of these metabolites will be dependent on the ATP/O quotient of the particular organism. Attempts to reduce the ATP demand for synthesis by feeding a mixture of preformed precursors has in some cases led to significant improvements in yields and specific rates of production. To illustrate, this the production of the biosurfactant so-

phorolipid and the antibiotics tyrocidine A, monensin B and penicillin G will be discussed.

The synthesis of a given metabolite may occur via a number of alternative routes. The precise biochemical pathways involved in the synthesis of tyrocidine, monensin and penicillin are not known. Moreover, the production conditions used are known only to commercial producers. Consequently, the discussion that follows is only an analysis of the energetic consequences of synthesis from single and mixed feedstocks. Nevertheless, it illustrates the importance of feedstock choice on carbon and energy flux to these metabolites and may explain why complex media are favored for their production. To avoid confusion all assumptions made regarding production stoichiometries will be given in Appendix 1.

Sophorolipid

Sophorolipid is a glycolipid containing the disaccharide sophorose and a b-hydroxy C-18 fatty acid. Each glucose of the sophorose is acylated with one acetate. Synthesis of sophorolipid from either glucose or an alkane has a high ATP demand due to the need to produce the carbohydrate moiety from the alkane and the lipid moiety from glucose, respectively:

From glucose

$$8.4 \text{ glucose} + 25.5 \text{ ATP} + O_2 = \\ \text{sophorolipid} + 16.4 \text{ } CO_2 + 25.5 \text{ ADP} + 25.5 \text{ Pi}$$

From a C-18 alkane

$$2.11 \text{ } C_{18}H_{38} + 3.11 \text{ } O_2 + 12.99 \text{ NAD} + 12.88 \text{ FAD} + \\ 19.11 \text{ ATP} = \text{sophorolipid} + 4 \text{ } CO_2 + 12.99 \text{ NADH}_2 \\ + 12.88 \text{ FADH}_2$$

There is a considerable reduction in the ATP demand for sophorolipid biosynthesis if a mixture of glucose and a C-18 alkane is used, and the demand is reduced further if the corresponding fatty acid is used instead of the alkane:

3.075 glucose + C-18 fattyacid + 3 NAD + O_2 + 6 ATP = sophorolipid + 3 $NADH_2$ + 2.45 CO_2

If, as suggested, the rate of ATP turnover (and associated metabolism) during sophorolipid synthesis has a consequence on production rate then production from mixed substrates may be advantageous. Theoretically mixed feedstocks also have advantages in terms of yield and oxygen demand Table 4. This has been found to be the case in practice; *Torulopsis bombicola* produced only trace amounts of sophorolipid when glucose or safflower oil were used as substrates, whereas, on a mixture of these substrates 70 gl^{-1} was produced. Moreover, the specific production rate was an order of magnitude higher than on a single substrate (Cooper & Paddock 1984). Similar results have been achieved in the author's laboratory in 200 l fermentations (Rye & Bailey unpublished). Not only were specific rates of sophorolipid production higher on a mixture of an alkane (C-18) and glucose but the final broth concentration increased from 5 gl^{-1} on glucose alone to 150 gl^{-1} on the mixture.

Tyrocidine A

A family of related peptide antibiotics known as tyrocidines is produced by *Bacillus brevis*. The synthesis of only one of these, tyrocidine A, will be considered. Tyrocidine A is composed of 10 amino acids (L orn-L leu-D phe-L pro-L phe-D phe-L asn-L gln-L tyr-L val) and synthesis occurs via a protein thiotemplate mechanism and not via synthesis on ribosomes.

If synthesis is assumed to occur from preformed amino acids, the overall process is energy requiring:

10 aminoacids + 20 ATP = Tyrocidine A + 20 ADP + 20 Pi

Assuming that the ATP demand is met via oxidation of glutamate to CO_2 and that *B brevis* has an ATP/O quotient of 2, an additional glutamate will be required to supply the ATP demand (10 ATP for synthesis and 1 ATP per amino acid transported

into the cell, Kiyoshi 1981). A fairly modest specific production rate of 0.05g tyrocidine A g^{-1} h^{-1} is equivalent to a specific ATP utilization rate (q_{ATP}) of 0.825 mmol g^{-1} h^{-1}. This is a significant rate of ATP turnover when it is remembered that production probably occurs in the stationary phase in the absence of growth. This q_{ATP} is of the same order as that required for cell growth at a specific growth rate of 0.012 h^{-1}.

In contrast, rapid synthesis of tyrocidine A from glucose poses the organism with rather different physiological problems, as it results in a significant net overproduction of ATP /mole tyrocidine. The magnitude of the overproduction will depend on the ATP/O quotient of the producing organism.

13.5 glucose + 25.75 ATP + 18 NAD + 10 NADP + 11 NH_3 = Tyrocidine A + 18 NADH2 + 10 NADPH2 + 25.75 ADP + 25.75 Pi

At an ATP/O quotient of 1 the energy generated in the production of tyrocidine A is sufficient to meet the demand for biosynthesis. At an ATP/O quotient of 2 there is an overproduction of 30.25 ATP/mole tyrocidine A synthesized and the overall q_{ATP} is 2.2 mmol g^{-1} h^{-1}. As discussed above (see Fig. 5) this is a very high rate of ATP turnover in the absence of growth and some means of energy dissipation will be required if rapid synthesis is to continue. The precise conditions used commercially are not known to the author, but it may be argued that a pathway leading to a moderate energy overproduction would probably be more stable, as synthesis would provide the cells 'maintenance energy' requirements and maintain metabolism in the absence of growth. A negative energetic balance for a given metabolite means that additional substrate has to be oxidized to make good the

Table 4. Effect of carbon source on yield of sophorolipid.

Carbon source	$Y_{substrate}$ (gg^{-1})	Y_{O_2} (gg^{-1})
Glucose	0.47	2.32
C-18 alkane	1.32	1.71
Glucose + C-18 fattyacid	0.81	11.10

shortfall. If the metabolite has no physiological value to the cell, it is difficult to understand how production can be stably maintained in strains exhibiting high specific production rates.

Monensin B

Streptomyces cinnamonensis produces a mixture of polyether antibiotics known as monensins (A-D). Although monensin A is the major metabolite in the mixture, for convenience the synthesis of monensin B will be considered and the synthesis of the methoxy group via methionine has not been included in the calculation. The basic skeleton of monensin B is derived from 5 acetate and 8 propionate units (Pospisil et al. 1988). These precursors can be generated from single or mixed substrate fermentations, a summary of the overall energetics of monemsin B synthesis from various carbon sources is as follows:

From acetate + propionate

13.5 acetate + 8 propionate + 63 ATP + 13.5 NAD + 18.5 FAD = monensin B + 13.5 NADH2 + 18.5 FADH2 + 37 CO_2 + 63 ADP + 63 Pi

From glucose

10.5 glucose + 13 ATP + 26 NAD + 5 HCO_3 + 5 H^+ = monensin B + 26 NADH2 + 21 CO_2

From glucose + acetate

---

5 acetate + 8.27 glucose + 39 ATP + 19.5 NAD = monensin B + 19.5 NADH2 + 25.74 CO_2

From glucose + acetate + propionate

12.8 acetate + 8 propionate + 2 glucose + 46.8 ATP + 17.88 NAD + 0.8 NADP + 0.54 FAD = monensin B + CO_2 + 17.88 NADH2 + 0.8 NADPH2 + 0.54 FADH2

From tripalmitin

0.84 tripalmatin + 42.04 ATP + 17.64 FAD + 4.16 NAD = monensin B + 17.64 FADH2 + 4.16 NADH2 + CO_2

In all cases except glucose, synthesis is ATP limited at an ATP/O quotient of 1. At an ATP/O quotient of 2, carbon and energy flux is balanced for synthesis from acetate + propionate and acetate + glucose Table 5. With glucose as the substrate, there is considerable over-production of energy at an ATP/O quotient of 2 Table 5. At an ATP/O quotient of 3, synthesis results in an over production of ATP in all cases. The analysis indicates that, although acetate and propionate are direct precursors in the synthesis of monensin B, the yield from both the carbon source and oxygen is as poor as it is from glucose. However, a combination of acetate and glucose leads to a significant increase in yield from oxygen and a corresponding reduction in the specific rate of ATP utilization for synthesis, Table 5. However, the yield from the carbon source is similar. By far the best substrate in terms of yield from both carbon source and oxygen is obtained with a plant oil as feedstock. This analysis is supported by studies to optimize monensin production where it has been found that a mixture of refined soybean oil and methyloleate produced the highest monensin titres (Stark et al. 1967). Nevertheless, as in the case of tyrocidine, a modest monensin production rate of 0.05 $gg^{-1}h^{-1}$ requires a substantial ATP turnover rate (Table 5), which is primarily due to the high demand for NADPH2. In physiological terms, this rate of ATP turnover in the absence of growth is a major event and would not be expected to be maintained unless there is some benefit to the organism.

Penicillin

The initial steps in the synthesis of penicillin occur via the condensation of a-aminoadipic acid, cysteine and valine to give the tripeptide (a-aminoadipyl cysteinylvaline) at the expense of 3 ATP. The lactam and thiazolidine rings are then closed to yield isopenicillin N + NADPH2 and FADH2. The a-aminoadipic moiety may be exchanged for phenylacetic acid or phenoxyacetic acid to give penicillin G or V, respectively. Phenyacetic acid is believed to enter the cell via diffusion and require activation (1 ATP) before exchange with a-aminoa-dipic acid can occur (Hersbach et al. 1984).

The large improvements in penicillin titre have been achieved by a combination of mutation / selection and media / process optimization. These improvements have been achieved empirically and the details are closely-guarded industrial secrets. In the absence of information concerning the nutrient regimes used and the nature of the mutations that have led to these significant increases in titre, it is difficult to analyze the energetic consequences of penicillin production on the producing organism. Nevertheless, using published information concerning penicillin biosynthesis, it is apparent that the overall energetics of production are significantly affected by the feedstocks used for production as well as the precise biochemical pathways involved in precursor synthesis. For a detailed discussion on the biochemical pathways involved in penicillin synthesis the reader is referred to the excellent review by Hersbach et al. 1984. The synthesis of penicillin from various feed stock is given below:

Preformed precursors

cysteine + valine + a-AAA + PAA + 7ATP + NAD + FAD = penicillin G + NADH2 + FADH2 + a-AAA

Total synthesis from glucose

2.5 glucose + 3NH$_3$ + SO$_4{}^{2-}$ + 3NAD + 2FAD + 14ATP = penicillin G + 3NADH2 + 2FADH2 + 4AMP + 11.5 ADP + 4 PPi + 11.5 Pi

Penicillin synthesis from preformed monomers is relatively energetically expensive, due to the ATP requirement for amino acid uptake, and no ATP is generated in the process. Synthesis from glucose contributes to the energetic cost of synthesis, and, at an ATP/O quotient of 2 for *P. chrysogenum* (Righelato et al. 1968) the overall synthesis is ATP limited and additional glucose will have to be oxidized to make good this shortfall. It is therefore not surprising to find that penicillin production is accompanied by high specific production rates of organic acids which would make a significant contribution to the cell's energy balance (Mason & Righelato 1976). It is apparent (Hersbach et al. 1984) that cysteine biosynthesis is energetically very expensive in terms of ATP requirement and therefore synthesis from a mixture of glucose and cysteine will significantly effect the overall energetics of synthesis and result in penicillin synthesis being an overall energy- generating process at an ATP/O quotient of 1:

2 glucose + cysteine + 2 NH$_3$ + PAA + FAD + 2.5 ADP + 4 ATP + 7NAD = penicillin G + a -AAA + 4 ADP + 2.5 AMP + 4Pi + 2.5 PPi + 7 NADH2 + FADH2

Assuming a specific penicillin production of 0.1 g g^{-1} h^{-1} and an ATP/O quotient = 2, specific ATP requirements of 2.1 and 4.2 mmol g^{-1} h^{-1} are necessary for synthesis from preformed monomers and from glucose, respectively, whereas there is a net ATP production rate of 3.6 mmol g^{-1} h^{-1} for synthesis from a combination of glucose and cysteine.

It is difficult to assess whether these energetic considerations are relevant to what has actually

Table 5. Effect of carbon source on the yield and rate of ATP turnover associated with monensin B production.

Carbon source	Yield from (gg^{-1} h^{-1})		Specific ATP utilisation rate (mmol g^{-1} h^{-1})	Net ATP balance (mol ATP/ mol monensin)
	carbon source	oxygen		
Acetate + propionate	0.32	1.27	4.90	0
Glucose	0.34	1.01	4.36	+ 39
Glucose + acetate	0.35	2.05	3.05	0
Glucose + acetate + propionate	0.37	1.71	3.65	− 8.3
Tripalmatin	1.84	1.90	3.27	+ 1.5

A specific monensin production rate of 0.05 g g^{-1} h^{-1} and an ATP/O quotient = 2 has been assumed.

been implemented to improve antibiotic titres. However, the continued use of complex media in antibiotic production points to the importance of the supply of mixtures rather than single substrates. It would be instructive to learn whether continual mutation and selection for increased titres on complex media have resulted in the deregulation of uptake systems to allow multiple substrate uptake simultaneously and the selection of pathways that offer the best energetic solution to the organism so that the desired pathway is a net generator rather than consumer of energy. Thus in the artificial environment of the industrial fermenter the production of the given secondary metabolite is not an energetic burden that is selected against but may actually contribute to the survival of the organism under these conditions.

References

Anderson A, Hacking AJ & Dawes E (1987) Alternative pathway for the biosynthesis of alginate from fructose and glucose in *Pseudomonas mendocina* and *Azotobacter vinelandii*. J. Gen. Microbiol. 133: 1045–1052

Andrew SPS (1983) Old and new feedstocks-chemistry and commerce. Inst Chemical Engineers Symposium. Series no. (79 pp 400–408)

Betlac MR, Campage MA Doherty DH Hassler RA Henderson NM Vanderslice RW Marelli JD & Ward MB (1987) Genetically engineered polymers: manipulation of xanthan biosynthesis. In: Yalpani M (Ed) Industrial Polysaccharides. Programme in Biotechnology, Vol 3 (pp 35–50). Elsevier, Amsterdam

Cooper DG & Paddock DA (1984) Production of a biosurfactant from *Torulopsis bombicola*. Appl. Environ. Microbiol. 47: 173–176

Cornish A, Linton JD & Jones CW (1987) The effect of growth conditions on the respiratory system of a succinoglucan-producing strain of *Agrobacterium radiobacter*. J. Gen. Microbiol. 133: 1971–2978

Cornish A, Greenwood JA, & Jones CW (1988a) Binding protein dependent transport glucose transport by *Agrobacterium radiobacter* grown in glucose limited continuous culture. J. Gen. Microbiol. 134: 3099–3110

–(1988b) The relationship between glucose transport and the production of succinoglucan exopolysaccharide by *Agrobacterium radiobacter*. J. Gen. Microbiol. 134: 3111–3122

Doherty D, Leigh JA Glazebrook J & Walker GC (1988) *Rhizobium meliloti* acid calcafluor-binding exopolysaccharide. J. Bacteriol. 170: 4249–4256

Dunnill P (1983) The future of Biotechnology. In: Biochem. Soc. Symp. No. 48, Biotechnol. 9–23

Farrand SG, Jones CW Linton JD & Stephenson RJ (1983) The effect of temperature and pH on the growth efficiency of the thermoacidophilic bacterium *Bacillus acidocaldarius* in continuous culture. Arch. Microbiol. 135: 276–283

Glazebrook J & Walker GC (1989) A novel exopolysaccharide can function in place of the calcafluor-binding exopolysaccharide in nodulation of Alfalfa by *Rhixobium meliloti*. Cell: 661–672

Grinberg AJ, Kosenko LV & Malashenko R Yu (1984) Formation of exopolysaccharide by methylotrophic micro-organisms. Mikrobiologischeskii Zhurnal 46: 22–26

Hersbach GJM, Van Der Beek CP & Van Dijk PWM (1984) The penicillins: Properties, biosynthesis and fermentation. In: Vandamme EJ (Ed) Biotechnology of Industrial Antiobiotics. Marcel Dekker, New York

Hueting S & Tempest DW (1977) Influence of acetate on the growth of *Candida utilis* in continuous culture. Arch. Microbiol. 115: 73–78

Hoischen C & Kramer SG (1990) Evidence for efflux carrier system involved in the secretion of glutamate by *Corynebacterium glutamicum*. Arch. Microbiol. 151: 342–347

Jarman TR & Pace GW (1984) Energy requirements for microbial exopolysaccharide synthesis. Arch. Microbiol. 137: 231–235

Jones CW (1988) Aerobic respiratory systems in bacteria. In: G Durand, L Bobichon & J Florent (Eds) 8th Int. Biotechnol. Symp. Vol 1 (pp 43–55). Soc. Franc. Microbiol. Paris

Kacser H & Burns JA (1973) The control of flux. In: DD Davies (Ed) Rate Control of Biological Processes. Symp. Soc. Gen. Micribiol. 27: 65–104. Cambridge

Kell DB (1989) Control analysis of microbial growth and productivity. In: Baumberg S, Hunter I & Rhodes M (Eds) Society for General Microbiology Symposium 44: 61–93

King PP (1982) Biotechnology: an industrial view. J. Chem. Tech. Biotechnol. 32: 2–8

Kiyoshi K (1981) Biosynthesis of peptide antibiotics. In: Corcovan JW (Ed) Antibiotics, Vol. iv. Springer-Verlag

Kula MR, Aharonowitz IY, Bulock JD, Chakrabarty AM, Hopwood DA, M athiasson B, Morris JG, Neijssel OM & Sahm H (1987) Microbiology and industrial products. Group report in biotechnology: potentials and limitations (Dahlem Konferenze 1986), University Press

Kristiansen B & Charley R (1981) Continuous process for the production of citric acid. In: M Moo-Young (Ed) Advances in Biotechnology 1. Proceed. 6th Int. Ferment Symp Pergamon Press, Toronto

Linton JD (1990) The relationship between metabolite production and growth efficiency of the producing organism. FEMS Microbiology Reviews 75: 1–18

Linton JD, Austin RN & Haugh DE (1984) The kinetics and physiology of stipitatic acid and gluconate production by carbon sufficient cultures of *Penicillium stipitatum* growing in continuous culture. Biotechnol. Bioeng. XXVI: 1455–1464

Linton JD, Griffiths J & Gregory M (1981) The effect of mixtures of glucose and formate on the yield and respiration of a chemostat culture of *Beneckea natriegens*. Arch. Microbiol. 129: 119–122

Linton JD, Evans, M Jones DS & Gouldney DN (1987a) Exocellular succinoglucan production be *Agrobacteriun radiobacter* NCIB 11883. J. Gen. Microbiol. 133: 2961–2969

Linton JD, Jones DS & Woodard S (1987b) Factors that control the rate of exopolysaccharide production by *Agrogacterium radiobacter* NCIB 11883 J. Gen. Microbiol. 133: 2979–2987

Linton JD, Gouldney D & Woodard S (1988) Efficiency and stability of exo-polysaccharide production from different carbon sources by *Erwinia herhicola*. J. Gen. Microbiol. 134: 1913–1921

Linton JD & Musgrove SG (1983) Product formation by a nitrogen limited culture of *Beneckea natriegens* in a chemostat in the presence of excess glucose. Env. J. Appl. Microbiol. Biotechnol. 18: 24–28

Linton JD, Watts PD, Austin RM, Haugh DE & Niekus HGD (1986) The energetics and kinetics of extracellular polysaccharide production from methanol by organisms possessing different pathways of C-1 fixation. J. Gen. Microbiol. 132: 779–788

Linton JD & Rye AR (1989) The relationship between the energetic efficiency in different microorganisms and the rate and type of metabolite over- produced. J. Indust. Microbiol. 4: 85–96

Mason HRS & Righelato RC (1976) Energetics of fungal growth: Effects of growth-limiting substrate on respiration of *Penicillium chrysogenum*. J Appl Chem. Biotechnol. 26: 145–152

Miall LM (1972) Stimulatory effect of organic acids in citric acid fermentation. UK Patent Spec. 1293786

Nakayama K (1976) The production of amino-acids. Proc Biochem 3: 4–9

(1986) Breeding of amino-acid mutants. In: Aida K, Chidala I, Nakayama H (Eds) Biotechnology of amino acid production. Progress in Industrial Microbiology (24: 3–33). Elsevier, Amsterdam

Neijssel OM (1977) The effect of 2, 4-dinitrophenol on the growth of *Klebsiella aerogenes* NCTC 418 in aerobic chemostat culture. FEMS Letts. 1: 47–50

Pospisil S, Sedmera P Matej J & Nohynek M (1988) Polyether Antibiotics: Monensin. Bioactive metabolites from microorganisms. In: Bushell ME & Grafe U (Eds) Progress in Microbiology 27. Elsevier, Amsterdam

Righelato RC, Trinci APJ & Pirt SJ (1968) The influence of maintenance energy and growth rate on the metabolic activity, morphology and conidiation of *Penicillium chrysogenum* J. Gen. Microbiol. 50: 399–412

Rye AJ, Drozd JW Jones CW & Linton JD (1988) Growth efficiency of *Xanthomonas campestris* in continuous culture. J. Gen. Microbiol. 134: 1055–1061

Stark WM Knox NG & Westhead JE (1967) Monensin, a new biologically active compound II. Fermentation studies, Antimicrobial Agents & Chemoth. 353–358

Stouthamer AH (1979) The search for correlation between theoretical and experimental growth yields. In: Quayle, JR (Ed) International Reviews of Biochemistry. Microbial Biochemistry, Vol 21 (pp 1–47). University Parks Press, Baltimore, MD

Stouthamer AH & Bettenhaussen CW (1973) Determination of efficiency of oxidative phosphorylation in continuous cultures of *Aerobacter aerogenes* Arch. Microbiol. 102: 187–195.

Stouthamer AH & van Verseveld HW (1985) In: Moo-Young M (Ed) Compehensive Biotechnology, Vol 1 (pp 215–238). Pergamon Press, Oxford

–(1987) Microbial energetics should be considered in manipulating metabolism for biotechnological purposes. Tr. Biotechnol. 5: 149–155

Tanaka K, Iwasaki T & Kinoshita S (1960) Studies on L-glutamic acid fermentation. Part 5. Biotin and L-glutamic acid accumulation by bacteria. J. Agri. Chem. Soc. (Japan) 34: 593–600

Tanaka K Machida-Shi K & Yamaguchi K (1969) Process for producing L-glutamic acid and alpha-ketoglutaric acid US Patent Office 3, 450, 599

Tong GE & Cannell PR (1983) The economics of organic chemicals from biomass. In: Wise DL (Ed) Organic chemicals from Biomass (pp 407–451). The Benjamin/Cummings Publishing Company Inc., London

Weenk G Olijve W & Harder W (1984) Ketogluconate formation by *Gluconobacter* species. Appl. Microbiol. Biotechnol. 20: 400–405

Appendix 1

The following assumptions were made in arriving at the stoichiometries for the production of:

EPS

For homopolymers: 1 mol ATP is required to to produce glucose 1 P, a second is required for the generation of ADP-glucose and finally a third is required for transport out of the cell and for addition to the growing chain. In the synthesis of complex polysaccharides composed of polymerised subunits of fixed composition, 1 ATP per repeat unit is required for transport out of the cell and addition to the growing chain.

Sophorolipid

1 ATP/glucose transported into the cell; 2 ATP / C-18 fattyacid coupled to sophorose; use of HMP pathway to produce NADPH or 1 ATP + NADH$_2$ + NADP = NADPH$_2$; ATP/O quotient = 1.5; shortfall in ATP supplied by complete oxidation of carbon source to CO_2; all reducing equivalents oxidised to H_2O via respiration. Alkane and fatty acid incorporation occurs directly into sophorolipid.

Monensin

The following assumptions have been made: The synthesis of manolyl CoA from acetyl CoA requires 1 ATP + HCO_3; methylmalonyl CoA formation from 2 acetyl CoA via succinate requires 1 ATP but generates 1 NADH2. The condensation of 5 malonyl CoA and 8 methylmalonyl CoA requires 24 NADPH2 and the reduction of NADP to NADPH2 via transhydrogenase requires 1 ATP + NADH2. ATP/O quotient = 2.

Antonie van Leeuwenhoek **60**: 313–323, 1991.
© 1991 *Kluwer Academic Publishers. Printed in the Netherlands.*

Determination of the maximum product yield from glucoamylase-producing *Aspergillus niger* grown in the recycling fermentor

Henk W. van Verseveld[1], M. Metwally[1,2], M. el Sayed[2], M. Osman[2], Jaap M. Schrickx[1] and Adriaan H. Stouthamer[1]
[1] *Vrije Universiteit, Biological Laboratories, Department of Microbiology, De Boelelaan 1087, NL-1081 HV Amsterdam, The Netherlands;* [2] *Tanta University, Faculty of Sciences, Botany Department, Tanta, Egypt*

Key words: Aspergillus, recycling fermentor, fungi, protein production, glucoamylase

Abstract

Aspergillus niger has been grown in glucose- and maltose-limited recycling cultures to determine the maximum growth yield, the maximum product yield for glucoamylase production, and the maintenance requirements at very slow specific growth rates. Using the linear equation for substrate utilization, and using the experimental data from both recycling experiments, both the maximum growth yield, Y_{xsm}, and the maximum product yield, Y_{psm}, could be determined. The values estimated were 157 g biomass per mol maltose for Y_{xsm} and 100 g protein per mol maltose for Y_{psm}. Expressed on a C_1-basis these values are 0.52 and 0.36 C-mole per C-mol for respectively Y_{xsm} and Y_{psm}. The found value for Y_{psm} is half the value found for alkaline serine protease production in *Bacillus licheniformis*, and it can be concluded that formation of extracellular protein is more energy consuming in filamentous fungi than in prokaryotic organisms. Maintenance requirements are no significant factor during growth of *Aspergillus niger*, and reported maintenance requirements are most probably due to differentiation.

Introduction

The filamentous fungi represent a physiologically diverse group of microorganisms which are characterized by a saprophytic, or less frequently a parasitic mode of nutrition. They grow slowly by comparison with many bacteria, and the morphology of their growth-form, usually as long filaments, as mats or as pellets, makes them rather difficult to handle and introduces problems of heterogeneity of the culture in study. However the economic and social importance of fungi as sources of food and biologically active metabolites has stimulated considerable interest in their growth physiology (Berry 1975; Righelato 1975).

Exploitation of enzymes has been carried out throughout the ages in leather tanning, cheese making, beer brewing etc. These processes used enzymes in the form of animal and plant tissues or microorganisms. The use of filamentous fungi and bacteria for commercial enzyme production has been developed in this century. In the 1960s commercial enzyme production was dramatically amplified by the introduction of enzyme-containing washing powders, that contain alkaline serine protease from *Bacillus licheniformis* and hydrolytic enzymes from *Aspergillus niger*. Nowadays the bulk of the world market for industrial enzymes comprises proteases and carbohydrases which together account for about 90% of the total market for enzymes. The production trend over the past 30 years has shifted from animal and plant sources towards microorganisms to the extent that new products are almost invariably derived from bacteria and fungi. There are several reasons for this phenomenon: (1) microorganisms grow rapidly and are ideal for intensive cultivation; (2) medium constituents are cheap and generally comprise agricultural products available in bulk and (3) the choice of producer organism is wide and they mostly can be improved by genetic manipulation (Priest 1984).

The extracellular enzyme glucoamylase cleaves a 1–4 and a 1–6 glucosidic linkages and converts amylopectin and glycogen almost quantitatively into D-glucose that can directly be used for food processing and preparation. Parameters from balance equations (Roels 1980) as well as the determination of growth parameters as maintenance requirements, maximal biomass yield and maximal product yield are of great importance for industrial production processes (Stouthamer & van Verseveld 1987), and can give strategies for an efficient and cheap production process. One of these parameters, the maximum product yield seldomly can be determined accurately. This paper describes a model in which the value of the maximum product yield can be estimated for glucoamylase producing *Aspergillus niger*.

Materials and methods

Organism and growth conditions

A wild type *Aspergillus niger* (ATCC 1015) from the Deutsche Sammlung von Mikroorganismen (Göttingen, Germany) was used. The organism was maintained on potato-dextrose agar slants. The medium composition was as described by Anderson & Smith (1971) with slight modifications. It contained per litre distilled water: $(NH_4)_2SO_4$, 1.98g; KH_2PO_4, 1g; $MgSO_4.7H_2O$, 0.25g; $CuSO_4.5H_2O$, 0.234mg; $FeSO_4.7H_2O$, 6.3mg; $ZnSO_4.7H_2O$, 1.1mg; $MnCl_2.4H_2O$, 3.5mg; and $CaCl_2$ 46.7mg. The medium was sterilized in the absence of carbon source by autoclaving. The carbon source was sterilized for 30 min. at 100 °C and added aseptically.

Aerobic carbon-limited recycling experiments with either maltose or glucose as limiting carbon- and energy-sources were carried out in a fermentor apparatus designed and manufactured by the electronics and mechanics workshops of the Faculty of Biology, Vrije Universiteit, Amsterdam, The Netherlands. A 2 litre fermentor vessel with a working volume of 1.5 l was used. The culture volume was controlled by means of a liquid-level indicator and a recycling unit both inserted through the top plate of the fermentor. The liquid level indicator regulated the speed of a peristaltic pump that kept the volume constant by withdrewing filtrate from the culture, while the biomass remained behind. The whole recycling system is designed and build by the electronics and mechanics workshops of the Faculty of Biology, Vrije Universiteit, Amsterdam, The Netherlands. Agitation was obtained by flat-blade, propeller-type impellers operating at about 700 rpm. When necessary wall growth was removed daily with a Teflon covered ring bar magnet which can be moved over the wall with an external horseshoe magnet. Culture pH was controlled at 4.5 ± 0.1 by addition of 1 N HCl or KOH, and the temperature was controlled at 30 °C.

In order to get good filamentous growth the fermentor vessel was inoculated with a spore suspension (10^5 spores/ml, final concentration), and a batch phase started, in which the pH initially was kept at 2.0 and air was led over the culture to get oxygen limitation. The pH was shifted to a value of 4.5 after batch growth was nearly completed, and growth in recycling culture started.

Analytical procedures

Dry weight was determined by filtration over Sartorius membrane filters of constant weight (catalogue no. 11307, Sartorius Membrane Filters Gmbh, Göttingen, Germany), washing twice with equal volumes of water and drying at 105 °C. The oxygen uptake and carbon dioxide production were measured with a mass spectrometer (VG, MM8–80, Single collector, Cheshire, England).

Extracellular products were measured with a Total Organic Carbon Analyzer (Type 915A, Beckman Instruments, Inc., Fullerton, Ca., USA). The glucoamylase activity was measured as described by Yamasaki et al. (1977) with slight modifications as follows: A reaction mixture containing 0.2 ml of 1% soluble starch, 0.5 ml 1 M acetate buffer (pH 5.3), and 0.3 ml enzyme solution (culture supernate sample), was incubated at 40 °C for 10 min. After incubation the reaction was stopped by heating in a boiling water bath. The formed glucose was measured with the GOD-Perid method (Boehringer, Mannheim, Germany). One Unit of glucoamylase is defined as the amount of enzyme which forms 1 μmol of glucose under the conditions described. Maltose was determined as described by Rauscher (1974), and protein as described by Lowry et al. (1951).

Data analysis of recycle cultures

The model and equations used are mentioned in the results section. For a complete treatment see van Verseveld et al. (1986).

Results

The model used for the description of growth and product formation in the recycling fermentor has been treated extensively by van Verseveld et al. (1986) and Stouthamer & van Verseveld (1987), and is based on carbon- and energy-balances (Roels 1980), the linear equation for substrate utilization (Pirt 1965) and a linear equation for product formation. The equations are given in Box 1 and a definition of terms in Box 2. From the derivation of equation 1 in Box 1 it is clear that it contains two unknown parameters, namely the maximum growth yield, Y_{xsm}, and the maximum product yield, Y_{psm}. Consequently it is not possible to estimate both parameters from one experiment. It is possible to estimate values for both maximum yield values from two or more experiments, which are sufficiently different and from which it is expected that both parameters will not be influenced by the different growth conditions. An example of such experiments is given here.

Growth and product formation in a maltose-limited recycling fermentor

Aspergillus niger ATCC1015 was grown in a maltose-limited recycling fermentor with 100% biomass feed back (Chesbro et al. 1979; van Verseveld et al. 1984) with a rate of maltose addition of 0.5 mmol/h, and the glucoamylase production was determined and corrected for glucoamylase washed out via the filtrate. During the course of recycling cultivation the glucoamylase production increased with increasing time, and in step with the protein production. The final production after 120 hours was 4666 units, and the specific activity was 4000 units glucoamylase per g protein.

Concerning the growth and the production pattern, experimental data for biomass accumulation (x_t) and product formation (P_t) as well as the rate of CO_2 evolution and O_2 consumption were used in a parameter optimization procedure as described by van Verseveld et al. (1986). This procedure uses equations 1 to 5 mentioned in Box 1 and all parameters can be fixed or floating.

The measured data have to comply with the terms of the model, especially to the product equation, $r_p = ax_t + br_x$. Consequently $P_t - P_0 = a\int x_t dt + b(x_t - x_0)$, and a plot of $\int x_t dt/(x_t - x_o)$ versus $(P_t - P_o)/(x_t - x_o)$ should yield a linear relationship with intercept b and slope a. From Fig. 1 it can be seen that the data do very reasonably fit to the product equation and that a is practically zero and b has a positive value. Since we have only b, in this case we assume that glucoamylase production is only dependent on the rate of biomass production and not on the biomass concentration as such. This yields a very reasonable fit of the data as can be seen in Fig. 2, in which b, and m_s were floating parameters and the others were fixed. The agreement between the fit and the experimental data points indicates that the carbon- and energy-balance is very good. Obviously $^rO_2$ and $^rCO_2$ were practically constant, already indicating that maintenance requirements are very low (0.4 μmol per g biomass per hour, see Table 1). Figure 3 shows the fitted values of Y_{xsm} and m_s for various values of Y_{psm}. Since the value of Y_{xsm} is dependent on Y_{psm}, it decreases with increasing Y_{psm}, while m_s is practi-

Box 1. Equations describing biomass production, product formation, oxygen uptake and CO_2 evolution.

The full derivation of these equations is given by van Verseveld et al (1986)
The basis of the model is formed by the carbon and energy balance for growth and product formation.

$$CH_mO_1 + aNH_3 + bO_2 \longrightarrow y_cCH_pO_nN_g + zCH_rO_sN_t + cH_2O + d\ CO_2$$

Substrate   biomass   product

and the linear equation for substrate utilization.

$$r_s = m_s x_t + \frac{r_x}{Y_{xsm}} + \frac{r_p}{Y_{psm}}$$

, and when $r_p = ax_t + br_x$, r_s becomes as in Eq. 1, when

$$m'_s = m_s + \frac{a}{Y_{psm}} \quad \text{and} \quad Y'_{xsm} = \left[\frac{1}{Y_{xsm}} + \frac{b}{Y_{psm}} \right]^{-1}$$

$$r_s = m'_s x_t + \frac{r_x}{Y'_{xsm}} \qquad (1)$$

Rearrangment leads to $r_x = (r_s - m_s' x_t)Y_{xsm}'$, and integration yields Eq.2, when $r_s = C$:

$$x_t = \frac{r_s}{m'_s} + \left[x_0 - \frac{r_s}{m'_s} \right] e^{-m'_s Y'_{xsm} t} \qquad (2)$$

The equation (3) for the rate of oxygen consumption is obtained from the energy balance ($b = 0.25[\gamma_s - y_c\gamma_x - z\gamma_p]$), and $rO_2 = br_sc$, in which c is the amount of carbon atoms in the substrate.

$$r_{O_2} = 0.25 \left[\left(\frac{\gamma_s}{Y_{xsm}} - \frac{\gamma_x}{M_x} \right) r_x c + \left(\frac{\gamma_s}{Y_{psm}} - \frac{\gamma_p}{M_p} \right) r_p c + \gamma_s m_s x c \right] \qquad (3)$$

The balance equation for CO_2 evolution is $d = 1 - y_c - z$, and consequently $rCO_2 = dr_sc$. After substitution we obtain Eq. 4:

$$r_{CO_2} = \left[\frac{1}{Y_{xsm}} - \frac{1}{M_x} \right] r_x c + \left[\frac{1}{Y_{psm}} - \frac{1}{M_p} \right] r_p c + m_s x c \qquad (4)$$

For the calculation of the amount of product formed the product equation
$r_p = ax_t + br_x$ must be integrated resulting in Eq. 5

$$P_t = P_0 + \frac{ar_s}{m'_s} t - \left[\frac{a}{m'_s Y'_{xsm}} - b \right] \left| x_t - x_0 \right| \qquad (5)$$

cally constant. The m_s found is much lower than the one calculated from chemostat cultures (Metwally et al. 1991).

Growth and product formation
in a glucose-limited recycling fermentor

Aspergillus niger ATCC1015 was also grown in a glucose-limited recycling culture with a rate of glucose addition of 1 mmol/h, and the glucoamylase production was determined. During the course of recycling cultivation the glucoamylase production also increased with increasing time, and was also in step with the protein production. The final production after 120 hours was 3344 units, and the specific activity was 2550 units glucoamylase per g protein. The glucoamylase production in units is only 72%

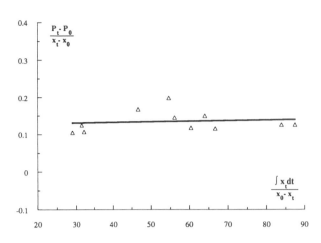

Fig. 1. Maltose-limited recycling experiment with *Aspergillus niger* (see Fig. 2). The biomass and product concentration data are verified to satisfy the model equation , $r_p = ax_t + br_x$. For further explanation see the text.

Box 2. Definition of terms/units.

r_s	rate of substrate consumption (mol h^{-1})
r_x	rate of biomass production (g h^{-1})
r_p	rate of product formation (g h^{-1})
rCO_2	rate of CO2 evolution (mol h^{-1})
rO_2	rate of O2 consumption (mol h^{-1})
γ_s	degree of reductance of substrate
γ_x	degree of reductance of biomass
γ_p	degree of reductance of product
x_t	biomass produced (g)
P_t	product formed (g)
m_s	maintenance coefficient (mol substrate (g biomass)$^{-1}$ h^{-1})
m_s'	maintenance coefficient (mol substrate (g biomass)$^{-1}$ h^{-1}) uncorrected for product formation
Y_{xsm}	maximum yield of biomass on substrate (g mol^{-1})
Y_{xsm}'	maximum yield of biomass on substrate (g mol^{-1}) uncorrected for product formation
Y_{psm}	maximum yield of product on substrate (g mol^{-1})
M_x	molecular weight of biomass (c C-atoms)
M_p	molecular weight of product (c C-atoms)

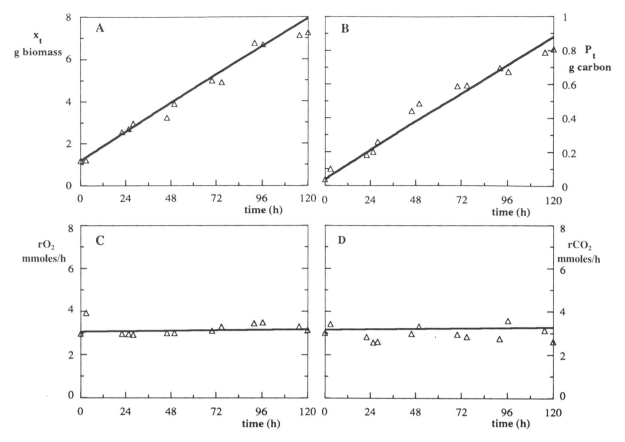

Fig. 2. Maltose-limited recycling experiment with Aspergillus niger. $r_s = 0.5$ mmol maltose/h. All data were fitted using the model equations mentioned in Box 1. Y_{xsm}, m_s and b were floating parameters, the others were fixed. *(A)* Biomass data (x_t, g dry weight) and fitted curve. *(B)* product data (P_t, g carbon) and fitted curve. *(C)* rate of oxygen consumption (r_{O2}, mmoles/h) data and fitted curve. *(D)* rate of carbondioxide production (r_{CO2}, mmoles/h) data and fitted curve.

Table 1. Growth parameters of maltose- and glucose-limited recycling experiments of *Aspergillus niger*.

Limitation	Y_{xsm}			m_s
	g DW/C-12 mol	g DW/mol	C-mol/C-mol	μmol.(g DW.h)$^{-1}$
Glucose	157.5	78.8	0.52	-0.8
maltose	157.5	157.5	0.52	0.4
	$Y_{psm}^{observed}$		$Y_{psm}^{protein}$	
	gC/C-12 mol	C-mol/C-mol	C-mol/C-mol	g/mol
Glucose	93	0.65	0.36	50
Maltose	52	0.36	0.36	100

Maximum growth yield, Y_{xsm}, maximum product yield, Y_{psm}, and maintenance requirements, m_s, are expressed as indicated. C-12 mol: molecular composition expressed on basis of the presence of 12 carbon atoms. C-mol: molecular composition expressed on basis of the presence of 1 carbon atom. DW: dry weight biomass.

of that produced under maltose-limited conditions, although the total protein production was 12% higher than under maltose-limitation. This can be explained by the following (i) glucoamylase produced under glucose-limited conditions in the recycling fermentor has a lower specific activity than when the enzyme is produced under maltose-limited conditions, or (ii) also other (or more other) enzymes are produced under glucose-limited conditions.

Concerning growth and production pattern, the experimental data for product concentration (P_t), CO_2 production and O_2 consumption were used in a parameter optimization procedure. From Fig. 4 it can be seen that most of the biomass is not submerged in the culture fluid, but settling on the wall, on the metal surface and on the baffle plate in the fermentor, causing the too low experimental x_t values. For the parameter optimization procedure, the biomass concentration data are not strictly necessary when all other data are measured very accurately. Since we should have a 100% carbon-balance, the remaining carbon should be formed as biomass, and this could be easily checked at the end of the experiment by measuring the total formed biomass including the settled part. The fitting procedure was applied, in which b, Y_{xsm}, and m_s were floating parameters and the rest was fixed, and again resulted in a very reasonable fit as can be seen in Fig. 4. Again rO_2 and rCO_2 values were practically time invariant, indicating very low maintenance requirements (practically zero, and at low Y_{psm} values negative). Figure 5 shows the fitted values of Y_{xsm} and m_s for various values of the maximal product yield (Y_{psm}), and of course Y_{xsm} decreases with increasing values of Y_{psm}. Again the m_s found is much lower than for the chemostat cultures. The difference in Y_{psm} between maltose- and glucose-limited cultures is due to the difference in protein production between the two cultures.

Calculation of the reliable value for
maximum product and maximum growth yields
in maltose- and glucose-limited recycling cultures

As mentioned before the maximal growth yield,

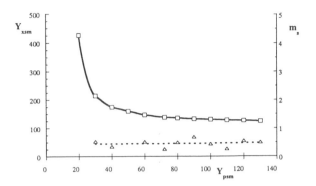

Fig. 3. Plot of the maximum growth yield ($\square$, Y_{xsm}, g dry weight/mol substrate) and the maintenance requirements ($\triangle$, m_s, μmol substrate (g dry weight.h)$^{-1}$) calculated at different values of the maximum product yield (Y_{psm}, g carbon/mol substrate) using the model equations mentioned in Box 1 of a maltose-limited recycling experiment of *Aspergillus niger*.

Y_{xsm}, and maximal product yield, Y_{psm}, are both unknown parameters in the model equations (see Box 1). Consequently, one parameter has to be known to calculate the other. If we can generate two equations from situations in which the amount of product differs strongly, the parameters Y_{xsm} and Y_{psm} can be determined. In this case we must be sure that in both situations the maximum yield of biomass corrected for product formation is equal (the same amount of energy, the same amount of carbon, is necessary to make one gram of biomass), and we have to assume that the efficiency of biomass production is not dependent on product formation. When beside protein other products such as overflow metabolites are produced, this will have consequences for the measured (or observed) value of Y_{psm} ($Y_{psm}^{observed}$). The percentage of protein of the total product can be calculated by time related correlation analysis. Taking into consideration that the molecular weight of 1 C-6 mol protein is equal to 138.48 ($C_6H_{9.48}N_{1.74}O_{2.04}$; Babel & Müller 1985) and since each standard mol protein contains 72 gram carbon, the protein formed (as gram carbon) can be calculated as follows:

$$\text{protein (gram carbon)} = \frac{72}{138.48} \times \text{protein (gram)}$$

Application of linear regression to both protein and total product formed expressed as gram carbon

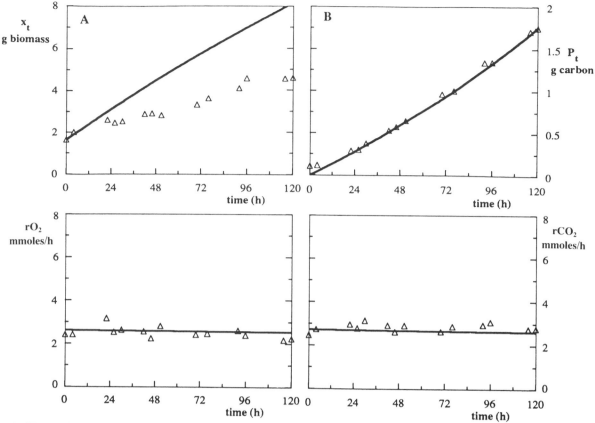

Fig. 4. Glucose-limited recycling experiment with *Aspergillus niger*. r_s = 1 mmol glucose/h. The biomass, product, rate of oxygen consumption and rate of carbondioxide production data were fitted using the model equations mentioned in Box1. Y_{xsm}, m_s and b were floating parameters, the others were fixed. *(A)* Biomass data (x_t, g dry weight) and fitted curve. *(B)* product data (P_t, g carbon) and fitted curve. *(C)* rate of oxygen consumption (r_{O2}, mmoles/h) data and fitted curve. *(D)* rate of carbondioxide production (r_{CO2}, mmoles/h) data and fitted curve.

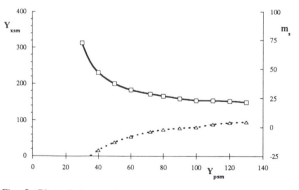

Fig. 5. Plot of the maximum growth yield (□, Y_{xsm}, g dry weight/mol C-12 substrate) and the maintenance requirements (△, m_s, μmol C-12 substrate (g dry weight.h)$^{-1}$) calculated at different values of the maximum product yield (Y_{psm}, g carbon/mol C-12 substrate) using the model equations mentioned in Box 1 of a glucose-limited recycling experiment of *Aspergillus niger*.

for both maltose and glucose-limited cultures resulted in 100% protein for maltose-limitation and 55% protein plus 45% overflow products for glucose limitation. Consequently the Y_{psm} for protein formation, $Y_{psm}^{protein}$, can be calculated at any value of $Y_{psm}^{observed}$ from the analysis of the nature of the products formed using the following equation.

$$Y_{psm}^{observed}(i) = \frac{\% \text{ protein}}{100}(i).Y_{psm}^{protein} + \frac{\% \text{ product}}{100}(i).MW_{product} \quad (6)$$

in which i should be $\geq$ 2, and MW is molecular weight. It is assumed that synthesis of other products than protein (overflow metabolites) does not consume energy (ATP) during their formation. For a good comparison between the glucose (C-6) and

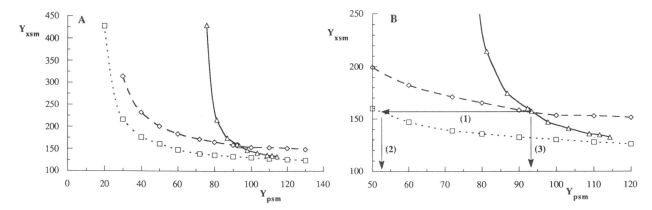

Fig. 6. Plot of the maximum growth yields (Y_{xsm}) of the maltose- ($\square \cdots$) and glucose- ($\diamond$--) limited recycling experiments with *Aspergillus niger* versus the observed Y_{psm} values, including the Y_{xsm} values of the maltose-limited experiment at the expected Y_{psm} values for glucose-limitation ($\triangle$ –), calculated with equation 6. For further explanation see the text. *(A)* all data. *(B)* enlargement of the relevant area. The arrows indicate: (1) the Y_{xsm} at which $Y_{psm}^{protein}$ is equal for both limiting conditions, (2) $Y_{psm}^{observed}$ for maltose-limitation and (3) $Y_{psm}^{observed}$ for glucose-limitation. See also Table 1.

maltose (C-12) experiments all parameters are expressed on a C-12-base.

From the analysis of the formed products it is clear that under maltose-limitation $Y_{psm}^{observed} = Y_{psm}^{protein}$, since only protein is produced, and for glucose-limitation $Y_{psm}^{observed} = 0.55 \cdot Y_{psm}^{protein} + 0.45 \times 144$.

The aim was now to find at what $Y_{psm}^{observed}$, Y_{xsm}(maltose) = Y_{xsm}(glucose) when expressed on a C-12 base, taking into consideration that 1 mole of maltose (C-12) can be converted stoichiometry into maximally 307.54 gram biomass. This can be made clear by plotting in one figure (see Fig. 6) the measured values of Y_{xsm} at different $Y_{psm}^{observed}$ values of both growth conditions and the values of Y_{xsm} of one of the growth conditions (maltose-limitation) at the expected different values of $Y_{psm}^{observed}$ for the other growth condition, calculated with equation 6. The calculation of the expected $Y_{psm}^{observed}$ proceeds as follows: Since the $Y_{psm}^{observed}$ of the maltose-limitation is equal to the $Y_{psm}^{protein}$, these values can be used in Eq. 6 and the expected $Y_{psm}^{observed}$ for glucose-limitation can be calculated. Then the Y_{xsm} values of the maltose-limitation are plotted versus the expected $Y_{psm}^{observed}$ values. As is clear from Fig. 6, the expected data do intercept the measured glucose-limited data at a point where Y_{xsm} and $Y_{psm}^{protein}$ are equal for both maltose- and glucose-limited conditions.

From Table 1 and Fig. 6 it can be seen that the maximum growth yield is 157 g biomass per C-12

substrate (or 0.52 C-mol per C-mol) and thus will be on glucose and maltose respectively 78.5 and 157 g biomass per mol substrate, the observed maximum product yield on glucose and maltose is respectively 93 and 52 g carbon per C-12 mol substrate, and the $Y_{psm}^{protein}$ is for both cultures 52 g carbon/mol substrate (on C-12 base), which is equal to 100 g protein per mol maltose or 50 g protein per mol glucose. With this kind of experiment the $Y_{psm}^{protein}$ is very low, 0.38 C-mol/C-mol, when compared with alkaline protease producing *Bacillus licheniformis* (0.7 C-mol/C-mol, Frankena et al. 1988).

Discussion

Microbial energetics and modelling of growth and product formation at very low specific growth rates in a recycling fermentor with 100% biomass fed back and fed batch cultures are of great importance for guiding and improving industrial processes, that are mostly carried out in fed-batch type fermentations. In the present study the glucoamylase production by *Aspergillus niger* grown in the recycling fermentor under maltose- and glucose-limited conditions is studied with the aim to obtain an estimate of the maximum yield of product on the used substrate, Y_{psm}. The choice of glucose and maltose is obvious, since both substrates are equal after glucoamylase converted maltose into two glucose molecules. Glucoamylase production by *Aspergillus niger* grown in the recycling fermentor under maltose-limited conditions is higher than under glucose-limited conditions. This is not a surprising observation, since glucoamylase production is not necessary for growth on glucose. Surprising is the higher level of glucoamylase production found at glucose-limitation (72% of that at maltose-limitation) at these low specific growth rates, when compared with the production in chemostat cultures at relatively high specific growth rate (17% of that found at maltose-limitation, see Metwally et al. 1991). It can be concluded that beside induction of glucoamylase production by substrate, also very slow specific growth rates seem to function in derepressing glucoamylase synthesis, which is a phenomenon we also observed in glucoamylase-overproducing transformants (J.M. Schrickx, unpublished results). The main objective of the presented experiments, the determination of the Y_{psm}, has been successful, since the two used carbon source-limiting experiments yielded two different sets of equations from which the two unknown parameters Y_{xsm} and Y_{psm} could be determined. However, it must be realized that the found Y_{psm} of 0.36 C-mol product per C-mol substrate is not necessarily the Y_{psm} for glucoamylase alone. Other enzymes and/or proteins can be produced, and consequently this Y_{psm} can better be considered as a Y_{psm} for extracellular protein formation. It still remains a fact that the found Y_{psm} for *Aspergillus niger* is only half the value of the Y_{psm} of 0.7 C-mol per C-mol found for alkaline serine protease producing *Bacillus licheniformis* (Frankena et al. 1988), strongly indicating that production of extracellular proteins by filamentous fungi is energetically very expensive as was already concluded from a comparison between glucose- and maltose-limited chemostat cultures (Metwally et al. 1991). From chemostat experiments it was concluded that the product yield was energetically, thus expressed as g protein produced per mol ATP utilized, only 10% of the theoretical expected yield.

A very notable observation is the extremely low maintenance requirements of $-$ 0.8 and 0.4 µmol substrate per g dry weight per hour for respectively glucose- and maltose-limitation. Analyzing chemostat data using the whole range of specific growth rates yielded maintenance requirements of about 0.1 mmol (100 µmol) per g dry weight per hour and Y_{xsm} values of about 210 g dry weight per C-12 mol substrate. However, when in chemostat experiments only those data were considered, where no differentiation in conidia occurred, the maintenance requirements and the Y_{xsm} dropped to respective values of about 20 µmol per g dry weight per hour and 170–190 g per C-12 mol substrate (Metwally et al. 1991). The results presented here are in accordance with these chemostat observations; differentiation was scarcely seen and both the maintenance requirements and Y_{xsm} were lowered to respectively practically zero and 157.5 g dry weight per C-12 mol substrate. The following conclusions can be drawn: (i) maintenance is not a significant factor during growth of filamentous fungi or (ii) maintenance requirements are linearly dependent on the specific growth rate and thus cannot be measured because then m_s becomes a third independent parameter in the linear equation of substrate utilization (see the derivation of equation 1 in Box 1) and (iii) reported maintenance requirements for fungi are most probably due to differentiation.

References

Anderson JG & Smith JE (1971) Synchronous initiation and maturation of *Aspergillus niger* conidiophores in culture. Trans. Br. Mycol. Soc. 56: 9–29

Babel W & Müller RH (1985) Correlation between cell composition and carbon conversion efficiency in microbial growth: a theoretical study. Appl. Microbiol. Biotechnol. 22: 201–207

Berry DR (1975) The environmental control of the physiology of filamentous fungi. In: Smith JE & Berry DR (Eds) The Filamentous Fungi Vol 1 (pp 16–32). London, Edward Arnold Publishers Ltd

Chesbro WR, Evans T & Eifert R (1979) Very slow growth of *Escherichia coli*. J. Bacteriol. 139(2): 625–638

Frankena J, van Verseveld HW & Stouthamer AH (1988) Substrate and energy cost of the production of exocellular enzymes by *Bacillus licheniformis*. Biotechnol. Bioeng. 32: 803–812

Lowry OH, Rosebrough NJ Farn AL & Randall RJ (1951)

Protein measurement with the Folin phenol reagent. J. Biol. Chem. 193: 265–275

Metwally M, el Sayed M, Osman M, Hanegraaf PPF, Stouthamer AH & van Verseveld H.W. (1991) Bioenergetic consequences of glucoamylase production in carbon-limited chemostat cultures of *Aspergillus niger*. A. van Leeuwenhoek 59: 35–43

Pirt S.J. (1965) The maintenance energy of bacteria in growing cultures. Proc. Roy. Soc. B. 163: 224–231

Priest FG (1984) Aspects of Microbiology 9: extracellular enzymes. Van Nostrand (UK) Go Ltd., Wokingham, Berkshire, England

Rauscher E (1974) In: Bergmeyer HW (Ed) Methods of Enzymiatic Analysis (p 890). Verlag Chemie, Weinhein and Academic Press, New York

Righelato RC (1975) Growth kinetics of mycelial fungi. In: JE Smith & DR Berry (Eds) The Filamentous Fungi, Vol 1 (pp 79–103). London, Edward Arnold Publishers Ltd

Roels JA (1980) Application of macroscopic principles to microbial metabolism. Biotechnol. Bioeng. 22: 2457–2514

Stouthamer AH, van Verseveld HW (1987) Microbial energetics should be considered in manipulating metabolism for biotechnological purposes. Tr. Biotechnol. 5: 149–155

Van Verseveld HW, Chesbro WR, Braster M & Stouthamer AH (1984) Eubacteria have three growth modes keyed to nutrient flow; consequences for the concept of maintenance energy and maximal growth yield. Arch. Microbiol. 137: 176–184

Van Verseveld HW, de Hollander JA, Frankena J, Braster M, Leeuwerik FJ & Stouthamer AH (1986) Modelling of microbial substrate conversion, growth and product formation in a recycling fermentor. A. van Leeuwenhoek 52: 325–342

Yamasaki Y, Suzuki Y & Ozawa J (1977) Purification and properties of two forms of glucoamylase from *Penicillium oxalicum*. Agr. Biol. Chem. 41: 755–762

Antonie van Leeuwenhoek **60**: 325–353, 1991.

Physiology of yeasts in relation to biomass yields

Cornelis Verduyn
*Department of Microbiology and Enzymology, Kluyver Laboratory of Biotechnology,
Delft University of Technology, Julianalaan 67, 2628 BC Delft, The Netherlands*

Key words: yeasts, cell yield, bioenergetics, Y_{ATP}, P/O-ratio, uncoupling

Abstract

The stoichiometric limit to the biomass yield (maximal assimilation of the carbon source) is determined by the amount of CO_2 lost in anabolism and the amount of carbon source required for generation of NADPH. This stoichiometric limit may be reached when yeasts utilize formate as an additional energy source. Factors affecting the biomass yield on single substrates are discussed under the following headings:

- Energy requirement for biomass formation (Y_{ATP}). Y_{ATP} depends strongly on the nature of the carbon source.
- Cell composition. The macroscopic composition of the biomass, and in particular the protein content, has a considerable effect on the ATP requirement for biomass formation. Hence, determination of for instance the protein content of biomass is relevant in studies on bioenergetics.
- Transport of the carbon source. Active (i.e. energy-requiring) transport, which occurs for a number of sugars and polyols, may contribute significantly to the calculated theoretical ATP requirement for biomass formation.
- P/O-ratio. The efficiency of mitochondrial energy generation has a strong effect on the cell yield. The P/O-ratio is determined to a major extent by the number of proton-translocating sites in the mitochondrial respiratory chain.
- Maintenance and environmental factors. Factors such as osmotic stress, heavy metals, oxygen and carbon dioxide pressures, temperature and pH affect the yield of yeasts. Various mechanisms may be involved, often affecting the maintenance energy requirement.
- Metabolites such as ethanol and weak acids. Ethanol increases the permeability of the plasma membrane, whereas weak acids can act as proton conductors.
- Energy content of the growth substrate. It has often been attempted in the literature to predict the biomass yield by correlating the energy content of the carbon source (represented by the degree of reduction) to the biomass yield or the percentage assimilation of the carbon source. An analysis of biomass yields of *Candida utilis* on a large number of carbon sources indicates that the biomass yield is mainly determined by the biochemical pathways leading to biomass formation, rather than by the energy content of the substrate.

Introduction

Biomass yield (amount of biomass produced per amount of substrate) may be an important factor in industrial processes. A high biomass yield is espe-cially important when the price of the raw material (e.g. the carbon source) makes up a large fraction of the costs of the final product. This, for example, holds for the cultivation of baker's yeast on molas-ses. Furthermore, any improvement in yield will decrease the oxygen demand and reduce heat production. This reduces the product price since supply of oxygen and cooling are also significant costs in large-scale processes.

Alternatively, in some processes it is necessary to keep biomass formation to a minimum in order to achieve a maximal conversion of the carbon source into extracellular products like ethanol or citric acid. It is evident, therefore, that more fundamental knowledge on the factors that govern the biomass yield is required for the optimization of large-scale fermentation processes.

In this work, the effect of various parameters on the biomass yield will be discussed in relation to the physiology. Although the data and discussions mainly concern yeasts, they in our opinion are applicable to heterotrophic organisms in general.

Practical aspects

In order to obtain reliable data on the relation between the physiology of an organism and its biomass yield, a number of practical aspects are of importance. These are summarized below.

Batch versus chemostat cultivation
Results on biomass yields of batch cultures are difficult to interpret. This is due to the fact that in batch cultures cells are exposed to a continuously changing environment, that exerts profound effects on the physiology. Therefore, in this contribution, mainly data obtained with chemostat cultivation are considered. In addition, biomass yields are nearly always expressed as weight of dry biomass obtained per amount of substrate utilized. However, the biomass composition (notably the protein content) needs consideration since comparison of growth yields of different yeasts requires knowledge on the biomass composition. Where possible, the protein and/or nitrogen contents of biomass will therefore be considered.

Medium composition
The importance of a correct medium composition should not be overlooked. Rigorous testing must be performed to determine the medium compo-

nent that limits the growth yield. Rieger et al. (1983), for example, showed that the physiology (and cell yields) of S. cerevisiae H1022 were strongly affected by a hidden manganese deficiency in the medium.

A number of theoretical studies have been published on the nutritional demands of yeasts (Egli 1980; Jones & Greenfield 1984; Jones & Gadd 1990). In anaerobic cultures, the provision of an adequate amount of sterols and unsaturated fatty acids is of primary importance. Low anaerobic growth yields and Y_{ATP} values (g cells.mol ATP formed^{-1}) as reported by Dekkers et al. (1981) can be explained (Verduyn et al. 1990a) by the fact that no ergosterol and oleic acid, which are required as vitamins during anaerobic growth (Andreasen & Stier 1953, 1954), had been added to the growth medium. All data discussed in this contribution refer to carbon- and energy-limited growth in chemostat cultures, unless mentioned otherwise.

Oscillations
Some yeasts, notably Saccharomyces and Schizosaccharomyces strains may exhibit oscillations in metabolic activity during sugar-limited growth due to spontaneous synchronization of the cell cycle (Parulekar et al. 1986; Novak & Mitchison 1990). These oscillations can result in variations in for instance dry weight (Rouwenhorst et al. 1991) and make yield studies with S. cerevisiae a difficult enterprise. Our data for S. cerevisiae presented herein were obtained with cultures that were not oscillating, as shown by measurements of O_2 consumption and CO_2 production.

Effluent removal
Selective effluent removal in chemostat cultures, especially when culture fluid is removed upwards from the culture surface by suction, can result in accumulation of biomass in the fermenter. This can be checked by measuring the biomass concentrations both in samples taken directly from the fermenter via a sample port and in samples taken from the effluent line. These data should be the same. We have found that, although a steady-state biomass concentration may be established, the difference between fermenter and outlet biomass con-

centration may in some cases be more than 10% (H. Noorman, pers. comm.). These problems can often be solved by placing the outlet tube in the centre of the fermenter fluid (rather than on the surface) and controlling the fluid removal rate by means of an electrical contact sensor at the surface of the fluid. The sensor then steers the effluent pump to maintain the required working volume.

Metabolite analysis and sampling
Assays of metabolites in culture supernatants may supply useful information on the presence of potentially inhibitory compounds, such as weak acids (to be discussed later). Simple HPLC methods can detect a wide range of short-chain acids via UV-detection at 210 nm with a high sensitivity (Favre et al. 1989; Verduyn et al. 1990b). Furthermore, by placing a refraction index meter in series with this equipment, (poly)alcohols and sugars can also be assayed. The detection limit for these latter compounds is, however, insufficient to assay residual substrate concentrations.

Sampling of culture fluid also deserves special attention when it concerns measurements of low residual substrate concentrations like, for example, glucose in glucose-limited chemostat cultures. Uptake of glucose will continue during sampling and thus leads to an underestimation of the residual concentration in the fermenter. Therefore, fast sampling methods have to be applied which either separate cells immediately from the medium or, alternatively, instantaneously stop metabolic activity. Methods to achieve this include rapid filtration (e.g. Rutgers et al. 1987), and the direct transfer of cells into liquid nitrogen (Postma et al. 1988).

Whatever the method applied, it should be checked that, as prescribed by chemostat mathematics, the residual concentration of the growth-limiting substrate is independent of the biomass density in the culture. To this end, the residual substrate concentration should be determined at different reservoir concentrations of the growth-limiting nutrient.

Carbon recovery
Finally, studies on biomass yields must include construction of carbon balances in order to check whether unexpected byproduct formation does occur. This should also be verified with a Total Organic Carbon (TOC) analysis of culture supernatants. The yield data reported herein refer to situations in which the carbon recovery was $100 \pm 5\%$.

Basic flows in metabolism: assimilation and dissimilation

The biomass yield is determined by two factors: the energy requirement of biomass formation and the efficiency of energy transduction in dissimilation. It is therefore convenient to make a general division of metabolism into two main processes: assimilation and dissimilation. Assimilatory processes involve the biosynthetic reactions leading to formation of biomass. Dissimilatory processes provide the energy (ATP) required for assimilation. Energy is also required for maintenance functions and transport of nutrients (Fig. 1). Finally a flow of carbon is directed towards the generation of NADPH, which is used as a reductant in anabolism. NADPH generation is considered here as an assimilatory process. The scheme shown in Fig. 1 is applicable to the carbon sources which support growth of yeasts. With very oxidized substrates, however, an additional dissimilatory flow is required to reduce the carbon source to the level of biomass, i.e. to provide reducing equivalents. The carbon sources for growth need not be the same as the energy source. Many yeasts can be grown on mixed substrates in such a way that ATP for assimilation of the carbon source is derived almost exclusively from an energy source which itself cannot be assimilated, such as formate (Fig. 1). Under these conditions the yield on the carbon source is at its maximum (i.e. all carbon in the carbon source is directed towards cell synthesis). This is discussed below.

Stoichiometric limits to the cell yield

An important issue in the context of yield studies is the theoretical limit to the yield that can be achieved on a given carbon source. Frequently it is

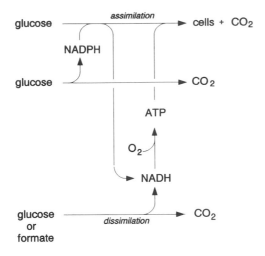

Fig. 1. Schematic representation of central metabolic processes. Formate can serve as an energy source during glucose-limited growth. That part of glucose which is normally dissimilated to provide ATP is then replaced by formate.

assumed that this is identical to complete incorporation of the substrate carbon into biomass, i.e. a carbon conversion of 100%. However, during assimilatory reactions, net production of CO_2 always occurs (Oura 1972). Therefore, the maximal carbon conversion is always lower than 100%. The fraction of CO_2 which is lost in assimilation depends on the carbon source and lies between approximately 10% for growth on, among others, glucose and methanol, and 29% for carbon sources such as succinic acid or oxalic acid (Gommers et al. 1988).

Another factor that must be taken into account in assessing the limits of growth yields is the NADPH requirement of biomass formation. In mammalian mitochondria (Rydström et al. 1970) and a number of prokaryotic organisms (Asano et al. 1967), NADPH can be generated from NADH by transhydrogenase (NADH + NADP$^+$ → NADPH + NAD$^+$). This enzyme has not been detected in yeasts (Bruinenberg et al. 1983a). In yeasts, NADPH is generated via NAD(P)$^+$-dependent isocitrate dehydrogenase and via the hexose monophosphate (HMP) pathway. It can be calculated (Bruinenberg et al. 1983b) that during growth on glucose, with ammonium as nitrogen source, insufficient NADPH can be produced via

the isocitrate dehydrogenase reaction alone. Therefore, the HMP route must be active and part of the glucose is oxidized to CO_2 via this route to generate NADPH. This means that, even in cases where the assimilation of a carbon source would result in the production of sufficient NADH to provide the ATP necessary for biomass formation, some dissimilation of the carbon source would still be necessary for the generation of NADPH. This is nicely illustrated with studies in which yeasts were grown under carbon and energy limitation on mixtures of glucose and formate. Formate can only be used as an energy but not as a carbon source by yeasts. Its oxidation is strictly NAD$^+$-dependent in yeasts (Veenhuis et al. 1983), hence formate cannot serve as a source of NADPH. Addition of formate to glucose-limited cultures of *Candida utilis* resulted in an increase in yield from 0.51 to 0.69 g.g glucose^{-1} (Bruinenberg et al. 1985). It can be calculated, however, that the theoretical maximum yield (not regarding the NADPH requirement) is 0.75 g.g glucose^{-1}. Indeed radiorespirometric studies on the contribution of the HMP pathway to glucose metabolism in *C. utilis* suggested that ca. 7% of the C-source is oxidized in the HMP pathway (Bruinenberg et al. 1986). Thus, during growth on mixtures of glucose and formate, formate oxidation can supply all NADH required for ATP formation. However, since formate cannot supply NADPH required for biomass synthesis, NADPH must be generated by oxidation of the carbon source (Fig. 1).

The amount of NADPH required for assimilation is dependent on the nitrogen source for growth. When nitrate serves as a nitrogen source, additional NADPH is required for reduction of nitrate to NH$_4^+$ (Bruinenberg et al. 1983). Indeed, it could be shown that during glucose-limited growth of *C. utilis* with nitrate, more glucose is directed to the HMP pathway (Bruinenberg et al. 1985). As a result, the maximum attainable yield on glucose in the presence of formate was lower during growth with nitrate than with NH$_4^+$ (Table 1).

Not only with glucose, but also with other carbon sources formate can serve as an energy source (Table 1). In most of these cases a considerable increase in yield (expressed as g biomass per g carbon source) was observed. A typical example of the effect of addition of formate to an ethanol-limited culture of *C. utilis* is shown in Fig. 2. In the ascending part of the yield profile, growth is still limited by the availability of energy (i.e. part of the carbon source is still dissimilated). Hence growth can be described as energy- and carbon-limited. When the energy supply is completely provided by reducing equivalents from formate oxidation, growth becomes carbon-limited (Fig. 2). Unlike growth on glucose, operation of the HMP pathway is, in theory, not necessary during growth on ethanol with NH_4^+ as nitrogen source. This is due to the fact that acetaldehyde can be converted into acetate via a $NADP^+$-dependent acetaldehyde dehydrogenase (Bruinenberg et al. 1983a). The maximal yield on ethanol observed in the presence of formate is similar to that calculated theoretically (Verduyn et al. 1991a).

Energy requirement for biomass formation: Y_{ATP}

In the classical study of Bauchop & Elsden (1960) it was shown that the biomass yield was related to the energy yield in the dissimilation. This led to the introduction of the term Y_{ATP} (g cells per mol ATP formed). This Y_{ATP} was calculated from product formation in anaerobic (batch) cultures and thus was an experimental value. As it was shown later that the observed Y_{ATP} may be affected by maintenance, notably in the case of bacterial growth, a

second term, Y_{ATP}^{max} (Stouthamer & Bettenhaussen 1973) was introduced, which is Y_{ATP} corrected for growth-independent maintenance energy. For yeasts no large differences between Y_{ATP} and Y_{ATP}^{max} are observed (discussed in Harder & van Dijken 1976).

It has also been attempted to calculate a theoretical Y_{ATP}^{max} value. This theoretical Y_{ATP}^{max} follows from a summation of all the ATP-consuming and -producing reactions leading to the formation of biomass (Gunsalus & Shuster 1961; Oura 1972; Stouthamer 1973; Verduyn et al. 1990a). The theoretical Y_{ATP}^{max} depends strongly on the carbon source, as can be shown by comparing the theoretical ATP requirements for growth on glucose, ethanol and lactic acid (Table 2). Lower values of Y_{ATP}^{max} are obtained if the carbon source is taken up by active (energy-requiring) transport.

Unfortunately, theoretical Y_{ATP}^{max} values cannot be used as such to calculate the cell yield. It has long been recognized that experimental Y_{ATP} values are considerably lower than theoretical values. For instance, the theoretical Y_{ATP}^{max} for anaerobic glucose-limited growth of *S. cerevisiae* at D = 0.10 h^{-1} was calculated to be 28, but the experimental Y_{ATP} was 14–16 g biomass.(mol ATP)$^{-1}$ for two strains (Verduyn et al. 1990a). Various models have been proposed that relate theoretical and experimental Y_{ATP} values (discussed in Verduyn et al. 1991a). It should be noted that part of the gap between theoretical and experimental ATP requirements could be due to processes which have

Table 1. Cell yields of yeasts (g cells · g carbon source^{-1}) in the absence of formate and maximum cell yield in the presence of formate with either ammonium or nitrate as nitrogen source.

Organism	C-source	N-source	Biomass yield	Max. yield with formate	Reference
Candida utilis	Glucose	Ammonium	0.51	0.68	Bruinenberg et al. (1985)
	Glucose	Nitrate	0.42	0.53	Bruinenberg et al. (1985)
	Glucose	Ammonium	0.51	0.66	Verduyn (unpublished)
	Glycerol	Ammonium	0.58	0.67	Verduyn (unpublished)
	Ethanol	Ammonium	0.69	0.84	Verduyn (unpublished)
Candida maltosa	Hexadecane	Ammonium	0.94	1.26	Müller & Babel (1988)
Hansenula polymorpha	Glucose	Ammonium	0.50	0.71	Babel et al. (1983)
	Acetic acid	Ammonium	0.37	0.47	Müller & Babel (1988)
	Acetic acid	Nitrate	0.25	0.40	Müller & Babel (1988)

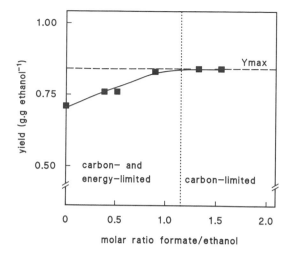

Fig. 2. Biomass yield of *Candida utilis* CBS 621 in ethanol-limited chemostat cultures (pH 5.0) in the presence of increasing concentrations of formate. The dilution rate was 0.20 h^{-1}. Ammonium served as nitrogen source. The horizontal broken line indicates the theoretical maximal yield when ethanol is completely assimilated (Verduyn, unpublished).

hitherto not been included in Table 2. These may include intracellular transport of proteins and metabolites, or underestimation of the costs of protein synthesis. It has therefore been proposed by Verduyn et al. (1991a) to use the theoretical ATP requirement for biomass formation and an additional, fixed, amount of ATP. The latter was obtained

by subtraction of the theoretical anaerobic ATP requirement from the experimental anaerobic ATP requirement for a given set of culture conditions of *S. cerevisae* at a dilution rate of 0.10 h^{-1} (Verduyn et al. 1991a). Calculated in this way, the difference between theoretical and observed ATP requirement amounted to 2700 mmol (g biomass)$^{-1}$.

Cell composition

The calculation of the theoretical ATP requirements for formation of yeast biomass as shown in Table 2 has been made for a given cell composition. If the results for glucose are evaluated, it is clear that the cost of protein polymerization makes up a major fraction of the theoretical ATP requirement. It is therefore relevant to obtain data on the macromolecular composition of the biomass, as this will influence the ATP requirement for biomass formation.

The composition of microorganisms can be broken down to four main polymers: protein, carbohydrate, nucleic acid and lipid. In addition, part of the biomass consists of inorganic metals ('ash'). The protein content of a number of yeasts grown in minimal media in carbon-limited chemostat cultures at a dilution rate of 0.1 h^{-1} is shown in Table 3.

Table 2. Calculation of the theoretical Y_{ATP}^{max} for growth of yeast on glucose, ethanol and lactic acid (cell composition: 52% protein, 28% carbohydrate, 7% RNA, 7% lipid and 6% ash).

	Glucose	Ethanol	Lactic acid
Amino acid synthesis	200	4264	1163
polymerization	1960	1960	1960
Carbohydrate synthesis	358	2329	1186
Lipid synthesis	179	651	407
RNA synthesis and polymerization	182	341	193
Turnover of m-RNA	71	71	71
NADPH generation	88	0	450
Transport: Ammonium	700	700	700
Potassium and phosphorus	240	240	240
Total	3978	10556	6371
Theoretical Y_{ATP}^{max}	25.1	9.5	15.7

Data are expressed as mmol ATP·(100 g cells)$^{-1}$.
It is assumed that the carbon source is taken up by passive (or facilitated) diffusion.

For some species the nitrogen content has also been determined (Verduyn 1991a; unpublished). A large difference is observed between the protein content of different species, with values ranging from 39 to 56%. Furthermore, even within one strain differences in the protein content are observed during growth on different carbon sources.

The protein content of biomass is not only influenced by the carbon source, but also by the growth rate. This has been studied in glucose-limited cultures of *S. cerevisiae* under anaerobic and aerobic conditions. In both cases a considerable increase (25%) in the protein content was observed with increasing growth rate (Fig. 3). We have not assayed the other major macromolecular components of biomass. However, literature data show that the carbohydrate content of *S. cerevisiae* is generally inversely related to the protein content, i.e. showing a considerable decrease with increasing growth rates (Küenzi 1970). Data on RNA, fat and ash contents of yeasts (Oura 1972; Dekkers et al. 1981; Atkinson & Mavituna 1983) indicate that the RNA content in different yeast species averages 6–8% (g.g cells^{-1}) at a dilution rate of 0.10 h^{-1}, but increases considerably with the growth rate to values between 10 and 15% of dry weight. Occasionally higher RNA contents (up to 20%) have been reported in batch cultures (e.g. Parada & Acevedo 1983). The lipid content of yeasts varies between 3 and 12% and averages 6–8% at a dilution rate of 0.10 h^{-1}. Reported ash contents fall in the range of 4–9%. To obtain relevant information on the biomass composition, it is probably sufficient to assay the protein content (see Verduyn et al. 1990a). Furthermore, protein is the most important macromolecular component of biomass in terms of bioenergetics. This can be seen from Table 4, in which the energy requirement for formation of 1 gram of the different polymers has been calculated from the data in Table 2. The energy requirement for the different polymers depends on the carbon source as is apparent from the large variations in theoretical Y_{ATP}^{max} calculated in Table 2. During growth on glucose, the synthesis costs of amino acids are lowest and small as compared to the polymerization costs. For growth on ethanol, synthesis costs of amino acids are high (Table 4). Note that data in Table 4 apply to synthesis and polymerization costs of the monomers only, and do not include uptake of carbon and nitrogen sources or NADPH formation. Due to the large variations in protein content and the high energy requirement of protein synthesis (Fig. 3, Table 4), it follows that the protein content will determine to a large extent the ATP requirement for biomass formation. Differences in RNA and lipid content will have less effect on the overal energy requirement since these compounds have a relatively low synthesis cost as

Table 3. Cell yield, and contents of protein and nitrogen (expressed as % of dry weight) of yeasts grown in carbon-limited chemostat cultures on minimal media at a dilution rate of 0.1 h^{-1}.

Organism	C-source	Yield (g·g^{-1})	Protein	N (%)	Reference
Candida utilis	Ethanol	0.72	56	nd	Shul'govskaya et al. (1988)
	Glucose	0.51	52	9.8	Verduyn et al. (1991)
	Glycerol	0.58	52	9.7	Verduyn et al. (1991)
	Acetic acid	0.39	53	9.7	Verduyn et al. (1991)
	Ethanol	0.69	52	9.8	Verduyn et al. (1991)
Hansenula nonfermentans	Glucose	0.51	39	nd	Van Urk et al. (1990)
Hansenula polymorpha	Glucose + methanol	–	40	7.3	Egli & Quayle (1986)
Saccharomyces cerevisiae	Glucose	0.10*	47	8.9	Verduyn et al. (1990b)
	Glucose	0.51	42	7.6	Verduyn et al. (1991)
	Acetic acid	0.29	47	8.9	Verduyn et al. (1991)
	Ethanol	0.61	46	8.6	Verduyn et al. (1991)
Schizosacch. pombe	Glucose	0.44	42	nd	van Urk et al. (1990)

* anaerobic growth   nd: not determined.

332

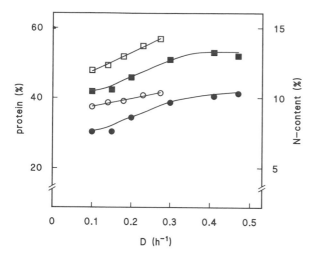

Fig. 3. Protein (□, ■) and nitrogen (○, ●) contents of *S. cerevisiae* CBS 8066 during anaerobic (open symbols) or aerobic (closed symbols) glucose-limited chemostat growth as a function of the dilution rate.

compared to protein, and furthermore make up a smaller part of the biomass. In addition, a high protein content also requires increased uptake of ammonium (which requires energy) and additional NADPH generation. The latter effect is augmented when nitrate is used as a source of nitrogen.

Cell composition is not only influenced by the growth rate but also by environmental factors, in particular temperature (e.g. Parada & Acevedo 1983). Similar observations have been made for bacteria (Harder & Veldkamp 1976).

Transport of carbon sources

A parameter which affects both the ATP requirement for assimilation and the ATP yield from dissimilation is the energy requirement for transport of the carbon source. A short summary on energetic aspects of substrate transport is presented below.

Transport of sugars

Uptake of sugars in yeasts may occur by facilitated diffusion and/or by active transport via a proton-

sugar symport mechanism (reviewed by Romano 1986). It is generally assumed that in *S. cerevisiae* uptake of glucose occurs via facilitated diffusion (Lang & Cirillo 1987, Romano 1982). Depending on the growth conditions, at least two glucose carriers with different K_m have been reported (Postma et al. 1989b) in glucose-limited *S. cerevisiae*. However, active uptake of fructose has been reported for some *S. cerevisiae* strains (Cason et al. 1986). In *C. utilis*, grown in glucose-limited cultures, two proton-symporters with different apparent K_m-values occur, as well as a diffusion carrier (Postma et al. 1988). At high glucose concentrations, the synthesis of these carriers is regulated by the growth conditions. When glucose is present in excess, synthesis of the symport mechanisms is repressed (Postma et al. 1988; Peinado et al. 1989).

The proton-sugar symport mechanisms described so far have a sugar/H^+ stoichiometry of 1. This has been shown for instance for uptake of galactose and xylose in *Rhodotorula glutinis* (Höfer & Misra 1978), maltose in *S. cerevisiae* (Serrano 1977) and glucose in *C. utilis* (Peinado et al. 1989). With an assumed H^+/ATP stoichiometry of the plasma membrane ATPase of 1, as has been reported for this enzyme from various eukaryotes (Malpartida & Serrano 1981; Nelson & Taiz 1989; Perlin et al. 1986), active uptake of a sugar molecule would require 1 ATP. Active uptake of a sugar will especially affect the biomass yield under anaerobic conditions, where the amount of energy obtained

Table 4. ATP requirements for biosynthesis of cell polymers during growth on glucose, ethanol and lactic acid.

Polymer	Carbon source		
	Glucose	Ethanol	Lactic acid
Protein: amino acid synthesis	3.8	82.0	22.4
polymerization	37.7	37.7	37.7
total	41.5	119.7	60.1
Carbohydrate	12.8	83.2	42.4
Nucleic acid (incl. polymerization)	26.0	48.7	27.5
Lipid	25.6	93.0	58.1

Data are expressed as mmol ATP·(g polymer)$^{-1}$.

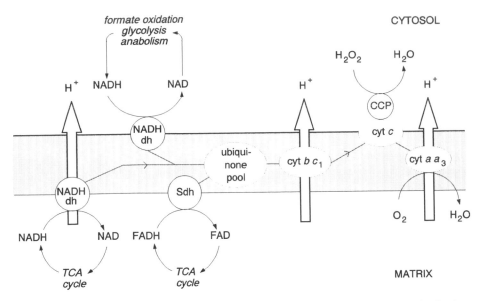

Fig. 4. Schematic representation of an electron transport chain with three proton translocating sites in the inner mitochondrial membrane of yeasts. In *S. cerevisiae*, the internal NADH dehydrogenase is located in the vicinity of succinate dehydrogenase (sdh), i.e. is not coupled to site I energy transduction (not shown in this figure).

from dissimilation of the sugar to ethanol is low. This point can be illustrated by comparing anaerobic chemostat data for growth of *S. cerevisiae* on maltose (active transport) and glucose (passive transport). Dissimilation of maltose (2 glucose units) results in a net yield of 3 ATP (4 via substrate phosphorylation minus 1 for transport), whereas dissimilation of 2 glucose yields 4 ATP. If the energy requirement for biomass formation is similar in both cases, the yield on maltose should be 75% of that on glucose. This has been confirmed experimentally (Table 5). The specific glycerol production (mmol.g cells^{-1}) is mainly dependent on the cell composition (Verduyn et al. 1990b) and was almost constant. The specific ethanol production should be higher during growth on maltose than on glucose, since additional ATP is required for transport of maltose. This was indeed found (Table 5).

Under aerobic conditions the effect of an energy requirement for sugar transport on the biomass yield is smaller than under anaerobic conditions. This is due to the much larger energy gain in aerobic dissimilation. For instance, with a biomass yield on glucose of 0.51 g.g glucose^{-1} (Table 3), 1089 mmol of glucose is required to form 100 g biomass (with an experimentally determined C, H, and N-content and a molecular weight of 100, including ash) according to (Verduyn et al. 1991a): 1089 $C_6H_{12}O_6$ + 720 NH_3 + 2269 $O_2 \rightarrow$ 1000 $C_{3.75}H_{6.6}O_{2.18}N_{0.70}$ (100g biomass) + 2784 CO_2 + 3324 H_2O. If it is further assumed that the effective P/O-ratio is 0.95 (Verduyn et al. 1991a) and that 1 ATP is consumed for each glucose taken up, the extra energy requirement for glucose uptake is 1089 mmol ATP.100 g biomass^{-1}. The additional amount of glucose that has to be dissimilated to provide this energy is (with formation of 12 reducing equivalents as well as 4 ATP by substrate phosphorylation during complete dissimilation of glucose to CO_2): 1089/(12 × 0.95 + 4) = 71 mmol glucose. The biomass yield will then become 0.47 g.g glucose^{-1}, a reduction of approximately 8%. In case of active transport of

Table 5. Growth and product formation during anaerobic chemostat growth of *S. cerevisiae* CBS 8066 on glucose and maltose.

	Glucose	Maltose
Yield (g·g sugar^{-1})	0.099	0.072
$Y_{glycerol}$ (mmol·g cells^{-1})	10.0	9.7
$Y_{ethanol}$ (mmol·g cells^{-1})	83	117

The dilution rate was 0.10 h^{-1} (Weusthuis, unpublished).

glucose, the theoretical ATP requirement for biomass formation as calculated in Table 2 would increase with 822 mmol ATP.(100 g biomass)$^{-1}$ (Verduyn et al. 1991a), an increase of 21%.

Transport of weak acids

Apart from sugars, many yeasts are also able to grow on various organic acids, including monocarboxylic acids like lactic and acetic acid. Little information is available on the mode of uptake of these acids. Transport of short-chain monocarboxylic acids, such as lactic acid, is claimed to be due to an electroneutral 'lactate-proton' symport in *S. cerevisiae* (Cassio et al. 1986) and *C. utilis* (Leao & van Uden 1986). This term is confusing as the term proton symport is associated with net uptake of protons, as in sugar-proton symport (see above). As pointed out previously (Verduyn et al. 1991a), electroneutral uptake of an acid cannot be considered an active (energy-requiring) transport mechanism since the proton taken up into the cytosol with the acid anion is subsequently consumed in metabolism and not extruded as in the case of active uptake of sugars. Eddy & Hopkins (1985) showed that the plasma membrane was not depolarized during uptake of lactate by *C. utilis*, in contrast to uptake of glucose (which is taken up by sugar/proton symport, see above), which also indicated that lactate uptake is electroneutral. If uptake of monocarboxylic acids occurs via active transport this will affect the bioenergetics and the cell yield. For instance, the biomass yield of *C. utilis* is 0.39 g.(g acetic acid)$^{-1}$ (Table 4), hence 3945 mmoles of acetate are required to form 100 g biomass. The theoretical ATP requirement for biomass formation from acetate is similar to that for ethanol (which is taken up by passive diffusion, Kotyk & Alonso 1985), i.e. 10556 mmoles ATP.100 g biomass (Table 2). This figure would thus increase with 37% when uptake of acetic acid requires 1 ATP.

Uptake of dicarboxylic acids has received even less attention than monocarboxylic acids, with the exception of malic acid. Most studies suggest that malic acid is taken up by diffusion in yeasts, including *Zygosaccharomyces bailii* (Baranoswki & Radler 1984) and anaerobically grown *S. cerevisiae* (Salmon 1987). However, for *Schizosaccharomyces pombe* active transport of malic acid has been claimed (Osothsilp & Subden 1986).

P/O-ratio

An important parameter in studies on aerobic biomass yields is the efficiency of respiration (P/O-ratio). In the sixties and seventies a number of studies have been published that dealt with the P/O-ratio of isolated yeast mitochondria. These studies (compiled by Lloyd 1974) indicated that the P/O-ratio of *S. cerevisiae* mitochondria was lower than that of, for instance, mitochondria of *C. utilis* or mammals. It was shown that *S. cerevisiae* lacks site I proton translocation (Ohnishi 1973). Furthermore, reducing equivalents are unable to cross the inner mitochondrial membrane (von Jagow & Klingenberg 1970). Therefore, cytosolic and mitochondrial pools of reducing equivalents are separated. Consequently, separate NADH dehydrogenases catalyze the oxidation of cytosolic and mitochondrial NADH in yeasts (reviewed by de Vries & Marres 1987). Oxidation of cytosolic NADH always by-passes site I proton translocation. Recent data on P/O-ratios (mainly obtained with rat mitochondria) suggest that the mechanistic P/O-ratio for oxidation of mitochondrial NADH is between 1.5 and 2.5 (with two or three proton-translocating sites, respectively, see Verduyn et al. 1991a for references). A P/O-ratio of 1.5 was reported for oxidation of ethanol (which is oxidized by a mitochondrial alcohol dehydrogenase) in mitochondria of *S. cerevisiae*, whereas a value of 1.25 was found for oxidation of cytosolic NADH or glycerol phosphate (Ouhabi et al. 1989). In view of the symmetrical electron transport chain of *S. cerevisiae*, the P/O-ratio is not dependent on the localization of the reducing equivalents. The P/O-ratio is thus probably not dependent on the carbon and energy source. For *C. utilis* a relation between the P/O-ratio and the carbon- and energy source is more complex because of the non-symmetrical electron transport chain (Fig. 4). Therefore, when

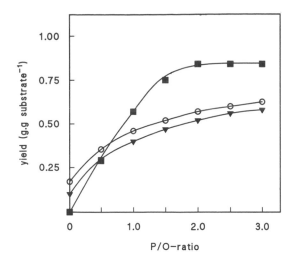

Fig. 5. Calculated biomass yield as a function of the effective P/O-ratio for growth on glucose ($\bigcirc$), ethanol ($\blacksquare$), and lactic acid ($\bigtriangledown$) assuming passive uptake of the carbon source. The cell composition is taken as presented in Table 2.

evaluating the mechanistic P/O-ratio, it is necessary to establish the localization of the redox reactions involved in catabolism. In this respect it should be noted that also during assimilation a net production of reducing equivalents occurs. For growth on ethanol, two extreme situations can be distinguished (Verduyn et al. 1991a), in which the first two steps of ethanol metabolism are taken either as mitochondrial or as cytosolic. In the former case, 77% of the total reducing equivalents used for energy transduction are generated in the mitochondria. If these enzymes are cytosolic, only 28% of the total number of reducing equivalents is generated in the mitochondria. The mechanistic P/O-ratio's calculated were 2.0 for growth on glucose and 1.8–2.3 for growth on ethanol (with alcohol and acetaldehyde dehydrogenases either both cytosolic or both mitochondrial, respectively) (Verduyn et al. 1991a).

Apart from the presence of either two or three proton-translocating sites, the electron transport chains of eukaryotes appear to be more uniform than those of prokaryotes, in which a large number of terminal oxidases are known to occur (see Jones et al. 1977). In some yeasts, a partial bypass of the electron transport chain has been shown, the so-called alternative respiration. Alternative respiration in yeasts is usually only coupled to site I proton-translocation (discussed in Alexander & Jeffries 1990). The occurrence of a significant rate of alternative respiration seems to be limited to a few yeasts and often it has to be induced, via for instance the addition of certain respiratory inhibitors. In *S. cerevisiae* only low rates of alternative respiration have been encountered (Goffeau & Crosby 1978).

Finally, the possibility should be considered that the stoichiometry of proton translocation is not constant. Experiments with rat and yeast mitochondria suggest that the H^+/e stoichiometry may be variable. For rat mitochondria, the change in H^+/e stoichiometry was located at the level of cytochrome *c* oxidase and it increased with decreasing values of the proton-motive force (Murphy & Brand 1988a,b). In contrast, mitochondria from *S. cerevisiae* showed an increased P/O-ratio at virtually constant proton-motive force (Ouhabi et al. 1989).

In summary, the magnitude of the P/O-ratio has an important effect on the biomass yield. The available data suggest that the carbon source has some effect on the P/O-ratio, but that large differences in P/O-ratios will be mainly caused by the number of proton-translocating sites as has also been documented for bacteria by Jones et al. (1977).

In order to evaluate the effect of the in vivo P/O-ratio on the biomass yield (g cells per g substrate), the latter has been calculated as a function of the P/O-ratio for growth on glucose, ethanol and lactic acid in Fig. 5. For a calculation of the yield, a value has to be set for Y_{ATP}. This has been done by adding a fixed amount of 2700 mmol.(100g biomass)$^{-1}$ to the theoretical Y_{ATP}^{max} of Table 2 (see previous paragraph; Verduyn et al. 1991a for a further discussion and for calculations). From Fig. 5 it can be seen that the biomass yield on ethanol becomes constant at a P/O-ratio of 2.0. At this point growth becomes limited by the availability of carbon. For growth on glucose and lactic acid the maximal theoretical growth yields, taken into account that part of the carbon source must be dissimilated to provide NADPH, are approximately 0.68 and 0.65 g.(g carbon source)$^{-1}$, respectively (calculations not shown). It can be concluded from Fig. 5 that, even if the effective (in vivo) P/O-ratio would be as high as 3, these values cannot be attained.

Maintenance energy and environmental factors

It has long been established that a fraction of the energy generated in catabolism is used in processes other than net biomass formation, the so-called maintenance energy (Pirt 1965). Processes involved may be, among others, turnover of macromolecules, futile cycles and maintaining cellular homeostasis (discussed by Tempest & Neijssel 1984). Maintenance is usually determined from plots of 1/Y versus 1/D, in which Y may represent the yield on carbon source, oxygen or ATP (the latter value usually being obtained from anaerobic chemostat experiments). It has been postulated by Neijssel & Tempest (1976) that also a growth-dependent maintenance energy may occur.

Generally speaking, the maintenance coefficients of yeasts on various carbon sources appear to be small as can be seen from listings by Roels (1983), Goldberg (1985) and Stouthamer & van Verseveld (1987). The effect of maintenance on the growth yield for *S. cerevisiae* during glucose-limited chemostat growth has been calculated by Roels (1981). It appears that the biomass yield of yeasts is less affected by maintenance than in bacteria, which generally have a much higher maintenance coefficient (discussed by Stouthamer & van Verseveld 1987).

In a number of cases it has not been possible to determine maintenance (e.g. Verduyn et al. 1990a). This is hardly surprising in view of the large changes in cell composition which may occur in, for instance, *S. cerevisiae* as a function of D (Fig. 3). With large changes in cell composition, maintenance may be masked by differences in Y_{ATP}^{max} values.

Maintenance energy requirement is probably affected by pH, temperature, osmotic value of the medium, etc. A clear interpretation of the reported effects of these environmental parameters is hampered by the fact that very few studies have been performed with chemostat cultures. Many studies only compare the maximal growth rates in batch cultures, and do not provide information on the biomass yield. In the following paragraphs a short, general overview is given of some of the environmental factors that affect the biomass yield, as well as their general mode of action. For convenience, all these factors are listed under the heading of maintenance. For a more detailed description the reader is referred to the various reviews cited.

Osmotic stress

Considerable attention has been focused recently on osmoregulation in microorganisms (reviewed by Vreeland 1987). When high concentrations of salt are added to the growth medium, water is lost from the cells and intracellular accumulation of solutes occurs in order to restore the osmotic equilibrium (Higgins et al. 1987). The intracellular salt concentration is maintained at a lower concentration than in the medium. The main osmoregulator in *S. cerevisiae* appears to be glycerol, whereas in some other yeasts both glycerol and arabinitol are produced intracellularly under salt stress (Reed et al. 1987; Meikle et al. 1988; Jovall et al. 1990). During batch cultivation of *S. cerevisiae* under osmotic stress, 29% (w/w) of the glucose was converted into glycerol (Larsson & Gustafsson 1987). Data on anaerobic growth of *S. cerevisiae* in the presence of NaCl have been presented by Watson (1970). It was shown that the maintenance energy requirement increased fourfold in the presence of 1 M NaCl. In contrast to many other earlier studies on energetics, the Y_{ATP} was calculated correctly here, i.e. it has been corrected for glycerol production which requires net input of 1 ATP/glycerol formed according to the equation: 0.5 glucose + ATP + NADH → glycerol + NAD⁺.

In some osmotolerant species, an active, sodium-driven, transport of glycerol has been shown (Lucas et al. 1990; van Zyl et al. 1990). These yeasts convert a smaller fraction of glucose to glycerol, and a major part of the glycerol is retained within the cell, unlike the situation with *S. cerevisiae*. Addition of uncouplers like 2,4-dinitrophenol (DNP) results in an efflux of glycerol (van Zyl et al.

Table 6. Pathways of oxygen radical generation and detoxification in yeasts.

Reaction	Enzyme	Localization
$O_2 + e \rightarrow O_2^-$	Various	Electron transport chain
$2 O_2^- + 2H^+ \rightarrow O_2 + H_2O_2$	Superoxide dismutases	Cytosol and mitochondria
$2 H_2O_2 \rightarrow O_2 + 2 H_2O$	Catalase	Peroxisomes
$2 \, Cyt \, c^{2+} + H_2O_2 \rightarrow 2 \, Cyt \, c^{2+} + H_2O$	Cytochrome c peroxidase	Mitochondria

1990). An active transport system for glycerol has not been found in *S. cerevisiae*.

Maiorella et al. (1984) showed that addition of a number of salts to the medium reservoir of glucose-limited chemostat culture of *S. cerevisiae* resulted in decreased yields, increased fermentation and increased glycerol production.

Oxygen and carbon dioxide pressures

High partial pressures of oxygen and carbon dioxide usually affect the physiology and yield of microorganisms (reviewed by Onken & Liefke 1989). Oxygen stress may be due to the formation of superoxide or hydroxyl radicals. Superoxide radicals are formed during respiration (see Cadenas et al. 1983 and Nohl 1986 for references). They can be converted to hydrogen peroxide via superoxide dismutases which, in different forms, occur in both the cytosol and mitochondria of yeasts (see Chang & Kosman 1989 for references). Hydrogen peroxide can subsequently be degraded by peroxidases and catalases. In yeasts, the mitochondrial enzyme cytochrome c peroxidase (CCP) probably has a major function in the detoxification of H_2O_2 (Verduyn et al. 1988a). Detoxification of H_2O_2 mediated by catalase is probably limited to hydrogen peroxide generated in peroxisomes by various oxidases. Catalase activity is not essential for the removal of H_2O_2 as was shown by the fact that catalase-negative mutants of *Hansenula polymorpha* are capable of growth on glucose/H_2O_2 mixtures in chemostat cultures (Verduyn et al. 1988a). Also catalase-negative mutants of *S. cerevisiae* can grow on oleic acid, which is metabolized via an H_2O_2-producing, peroxisomal oxidase (Veenhuis et al. 1987). An overview of some of the important oxygen-linked

radical- and H_2O_2-generating reactions and associated defence mechanisms is shown in Table 6.

It has been shown that yeast mitochondria can produce H_2O_2, probably with O_2^- as an intermediate, albeit at a low rate of 2 nmol.min^{-1}.(mg mitochondrial protein)$^{-1}$ (Boveris 1978). Because of the presence of CCP in mitochondria, this rate may have been underestimated. Decomposition of H_2O_2 by CCP affects the biomass yield as electrons are accepted at the level of cytochrome c and therefore bypass the cytochrome c oxidase complex, thus reducing the P/O-ratio (Fig. 4). In Table 7 (adapted from Verduyn et al. 1991b) the effect of H_2O_2 on the biomass yield of glucose-limited *H. polymorpha* is shown. By adding increasing amounts of H_2O_2 to the medium reservoir, progressively more oxygen is replaced as an electron acceptor by H_2O_2. The residual concentration of H_2O_2 in the fermenter was low (< 0.5 mM). Nevertheless it was calculated that the negative effect of H_2O_2 on the biomass yield was larger than expected if only the P/O-ratio was decreased, due to a partial bypass of site III proton translocation. Hence, despite its low concentration in the fermenter, H_2O_2 affected the physiology (Verduyn et al. 1991b). H_2O_2 reduction can be coupled to anaerobic ox-

Table 7. Effect of H_2O_2 on cell yield in glucose-limited chemostat cultures of *Hansenula polymorpha* ATTCC 46059 at a dilution rate of 0.10 h^{-1}.

H_2O_2 reservoir concentration (mM)	Yield (g·g glucose^{-1})
0	0.44
100	0.39
170	0.34
220	0.30

Adapted from Verduyn et al. 1991b.

idation of ethanol, demonstrating that H_2O_2 can indeed function as an electron acceptor (Verduyn et al. 1991). Whether endogenous H_2O_2 production may affect the bioenergetics is not clear. This is due to the fact that the exact H_2O_2 production rate is not known.

The susceptibility of microorganisms to high oxygen pressures and/or oxygen radicals is usually studied by aeration with pure oxygen or the addition of oxygen-radical-generating compounds, such as paraquat (Carr et al. 1986). Flushing of chemostat cultures of *S. cerevisiae* with 100% oxygen resulted in decreases in yield of 25–40% (depending on D) and an increase in total superoxide dismutase activity (Lee & Hassan 1987). The biomass yield of glucose-limited *S. cerevisiae* CBS 8066 decreased by 24% during growth at an oxygen tension of 60% at a dilution rate of $0.10h^{-1}$ (Verduyn, unpublished results). Also *Candida* species are sensitive to increased oxygen pressures. The yield of ethanol-grown *C. utilis* strongly declined at $pO_2 > 350$ mbar (Paca & Gregr 1979). In another *Candida* strain (cited in Onken & Liefke 1989), the yield declined to zero between 210 mbar (i.e. saturation at 1 bar air pressure) and 500 mbar. These data indicate that dissolved-oxygen tensions which are only slightly higher than 210 mbar can lead to considerable decreases in biomass yield.

The effects of carbon dioxide on yeast physiology and biomass yields have been reviewed by Jones & Greenfield (1982) and Onken & Liefke (1989). Part of the toxic effect may be due to dissociation of CO_2 to HCO_3^-. This compound inhibits numerous enzymes including TCA-cycle enzymes like succinate dehydrogenase (listed in Jones & Greenfield 1982) and may affect the plasma membrane permeability. Few data are available on the effect of CO_2 on yields of yeasts, most studies having centered on the reduction in μ_{max} in batch cultures. Aerobic fed-batch cultures of *S. cerevisiae* had a reduced yield at CO_2 pressures higher than 350 mbar (Chen & Gutmanis 1976), but the reduction in yield was only 10% at a pCO_2 of 500 mbar (for comparison: 1 bar air pressure equals 0.35 mbar CO_2).

High cell density cultures, as used in many commercial processes, may give rise to gas pressure related problems. Aeration with air is sometimes insufficient to ensure aerobic conditions. In this case, oxygen is mixed with, or replaces the influent air stream. Under these conditions, local high O_2 concentrations may give rise to oxygen radicals. CO_2 effects can be expected when CO_2 is not stripped, for instance in non-aerated cultures as during beer brewing. In fact, factors negatively affecting biomass yields have sometimes been used successfully to increase product conversion. For instance, the yield of citric acid production by *Aspergillus* strains and *C. tropicalis* could be improved by high oxygen pressures (these and further examples are listed in Onken & Liefke 1989).

Temperature

It is often inferred, for instance in textbooks on microbiology, that the optimum growth temperature is that temperature at which the growth rate is highest (e.g. Stanier et al. 1987). However, the biomass yield may not be maximal at the 'optimal' growth temperature. For instance, in batch cultures of *K. marxianus* a maximal growth rate was observed at 40 °C (Rouwenhorst et al. 1988). However, during chemostat cultivation at D = 0.10 h^{-1} the biomass yield at 38 °C was lower than at 30 °C, 0.42 and 0.48 g.(g glucose)$^{-1}$, respectively (Verduyn, unpublished). A similar effect of temperature on growth yields was reported for *Candida pseudotropicalis* (Gomez & Castillo 1983).

At high temperatures the maintenance coefficient of *Escherichia coli* on various carbon sources, increased 7 to 10-fold when the growth temperature was increased from 30 to 40 °C. It appeared that increased turnover of cell material contributed to, but was not the main cause of the observed increase in maintenance (Wallace & Holms 1986). Also for methanol-grown *Hansenula polymorpha* increases in the cultivation temperature resulted in an increased maintenance coefficient (cited in Heijnen & Roels 1981).

A detailed analysis of the effects of temperature on yeast growth has been given by van Uden (1984). High temperatures often give rise to petite mutants, and may affect the permeability of the plasma membrane. Thus, an effect on the yield may be expected as a result of a decrease in the efficiency of energy generation and an increase in maintenance. Furthermore, it has been suggested that a decrease in yield at high temperatures may be due to glucose dissimilation by a non-viable fraction of the culture (van Uden & Madeira-Lopes 1976). The negative effects of high temperatures on growth are enhanced in the presence of alcohols (see van Uden 1984) and acids, including octanoic acid (Sá-Correia 1986) and acetic acid (Pinto et al. 1989).

The biomass composition can vary significantly with the growth temperature. Data of Harder & Veldkamp (1967) for a bacterium, which may be typical for microorganisms in general, showed that the protein and RNA contents of biomass were minimal at an optimum growth temperature, but increased considerably at both lower and higher temperatures. Also for *C. utilis*, it has been shown that the RNA and protein contents increased when the cultivation temperature was decreased from 30 to 20 °C (Brown & Rose 1969). In a study on *S. cerevisiae*, Parada & Acevedo (1983) showed that the RNA content increased with increasing cultivation temperatures. In *Trichosporon cutaneum*, RNA and protein contents also went through a minimum at an optimum temperature (cf. Fiechter et al. 1989). As can be seen from Table 4, the biomass composition, in particular the protein content, influences the ATP requirement for biomass and hence will affect the biomass yields.

pH

Yeasts are able to maintain a more or less constant cytosolic pH over an extracellular pH range of 3.5–9 (Höfer et al. 1985; Warth 1988; Viegas & Sá-Correia 1991). The plasma membrane ATPase plays a central role in pH-homeostasis (reviewed by Serrano 1988). During growth in media buffered at pH 3.5 a 2–3 fold increase in the plasma membrane ATPase activity was observed as compared to growth at pH 5–6 (Eraso & Gancedo 1987). Mutants of *S. cerevisiae* with a reduced plasma membrane ATPase activity were more sensitive to growth at low pH as well as to the effect of added

acetic acid (which may lower the cytosolic pH) than the wild type (Vallejo & Serrano 1989).

The biomass yield in chemostat cultures is often independent of the medium pH over a certain range as shown for instance for *K. marxianus* (Rouwenhorst et al. 1988) and *S. cerevisiae* (Verduyn et al. 1990b). At lower pH values the yield decreases. The protein content of *S. cerevisiae* was independent of the pH, therefore the decline in yield was probably not due to a changed biomass composition. A decline in yield may be due to a greater passive influx of protons at lower pH. Little is known about the passive proton permeability of the plasma membrane of yeasts. A value of 0.12 $mmol.g^{-1}.h^{-1}$ was reported by Leao & van Uden (1984). If this influx is to be countered by the plasma membrane ATPase, this would require a similar ATP hydrolysis rate. This loss of ATP is low as compared to the total qATP (approximately 2% at $D = 0.10\ h^{-1}$ with a Y_{ATP} of 16). However, Verduyn et al. (1990a) have shown that an apparent correlation exists between culture pH and maintenance for anaerobic glucose-limited chemostat cultures of *S. cerevisiae* CBS 8066 and H1022. This correlation was calculated assuming that the energy requirement for biomass formation (i.e. experimental Y_{ATP}^{max}) is constant. The maintenance energy thus calculated increased strongly with decreasing culture pH and reached a maximum of approximately 12 $mmol\ ATP.g^{-1}.h^{-1}$. Plotting of the maintenance energy as a logarithm versus culture pH resulted in a straight line (Verduyn et al. 1990a). This suggests that the extent of uncoupling is a linear function of the proton concentration. Some peculiar observations have been made for growth of *Trichosporon cutaneum* as a function of culture pH (Fiechter et al. 1989). This yeast appeared to be very sensitive to external pH: the yield decreased linearly between pH 6 and 3.5 during aerobic glucose-limited growth and wash-out occurred at pH 3.0. Interestingly, the cell composition was significantly affected by decreasing medium pH, with a 25% increase in protein and RNA-contents (Fiechter et al. 1989). These large, as yet unexplained, changes in cell composition complicate an analysis of the effect of the medium pH, as the ATP requirement for biomass formation is dependent on the cell composition (see above). The combined effects of temperature and pH on the biomass yield of *S. cerevisiae* TBPM 14 in glucose-limited chemostat cultures were examined by Eroshin et al. (1976). Over a pH range of 2–7 and a temperature range of 22–35 °C, a maximum yield for this strain was found at pH 4.1 and 28.5 °C.

Effect of byproducts on the biomass yield

Alcohols

Production of ethanol is not desirable in commercial processes in which a maximal biomass yield is important, e.g. the production of baker's yeast. *S. cerevisiae* may exhibit the so-called Crabtree effect: the occurrence of alcoholic fermentation under aerobic conditions (see van Dijken & Scheffers 1986). This phenomenon occurs in sugar-limited chemostat cultures above a 'critical' dilution rate, the value of which is strongly dependent on the strain (Alexander & Jeffries 1990). In *S. cerevisiae* CBS 8066 the critical D was as high as 0.38 h^{-1} (Postma et al. 1989), whereas for many other strains ethanol formation already started between dilution rates of 0.25 and 0.30 h^{-1} (data compiled by Alexander & Jeffries 1990).

Apart from the loss of carbon, ethanol can also interfere with cellular metabolism. The inhibition of yeast growth and fermentation by ethanol has been a major subject in the literature (reviewed by D'Amore & Stewart 1987). Most observations on the effect of ethanol suggest an increased membrane permeability, leading to leakage of amino acids (Salgueiro 1988), enhancement of the uncoupling effect of acetate (Pampulha & Loureiro 1989), dissipation of the proton-motive force (pmf) across the plasma membrane (Cartwright et al. 1986) and inhibition of some enzymes, notably the plasma membrane ATPase (Cartwright et al. 1987). The proposed inhibition of the plasma membrane ATPase (Cartwright et al. 1987) has been questioned by Petrov & Okarokov (1990). Experiments with plasma membrane vesicles of *S. carlbergensis* indicated that the pmf across the plasma membrane decreased in the presence of ethanol but that this was not due to inhibition of the AT-Pase. It was suggested that the permeability of the vesicles to protons was increased in the presence of ethanol. Rosa and Sá-Correia (1991) found inhibition of the plasma membrane ATPase at ethanol concentrations higher than 3 vol%. However, higher concentrations of ethanol (6–8% v/v) raised the level of the plasma membrane ATPase activity up to 3-fold, thereby compensating for the inhibition of the enzyme by ethanol. Furthermore, there seemed to be a clear difference between the plasma membrane permeability (tested as uptake rate of acetic acid) of batch-grown and chemostat-grown *S. cerevisiae*. In the latter case, permeability appeared not to be affected by exposure to ethanol up to 10% v/v (Jones & Greenfield 1987). Direct inhibition or inactivation of cytosolic enzymes only seems to occur at high ethanol concentrations, in the order of 15% v/v (Pascual et al. 1988). This concentration is approximately the limit which can be tolerated by most *S. cerevisiae* strains. It has recently been shown that the apparent increase in passive proton influx (as suggested by a drop in the intracellular pH) after an ethanol pulse to *S. cerevisiae* is due (at least in part) to the formation of acetic acid, which results in a transient decrease in the intracellular pH (Loureiro-Dias & Santos 1990).

Unfortunately, few quantitative data on the effect of ethanol on biomass yield are available. In batch cultures of *S. cerevisiae* grown on glucose supplemented with ethanol, the biomass yield began to decrease at ethanol concentrations higher than 4% (Rosa & Sá-Correia 1991). Further effects of ethanol, notably on transport systems (reviewed by van Uden 1989) will not be discussed here.

Some reported toxic effects of ethanol, notably the inhibition of fermentation, have in fact been shown to be due to exhaustion of medium components other than the carbon source. For instance it has been shown that addition of biotin (Winter et al. 1989) or magnesium (Dasari et al. 1990) can relieve the effect of apparent ethanol inhibition. Furthermore, the observation that added ethanol is less toxic than endogenously produced ethanol (Novak et al. 1981) has led to suggestions that many of the effects orginally ascribed to ethanol are due to other metabolites, such as longer alcohols (Okolo et al. 1987) or weak acids (Viegas et al. 1985).

Weak acids

Production of weak acids is a normal event during growth of yeasts, even under carbon-limited conditions. Acetic acid is of the best-known acidic products formed during cultivation of yeast. Formation of acetic acid has been shown in aerobic and anaerobic glucose-limited cultures of S. cerevisiae (Postma et al. 1989; Verduyn et al. 1990b). Acetic acid is formed from pyruvate via pyruvate decarboxylase and acetaldehyde dehydrogenases (Fig. 6). In principle, acetate can be converted into acetyl-CoA by acetyl-CoA synthetase. However, this latter enzyme is subject to glucose repression. This results in an increasing production rate of acetate as a function of D in glucose-limited chemostat cultures of S. cerevisae CBS 8066 (Postma et al. 1989). A relation has been shown between the amount of acetate produced in two S. cerevisiae strains during anaerobic chemostat cultivation and the amount of acetaldehyde dehydrogenase(s) and acetyl-CoA synthetase (Verduyn et al. 1990b). A common observation in fermenting cultures is that the presence of acetic acid coincides with an increase in fermentation rate (Pons et al. 1986; McDonald et al. 1987). In aerobic glucose-limited chemostat cultures of S. cerevisiae, the appearance of acetic acid coincided with an increase in the respiration rate (Postma et al. 1989a). This also suggests that weak acids may uncouple energy generation from biomass formation.

The mechanism of uncoupling by weak acids is schematically shown in Fig. 7. Non-metabolizable weak acids enter the cell by passive diffusion in the undissociated form as indicated by the fact that the uptake rate is pH-dependent (Pampulha & Loureiro 1989; Viegas et al. 1989). Once inside the cell, the acid will dissociate due to the relatively high pH of the cytosol. This means that, in effect, weak acids act as proton conductors. If this process continues unabated, collapse of ΔpH across the plasma membrane will follow and the intracellular pH will become similar to the external pH. Therefore, protons have to be removed via the plasma membrane ATPase, which requires hydrolysis of ATP. In order to provide this ATP, increased respiration and/or fermentation (depending on the growth conditions) is necessary. The fate of the anion is not fully clear. It is generally assumed that the membrane is impermeable to anions. Therefore, the action of weak acids is different from that of 'real' uncouplers like 2,4-dinitrophenol (DNP), which are permeable both in the undissociated and dissociated forms. This permits rapid cycling of these compounds across membranes, leading to dissipation of both the chemical (ΔpH) and electrical ($\Delta\psi$) component of the proton-motive force. However, it has been suggested that efflux of some acids, including benzoate and sorbate, may occur via an as yet unresolved active transport system (Warth 1988, 1989).

According to the mechanism described above, uptake of, and uncoupling by, weak acids should be minimal at high external pH values, at which weak acids are nearly completely dissociated. This was confirmed in an experiment (Verduyn, unpublished) in which 5 mM benzoate (a non-metabolizable weak acid) was added to the medium reservoir of a glucose-limited aerobic chemostat culture of Hansenula polymorpha. The biomass yield in the presence of benzoate was dependent on the culture pH. At pH 7.1 the yield of a control culture (no benzoate) and the culture with benzoate were virtually similar. The lowest pH at which a steady state could be attained in the presence of benzoate was 5.4. At this pH, benzoate caused a 50% decrease of the biomass yield. Below pH 5.4, the culture with benzoate washed out. The yield of the culture without benzoate was not affected in this pH range (Fig. 8).

The uptake rate of weak acids depends on the chain length of the acid, as shown by Warth (1989). Apart from an effect on the proton-motive force, accumulation of the acids within the cell might also inhibit various cellular processes. Pampulha & Loureiro-Dias (1990) have studied the effect of acetic acid on the activity of glycolytic enzymes in S. cerevisiae. The concentrations of acetic acid required to inhibit glycolytic enzymes by 50% differed from 0.12M to more than 1 M, depending on the enzyme assayed.

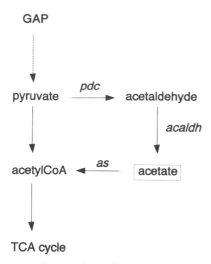

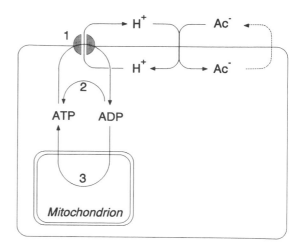

Fig. 6. Pathway of acetate formation and consumption in yeasts. Enzymes: *pdc*: pyruvate decarboxylase; *acaldh*: acetaldehyde dehydrogenases; *as*: acetylCoA synthase.

Fig. 7. Schematic representation of uncoupling of the plasma membrane by weak acids. (1): plasma membrane ATPase. (2): ATP formation by substrate phosphorylation and (3): ATP formation via respiration.

We have attempted to quantify the effect of weak acids on the bioenergetics of two strains of *S. cerevisiae* by adding various weak acids to the medium reservoir of anaerobic glucose-limited cultures. Anaerobic cultures offer some advantages over aerobic cultures in this respect since

- Y_{ATP} can be calculated simply from biomass and metabolite formation,
- uncoupling of the mitochondrial membrane (thus influencing the P/O-ratio) will not play a role under these conditions, and
- acids which can be metabolized together with glucose (like acetate) under aerobic conditions, cannot be metabolized under anaerobic conditions.

It has been established (Verduyn et al. 1990b) that anaerobic growth of *S. cerevisiae* in glucose-limited chemostat cultures in the presence of acetate, propionate, butyrate, or benzoate resulted in a decrease in biomass yield, accompanied by an increased fermentation rate. Assuming that the ATP requirement for biomass formation (which had a constant composition in these experiments) does not change, it could be calculated that the effect of a weak acid can be described as a growth rate-independent maintenance effect, i.e. follows the equation $(1/Y_{ATP})\text{observed} = m_e/\mu + (1/Y_{ATP}^{max})$. This is visualized in Fig. 9 in which the effect of a

fixed concentration of 20 mM propionate on the biomass yield and observed Y_{ATP} of an anaerobic glucose-limited chemostat culture of *S. cerevisiae* H1022 is shown as a function of the dilution rate (Verduyn, unpublished). In cultures without propionate, a small decrease in biomass yield and Y_{ATP} was observed with increasing D. This is probably due to changes in the protein content (Fig. 3) which results in an increased ATP requirement for biomass formation (Table 2). In the cultures in which propionate had been added to the medium reservoir, the biomass yield and Y_{ATP} increased with D, but were much lower than in the culture without propionate. With the aid of the formulae shown above, the m_e can then be calculated, which results in an approximately constant maintenance energy value (broken line in Fig. 9). The μ_{max} in the presence of 20 mM propionate was only 0.21 h⁻¹ whereas the μ_{max} in the absence of propionate was 0.30 h⁻¹. It is possible that the fermentation rate, which determines the maximal ATP production rate, becomes limiting in the presence of propionate at D = 0.21 h⁻¹ (for a further discussion see Verduyn et al. 1990a).

The maintenance energy requirement is linearly related to the concentration of the acid (Verduyn et al. 1990a). The maximal maintenance energy calculated (i.e. just before the culture washed out) in the

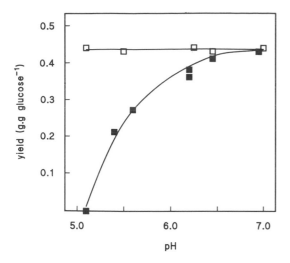

Fig. 8. Biomass yields of glucose-limited chemostat cultures of *Hansenula polymorpha* ATCC 46059 without benzoate (□) and with 5 mM benzoate (■) as a function of the culture pH. The dilution rate was 0.10 h^{-1} (Verduyn, unpublished).

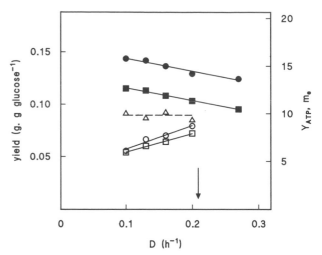

Fig. 9. Biomass yields (g·g glucose^{-1}) (□) and Y_{ATP} (g cells·mol ATP formed^{-1}) (○) in anaerobic glucose-limited chemostat cultures of *S. cerevisiae* H1022 in the absence (closed symbols) and presence (open symbols) of 20 mM propionate, and calculated maintenance energy (m_e, mmol ATP·g^{-1}·h^{-1}) due to the presence of propionate (broken line) as a function of the dilution rate. The arrow indicates the D at which the culture with propionate washed out (Verduyn, unpublished).

presence of weak acids was 17 ± 2 mmol ATPeq.g^{-1}.h^{-1}. This is a considerable amount as can be seen from the following calculation: if the Y_{ATP} for biomass formation is assumed to be 16, qATP (with qATP = μ/Y_{ATP}) is 6.25 mmol ATP.g^{-1}.h^{-1}. The total qATP (qATP biomass + qATP acid) is 23 mmol ATP.g^{-1}.h^{-1}, hence almost 75% of the total energy generated is lost due to acid uncoupling.

Uncoupling by weak acids can also be studied in aerobic cultures, provided that the acid is not metabolized. Benzoate, a well-known food preservative (Jay 1978; Lueck 1980), was used to evaluate the effect of acid uncoupling in aerobic glucose-limited cultures of *S. cerevisiae* CBS 8066. Addition of benzoate to the medium reservoir resulted in a decrease of the biomass yield (g biomass per g glucose) and an increase in specific oxygen uptake rate (qO$_2$). Alcoholic fermentation did not occur at residual benzoate concentrations up to 10 mM (Fig. 10). At this benzoate concentration, the biomass yield and qO$_2$ were 0.15 g.g glucose^{-1} and 19.5 mmol O$_2$.g^{-1}.h^{-1}, respectively. A further increase of the benzoate concentration resulted in ethanol formation and a drastic reduction of the yield. Furthermore, the qO$_2$ declined from 19.5 to 12–13 mmol O$_2$.g^{-1}.h^{-1} (Fig. 10). Apparently, repression of respiration occurred. The maximal qO$_2$ of 19.5

mmol g^{-1}.h^{-1} as measured in these experiments is remarkably high and significantly exceeds the qO$_2$ of this strain at high dilution rates (11–12 mmol.g^{-1}.h^{-1} between 0.38 and 0.47 h^{-1}, Postma et al. 1989a). The amount of ATP lost as a result of acid uncoupling can be calculated from the difference in qO$_2$ of cultures grown with and without benzoate. With an assumed P/O-ratio of 0.95 (Verduyn et al. 1991a), it can then be calculated (Verduyn 1992) that the extent of uncoupling by benzoate (and by a number of other acids, not shown) is similar under anaerobic and aerobic conditions.

From these experiments it can be concluded that weak acids may have a considerable affect on the cell yield, even when present at relatively low concentrations. Since the concentration of weak acids depends on the biomass concentration, acid uncoupling could be especially important in high-density cultures. Medium-chain fatty acids, which are intermediates in lipid synthesis (Taylor & Kirsop 1977), cannot be detected easily and require extensive extraction and concentration procedures (e.g. Viegas et al. 1989). Nevertheless, such fatty acids could play an important role in acid uncoupling

344

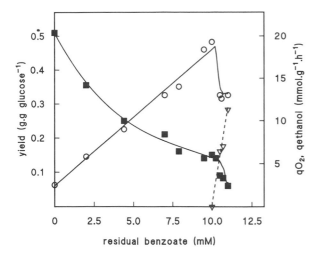

Fig. 10. Biomass yields (■), specific oxygen uptake rate (○) and specific ethanol production rate (▽) in glucose-limited cultures of *S. cerevisiae* CBS 8066 as a function of the residual benzoate concentration in the fermenter. In these experiments, benzoate was added to the medium reservoir. The dilution rate was 0.10 h^{-1} (Verduyn, unpublished).

under certain conditions. It has been shown that octanoic and decanoic acids are extremely potent uncouplers, and already affect biomass yields when present in micromolar concentrations (Lafon-Lafourcade et al. 1984; Viegas et al. 1989). They presumably only function as uncouplers if in soluble state, i.e. if sufficient ethanol is present. Apart from acting as proton conductors, medium-chain length fatty acids may also alter the permeability of the membrane by being inserted into it (Ingram & Buttke 1984; see Viegas et al. 1989 for further references).

As the formation of toxic byproducts, such as acetic acid, is usually proportional to the biomass concentration, problems arising from accumulation of the compounds are most likely to occur in high-density cultures. Furthermore, acid uncoupling is not necessarily due to endogenously-produced compounds. Complex growth substrates may contain acids. For instance molasses, used as a substrate in the commercial production of baker's yeast, may contain acetic, lactic, and butyric acids (Essia Ngang et al. 1989; Beudeker et al. 1990). Whether these acids indeed act as uncouplers probably depends to a major extent (apart from culture pH) on the imposed growth rate. At higher dilution

rates (high glucose fluxes) enzymes involved in the oxidation of for instance acetic acid may be repressed (Postma et al. 1989a) and hence acetic acid will not be co-utilized with the sugar(s) anymore. The uncoupling effects of organic acids, often in combination with a low pH, may be useful in order to improve product yields in processes in which a high biomass yield is not wanted, for instance in citric acid production (cf. Linton & Rye 1989).

A relation between the energy content of the substrate and the biomass yield?

It has long been attempted to predict the yield of microorganisms on different carbon sources by relating the energy content of the carbon source (usually expressed as the degree of reduction) to a number of parameters. These parameters include the yield on available electrons (Payne 1970) or the heat production per amount of biomass formed (Birou et al. 1987; von Stockar & Marison 1989). These various approaches are discussed in the contribution of Heijnen (this issue). Linton & Stephenson (1978) have shown that a correlation exists between the heat of combustion of the substrate and the maximal biomass yield expressed as g biomass.(g substrate carbon)$^{-1}$. A more refined model was presented by Gommers et al. (1988). In this model the percentage assimilation of the carbon source was related to its heat of combustion. This avoids the problem of the loss of CO_2 in anabolic reactions as outlined previously. In both studies, the highest yields reported in the literature for both bacteria and yeasts during carbon- or energy-limited growth in minimal media were employed. The general trend of these studies was a figure similar in shape to Fig. 2, i.e. an ascending part (up to a degree of reduction of approximately 4.5–5) and a horizontal part at higher degrees of reduction in which the assimilation was close to 100% (i.e. possibly carbon-limited growth). However, there are a number of exceptions to the apparent rule that the yield on a carbon source with a high degree of reduction will automatically be higher than on one with a lower degree of reduction. For instance, methane has a degree of reduction of 8. It is metab-

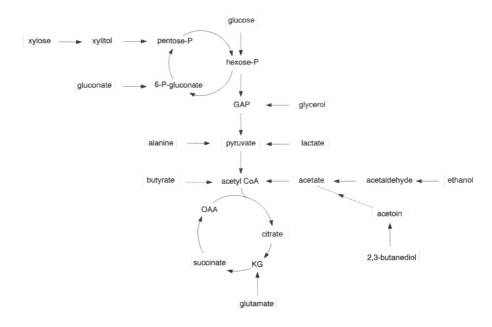

Fig. 11. Schematic representation of metabolic pathways during growth of *C. utilis* on various carbon sources (indicated in boxes). Broken lines indicate unidentified routes.

olized by bacteria via methanol (degree of reduction 6) by a mixed-function oxygenase according to: $CH_4 + O_2 + NADH \rightarrow CH_3OH + NAD^+ + H_2O$. It is clear from this equation that the yield on methane will be lower than that on methanol due to the required input of reducing equivalents. This has been confirmed experimentally (for data see van Dijken & Harder 1975).

In order to establish whether general data on the maximum growth yield of a number of microorganisms have practical use for the estimation of the biomass yield of a given organism, we decided to cultivate one yeast on a wide variety of carbon sources. For this purpose we chose *C. utilis*, as it has a broad substrate specificity. The organism was cultivated in aerobic carbon- and energy-limited chemostat cultures in minimal media as described by Verduyn et al. (1991a). The general metabolic pathways employed during metabolism of the various carbon sources are schematically depicted in Fig. 11. For all carbon sources, the biomass yield (Table 8) as well as the protein content were determined. Biomass and CO_2 were the only products. The protein and carbon contents were almost constant at values of $52 \pm 3\%$. and $47 \pm 1.5\%$, respec-

tively. The average cell formulae (with a molecular weight of 100, including ash) can be represented by $C_{3.9}H_{6.5}N_{0.72}O_{2.03}$. The yield data can also be expressed as g biomass per g substrate carbon (according to Linton & Stephenson 1978). When plotted against the degree of reduction, a considerable scatter is observed (Fig. 12A). It is also possible to plot the yield data as the percentage assimilation of the carbon source (Fig. 12B). This, however, presents an unusual problem as the exact metabolic pathways of some of the carbon sources listed in Table 8 are not known (indicated by broken lines in Fig. 11). For calculation of the percentage assimilation the procedures of Gommers et al. (1988) have been followed. These assume that, for instance, xylose and gluconic acid are metabolized via the most 'simple' route, which is conversion to hexose-phosphates and subsequent metabolism via the normal glycolytic pathways. Thus, xylose metabolism can be represented in terms of glucose metabolism according to 1.2 xylose + 1.2 NADH $\rightarrow$ 1 glucose (= 2 glyceraldehyde-3-phosphate) + 1.2 NADPH. The loss of carbon in anabolism is thus similar to that on glucose. However, two different pathways can be envisaged: 1) via transketolase/

346

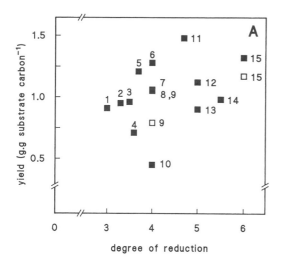
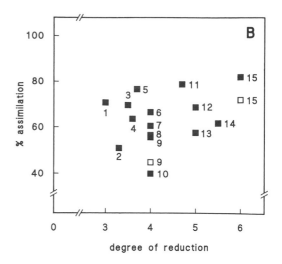

Fig. 12A. Biomass yields of *C. utilis* CBS 621 (■), and *S. cerevisiae* CBS 8066 (□) (expressed as g biomass.g substrate carbon^{-1}) as a function of the degree of reduction of various carbon sources during carbon- and energy-limited chemostat cultivation at a dilution rate of 0.10 h^{-1}. The nitrogen source was ammonium. 1: citric acid; 2: pyruvic acid; 3: succinic acid; 4: glutamic acid; 5: gluconic acid; 6: glucose; 7: xylose; 8: lactic acid; 9: acetic acid; 10: alanine; 11: glycerol; 12: butyric acid; 13: acetoin; 14: 2,3-butanediol, and 15: ethanol. The nitrogen source was ammonium (not added during growth on alanine and glutamic acid).

Fig. 12B. Percentage assimilation of the carbon source as a function of its degree of reduction Legend as in *(A)*.

transaldolase, which results in formation of GAP, but in which carbon is lost via decarboxylation in the HMP pathway and 2) via phosphoketolase, which results in formation of GAP and acetylphosphate. A low phosphoketolase activity has been found in some yeasts (Whitworth & Ratledge 1977). The relative contribution of the three different routes to xylose metabolism is not known. Butyrate and 2,3-butanediol metabolism have not been studied extensively in yeasts. It has been assumed that butyrate is assimilated via acetyl CoA, without a net loss of CO_2. 2,3-Butanediol is first converted into acetoin by an NAD$^+$-dependent butanediol dehydrogenase (Verduyn et al. 1988b). The next steps are not known, but since cells grown on butanediol have a high isocitrate lyase activity (Verduyn et al. 1988c), it is likely that acetate is an intermediate. In view of these uncertainties, the results of the calculation on the percentage assimilation of some of the carbon sources as shown in Fig. 12B should be viewed with care. In terms of energetics it is most easy to compare several substrates which are interconverted via a simple linear reaction sequence. For instance, lactate is converted into pyruvate with production of FADH. Thus

(if the uptake system of these acids is similar), the percentage assimilation of lactate should be larger than for pyruvate. This was confirmed in practice (Fig. 12B). D-alanine is also converted into pyruvate by a peroxisomal oxidase according to (Zwart et al. 1983): D-alanine + O_2 → pyruvate + H_2O_2 + NH_3. Although D-alanine has a higher degree of reduction than pyruvate, its oxidation to pyruvate does not yield reducing equivalents, but rather hydrogen peroxide (which is subsequently degraded to oxygen and water by catalase, Table 7). The percentage assimilation of D-alanine would therefore be expected to be approximately similar to that of pyruvate. It should be realized that alanine is both a nitrogen and a carbon source and that excess ammonia is produced. No addition and uptake of ammonium is required (which should theoretically increase the Y_{ATP}^{max}, cf. Table 2). However, the percentage assimilation of alanine is significantly lower than that of pyruvate (40 and 51%, respectively, Fig. 12B). This suggests that energy is required in the uptake of alanine and/or its conversion into pyruvate. Indeed, amino acids are taken up by yeasts via active transport mechanisms (Cooper 1982). Therfore, the fact that alanine exhibits a much lower percentage assimilation than pyruvate is most likely due to the transport costs of alanine.

Also noteworthy are the low yields on 2,3-butanediol and acetoin, two compounds with a high degree of reduction. A number of oxidative cleavage reactions for the breakdown of acetoin, all to C_2-units, have been presented by Fründ et al. (1989). From enzymic studies it was concluded that, after conversion of butanediol to acetoin (net formation of 1 NADH), breakdown of acetoin occurs via a CoASH-dependent 'acetoin-dichlorophenol oxidoreductase' according to the reaction: acetoin + CoASH → acetate + acetylCoA + 'H_2' (reducing equivalents). It is expected that the percentage assimilation would then be considerably higher than for acetic acid as 1) for each acetate (or acetyl CoA) formed an additional reducing equiv-

Table 8. Cell yield (g biomass·g carbon source^{-1}) of *Candida utilis* CBS 621 and *Saccharomyces cerevisiae* CBS 8066 in aerobic carbon-limited chemostat cultures (minimal media) on various carbon sources as a function of the degree of reduction.

C-source	Degree of reduction	Yield $(g \cdot g^{-1})$
C. utilis:		
Citric acid	3.0	0.34
Pyruvic acid	3.3	0.39
Succinic acid	3.5	0.39
Glutamic acid	3.6	0.29
Gluconic acid	3.7	0.44
Acetic acid	4.0	0.39
D-alanine*	4.0	0.31
Glucose	4.0	0.51
Lactic acid	4.0	0.42
Xylose	4.0	0.42
Glycerol	4.7	0.58
Acetoin	5.0	0.49
Butyric acid	5.0	0.61
2,3-Butanediol	5.5	0.52
Ethanol	6.0	0.69
S. cerevisiae:		
Acetic acid	4.0	0.29
Ethanol	6.0	0.61

* Zwart et al. (1983).
The dilution rate was 0.10 h^{-1}. The nitrogen source was ammonium (not added during growth on alanine and glutamic acid). All results from Verduyn et al. (1991a; unpublished) unless indicated otherwise.

alent is produced and 2) the high activation cost of acetate (acetate + CoA + 2 ATP → acetylCoA + AMP + PP$_i$) are partially avoided through direct formation of acetylCoA. It must be concluded that either somewhere in the metabolic routes between acetoin and acetate, ATP is required, or active transport of the carbon source is involved, or both. However, it seems unlikely that a compound like 2,3-butanediol is taken up by active transport. Glycerol uptake in *C. utilis*, for instance, is by passive or facilitated diffusion (Gancedo et al. 1968). It is also possible that reoxidation of the reducing equivalents formed during the oxidation of acetoin is not coupled to proton translocation. Thus far, the energetics of 2,3-butanediol and acetoin metabolism remain unclear.

From these various examples it can be concluded that, in general, carbon sources with a high degree of reduction have a higher percentage assimilation than those with a low degree of reduction, but that exceptions are common. It is thus not surprising that no clearcut pattern is observed between the percentage of the carbon source that is assimilated and the degree of reduction (Fig. 12B). Furthermore, if results for *S. cerevisae* CBS 8066 are compared to those for *C. utilis*, the yield is similar to that of *C. utilis* on glucose, but much lower on ethanol and acetic acid (Table 4, Fig. 12A). This seems peculiar as it is well known that *C. utilis* has three proton-translocating sites as compared to only two in *S. cerevisiae* (see section on P/O-ratio). The observation that the yield is similar during growth on glucose is due to the fact that the ATP requirement for biosynthesis in *C. utilis* is larger than for *S. cerevisiae* since 1) *C. utilis* has an active transport system for glucose and 2) the protein content of *C. utilis* is 25% larger at this dilution rate (Table 3). Thus, during growth on glucose the higher P/O-ratio of *C. utilis* is masked by a low Y_{ATP}. The larger P/O-ratio of *C. utilis* is clearly reflected in the higher yields during growth on ethanol or acetic acid (Table 8). With these carbon sources the theoretical ATP requirement for biomass formation is similar for both yeasts (Verduyn et al. 1991). Thus Fig. 12A and 12B have a limited usefulness in predicting biomass yields. It appears that the biochemistry of the pathways leading to central metabolic precursors (like pyruvate, phosphoglycer-ate, oxaloacetic acid, etc.) rather than the energy content of the carbon- and energy source deter-mines the biomass yield to a major extent.

Acknowledgements

The experimental work of the author that is dis-cussed in this contribution was supported by The Netherlands Organization for Scientific Research (N.W.O.). I am indebted to R.A. Weusthuis for providing unpublished results and assistance with the preparation of the figures, and to J.P. van Dijken, J.T. Pronk, and W.A. Scheffers for valua-ble discussions and a critical reading of the manu-script.

References

Alexander B, Leach S & Ingledew JW (1987) The relationship between chemiosmotic parameters and sensitivity to anions and organic acids in the acidophile *Thiobacillus ferrooxidans*. J. Gen. Microbiol. 133: 1171–1179

Alexander MA & Jeffries TW (1990) Respiratory efficiency and metabolite partitioning as regulatory phenomena in yeasts. Enzyme Microb. Technol. 12: 2–19

Andreasen AA & Stier TJB (1953) Anaerobic nutrition of *Saccharomyces cerevisiae*. I. Ergosterol requirment for growth in a defined medium. J. Cell. Comp. Physiol. 41: 23–26

— (1954) Anaerobic nutrition of *Saccharomyces cerevisiae*. II. Unsaturated fatty acid requirement for growth in a defined medium. J. Cell. Comp. Physiol. 43: 271–281

Asano A, Imai K & Sato R (1967) Oxidative phosphorylation in *Micrococcus denitrificans*. II. The properties of pyridine nu-cleotide transhydrogenase. Biochim. Biophys. Acta 143: 477–486

Atkinson B & Mavituna F (1983) Biochemical Engineering and Biotechnology Handbook (pp 120–125). The Nature Press, New York

Babel W, Müller RH & Markuske KD (1983) Improvement of growth yield on glucose to the maximum by using an addition-al energy source. Arch. Mcrobiol. 136: 203–208

Baranowski K & Radler F (1984) The glucose dependent trans-port of L-malate in *Zygosacchromyces bailii*. A. v. Leeuwen-hoek 50, 329–340

Bauchop T & Elsden SR (1960) The growth of micro-organisms in relation to their energy supply. J. Gen. Microbiol. 23: 457–469

Beudeker RF, van Dam HW, van der Plaat JB & Vellenga K (1990) Developments in baker's yeast production. In: Ve-

rachtert H & de Mot R (Eds) Yeast biotechnology and Bioca-talysis (pp 103–145). Marcel Dekker Inc., New York & Basel

Birou B, Marison IW & Von Stockar U (1987) Calorimetric investigation of aerobic fermentations. Biotechnol. Bioeng. 30: 650–660

Boveris A (1978). Production of superoxide anion and hydrogen peroxide in yeast mitochondria. In: Bacila M, Horecker BL & Stoppani AOM (Eds) Biochemistry and Genetics of Yeasts: Pure and Applied Aspects (pp 65–80). Academic Press, New York, San Francisco and London

Brown CM & Rose AH (1969) Effects of temperature on com-position and cell volume of *Candida utilis*. J. Bacteriol. 97: 262–272

Bruinenberg PM, van Dijken JP & Scheffers WA (1983a) An enzymic analysis of NADPH production and consumption in *Candida utilis*. J. Gen. Microbiol. 129: 965–971

— (1983b) A theoretical analysis of NADPH production and consumption in yeasts. J. Gen. Microbiol. 129: 953–964

Bruinenberg PM, Jonker R, van Dijken JP & Scheffers WA (1985) Utilization of formate as an additional energy source by glucose-limited chemostat cultures of *Candida utilis* CBS 621 and *Saccharomyces cerevisiae* CBS 8066. Evidence for the absence of transhydrogenase activity in yeasts. Arch. Micro-biol. 142: 302–306

Bruinenberg PM, Waslander GW, van Dijken JP & Scheffers WA (1986) A comparative radiorespirometric study of glu-cose metabolism in yeasts. Yeast 2: 117–121

Cadenas E, Brigelius R, Akerboom Th & Sies H (1983) Oxygen radicals and hydroperoxides in mammalian organs: aspects of redox cycling and hydrogen peroxide metabolism. In: Sund H & Ullrich V (Eds) Biological Oxidations (pp 288–310). Springer-Verlag, Berlin, Heidelberg

Carr RJG, Bilton RF & Atkinson T (1986) Toxicity of paraquat to microorganisms. Appl. Env. Microbiol. 52: 1112–1116

Cartwright CP, Juroszek JR, Beavan MJ, Ruby FMS, Morais SMF & Rose AH (1986) Ethanol dissipates the proton-motive force across the plasma membrane of *Saccharomyces cere-visiae*. J. Gen. Microbiol. 132: 369–377

Cartwright CP, Veazey FJ & Roase AH (1987) Effect of ethanol on activity of the plasma membrane ATPase in, and accumu-lation of glycine by, *Saccharomyces cerevisiae*. Appl. Envi-ron. Microbiol. 53: 509–513

Cason DT, Spencer Martins I & van Uden N (1986) Transport of fructose by a proton symport in a brewing yeast *Saccharomyc-es cerevisiae*. FEMS Microbiol. Lett. 36: 307–310

Cassio F, Leao C & van Uden N (1986) Transport of lactate and other short-chain monocarboxylates in the yeast *Saccharo-myces cerevisiae*. Appl. Env. Microbiol. 53: 509–513

Chang EC & Kosman DJ (1989) Intracellular Mn(II)-associated superoxide scavenging activity protects Cu,Zn superoxide dismutase deficient *Saccharomyces cerevisiae* against dioxy-gen stress. J. Biol. Chem. 264: 12172–12178

Chen SL & Gutmanis F (1976) Carbon dioxide inhibition of yeast growth in biomass production. Biotechnol. Bioeng. 18: 1455–1462

Cooper TG (1982) Transport in *Saccharomyces cerevisiae*. In:

Strathern JN, Jones EW & Broach JR (Eds) The Molecular Biology of the Yeast *Saccharomyces*. Metabolism and Gene Expression (pp 400–461). Cold Spring Harbor Lab., New York

D'Amore T & Stewart GG (1987) Ethanol tolerance of yeast. Enzyme Microb. Technol. 9: 322–330

Dasari G, Worth MA, Connor MA & Pamment NB (1990) Reasons for the apparent difference in the effects of produced and added ethanol on culture viability during rapid fermentations by *Saccharomyces cerevisiae*. Biotechnol. Bioeng. 35: 109–122

Dekkers JGJ, de Kok HE & Roels JA (1981) Energetics of *Sacchromyces cerevisiae* CBS 426: comparison of anaerobic and aerobic glucose limitation. Biotechnol. Bioeng. 23: 1023–1035

De Vries S & Marres CAM (1987) The mitochondrial respiratory chain of yeast. Structure and biosynthesis and the role in cellular metabolism. Biochim. Biophys. Acta 895: 205–239

Eddy AA & Hopkins PG (1985) The putative electrogenic nitrate-protonsymport of the yeast *Candida utilis*. Comparison with the systems absorbing glucose or lactate. Biochem. J. 231: 291–297

Egli T. (1980) Wachstum von Methanol assimilierenden Hefen. Diss. ETH Nr.6538, Zürich, Switzerland

Egli T & Quayle JR (1986) Influence of the carbon-nitrogen ratio of the growth mdium on the cellular composition and the ability of the methylotrophic yeast *Hansenula polymorpha* to utilize mixed carbon sources. J. Gen. Microbiol. 132: 1779–1788

Eraso P & Gancedo C (1987) Activation of yeast plasma membrane ATPase by acid pH during growth. FEBS 224: 187–192

Eroshin VK, Utkin IS, Ladynichev SA, Samoylov VV, Kuvshinnikov VD & Skryabin GK (1976) Influence of pH and temperature on the substrate yield coefficient of yeast growth in a chemostat. Biotechnol. Bioeng. 18: 289–295

Essia Ngang JJ, Letourneau F & Villa P (1989) Alcoholic fermentation of beet molasses: effects of lactic acid on yeast fermentation parameters. Appl. Microbiol. Biotechnol. 31: 125–128

Favre E, Pugeaud P, Raboud JP & Peringer P (1989) Automated HPLC monitoring of broth components on bioreactors. J. Auto. Chem. 11: 280–283

Fiechter A, Käppeli O & Meussdoerffer F (1989) Batch and continuous cultures. In: Rose AH & Harrison JS (Eds) The Yeasts, Vol 2 (pp 99–129). Academic Press, London

Fründ C, Priefert H, Steinbüchel A & Schlegel H (1989) Biochemical and genetical analysis of acetoin catabolism in *Alcaligenes eutrophus*. J. Bacteriol. 171: 6539–6548

Gancedo C, Gancedo JM & Sols A (1968) Glycerol metabolism in yeasts. Pathways of utilization and production. Eur. J. Biochem. 5: 165–172

Goffeau A & Crosby B (1978) A new type of cyanide-insensitive, azide-sensitive respiration in the yeasts *Schizosaccharomyces pombe* and *Saccharomyces cerevisiae*. In: Bacile M, Horecker BL & Stoppani AOM (Eds) Biochemistry and Genetics of Yeasts (pp 81–96). Academic Press, New York

Goldberg I (1985) Single cell protein (p 79). Springer Verlag, Berlin Heidelberg

Gomez A & Castillo FJ (1983) Production of biomass and β-D-galactosidase by *Candida pseudotropicalis* grown in continuous culture on whey. Biotechnol. Bioeng. 25: 1341–1357

Gommers PJF, van Schie BJ, van Dijken JP & Kuenen JG (1988) Biochemical limits to microbial growth yields: An analysis of mixed substrate utilization. Biotechnol. Bioeng. 32: 86–94

Gunsalus IC & Shuster CW (1961) Energy-yielding metabolism in bacteria. In: Gunsalus IC and Stanier RY (Eds) The Bacteria, Vol 2 (pp 1–58). Academic Press, New York and London

Harder W & Veldkamp H (1967) A continous culture of an obligately psychrophilic *Pseudomonas* species. Archiv für Mikrobiologie 59: 123–130

Harder W & van Dijken JP (1976) Theoretical culculations on the realtion between energy production and growth of methane-utilizing bacteria. In: Microbial Production and Utilization of Gases (pp 403–418). E. Goltze Verlag, Göttingen

Heijnen JJ & Roels JA (1981) A macroscopic model describing yield and maintenance relationships in aerobic fermentation processes. Biotechnol. Bioeng. 23: 739–763

Higgins CF, Cairney J, Stirling DA, Sutherland L & Booth IR (1987) Osmotic regulation of gene expression: ionic strength as an intracellular signal. Trends Biochem. Sci. 12: 339–344

Höfer M & Misra PC (1978) Evidence for a proton/sugar symport in the yeast *Rhodotorula gracilis* (*glutinis*). Biochem. J. 172: 15–22

Höfer M, Nicolay K & Robillard G (1985) The electrochemical H^+ gradient in the yeast *Rhodotorula glutinis*. J. Bioenerg. Biomembr. 17: 175–182

Ingram LO & Buttke T (1984) Effects of alcohol on microorganisms. Adv. Microbial Physiol. 25: 254–290

Jay MJ (1978) In: Modern Food Microbiology (pp 163–165) D. van Nostrand Comp., New York, Cincinnati, Toronto, London, Melbourne

Jones CW, Brice JM & Edwards C (1977) The effect of respiratory chain composition on the growth efficiencies of aerobic bacteria. Arch. Mikrobiol. 115: 85–93

Jones RP & Greenfield PF (1982) Effect of carbon dioxide on yeast growth and fermentation. Enz. Microb. Technol. 4: 210–223

— (1984) A review of yeast ionic nutrition. Part I: growth and fermentation requirements. Process Biochem. 19: 48–60

— (1987) Ethanol and the fluidity of the yeast plasma membrane. Yeast 3: 223–232

Jones RP & Gadd GM (1990) Ionic nutrition of yeast – physiological mechanisms involved and implications for biotechnology. Enzyme Microb. Technol. 12: 402–418

Jovall P-A, Tunblad-Johansson I & Adler L (1990) ^{13}C NMR analysis of production and accumulation of osmoregulatory metabolites in the salt-tolerant yeast *Debaromyces hansenii*. Arch. Microbiol. 154: 209–214

Kotyk A & Alonso A (1985) Transport of ethanol in baker's yeast. Folia Microbiol. 30: 90–91

Küenzi M (1970) Uber der Reservekohlenhydratstoffwechsel

350

von *Saccharomyces cerevisiae*. Diss. Nr. 4544, ZÜrich, Switzerland

Lafon-Lafourcade S, Geneix C & Ribereau-Gayon P (1984) Inhibition of alcoholic fermentation of grape must by fatty acids produced by yeasts and their elimination by yeast ghosts. Appl. Env. Microbiol. 47: 1246–1249

Lang JM & Cirillo VP (1987) Glucose uptake in a kinaseless *Saccharomyces cerevisiae* mutant. J. Bacteriol. 169: 2932–2937

Larsson C & Gustafsson (1987) Glycerol production in relation to the ATP pool and heat production rate of the yeasts *Debaromyces hansenii* and *Saccharomyces cerevisiae* during salt stress. Arch. Microbiol. 147: 358–363

Leao C & van Uden N (1984) Effects of ethanol and other alkanols on passive proton influx in the yeast *Saccharomyces cerevisiae*. Biochim. Biophys. Acta 774: 43–48

— (1986) Transport of lactate and other short-chain monocarboxylates in the yeast *Candida utilis*. Appl. Microbiol. Biotechnol. 23: 389–393

Lee FJ & Hassan HM (1987) Biosynthesis of superoxide dismutase and catalase in chemostat cultures of *Saccharomyces cerevisiae*. Appl. Microbiol. Biotechnol. 26: 531–536

Linton JD & Stephenson RJ (1978) A preliminary study on the growth yields in relation to the carbon and energy content of various organic growth substrates. FEMS Microbiol. Lett. 3: 95–98

Linton JD & Rye AJ (1989) The relationship between the energetic efficiency in different microorganisms and the rate and type of metabolite overproduced. J. Indust. Microbiol. 4: 85–96

Lloyd D (1974) In: The mitochondria of microorganisms, pp 89–90. Academic Press, London

Loureiro-Dias MC & Santos H (1990) Effects of ethanol on *Saccharomyces cerevisiae* as monitored by in vivo ^{31}P and ^{13}C nuclear magnetic resonance. Arch. Microbiol. 153: 384–391

Lucas C, da Costa M & van Uden N (1990) Osmoregulatory active sodium-glycerol co-transport in the halotolerant yeast *Debaromyces hansenii*. Yeast 6: 187–191

Lueck E (1980) In: Antimicrobial food additives (pp 210–217). Springer Verlag, Berlin, Heidelberg, New York

Maiorella B, Blanch HW & Wilke CR (1983) By-product inhibition effects on ethanolic fermentation by *Saccharomyces cerevisiae*. Biotechnol. Bioeng. 25: 103–121

— (1984) Feed component inhibition in ethanolic fermentation by *Saccharomyces cerevisiae*. Biotechnol. Bioeng. 26: 1155–1166

Malpartida F & Serrano R (1981) Proton translocation catalyzed by the purified yeast plasma membrane ATPase reconstituted in liposomes. FEBS Lett. 131: 351–354

McDonald IJ, Walker T & Johnson BF (1987) Effects of ethanol and acetate on glucose-limited chemostat cultures of *Schizosaccharomyces pombe*, a fission yeast. Can. J. Microbiol. 33: 598–601

Meikle AJ, Reed RH & Gadd GM (1988) Osmotic adjustment and the accumulation of organic solutes in whole cells and protoplasts of *Saccharomyces cerevisiae*. J. Gen. Microbiol 134: 3049–3060

Mishra P & Kaur S (1991) Lipids as modulators of ethanol tolerance in yeast. Appl. Microbiol. Biotechnol. 34: 697–702

Müller RH, Markuske KD & Babel W (1985) Formate gradients as a means for detecting the maximum carbon conversion efficiency of heterotrophic substrates: correlation between formate utilization and biomass increase. Biotechnol. Bioeng. 27: 1599–1602

Müller RH & Babel W (1988) Energy and reducing equivalent potential of C_2-compounds for microbial growth. Acta Biotechnol. 8: 249–258

Murphy MP & Brand MD (1988a) Membrane-potential-dependent changes in the stoichiometry of charge translocation by the mitochondrial electron tranpsort chain. Eur. J. Biochem. 173, 637–644

— (1988b) The stoichiometry of charge translocation by cytochrome oxidase and the cytochrome bc_1 complex of mitochondria at high membrane potential. Eur. J. Biochem. 173: 645–651

Neijssel OM & Tempest DW (1976) Bioenergetic aspects of aerobic growth of *Klebsiella aerogenes* NCTC 418 in carbon-limited and carbon-sufficient chemostat culture. Arch. Microbiol. 107: 215–221

Nelson N & Taiz L (1989) The evolution of H^+-ATPases. TIBS 14: 113–116

Nohl H (1986) A novel superoxide radical generator in heart mitochondria. FEBS 214: 269–273

Novak M, Strehaiano P, Moreno M & Goma G (1981) Alcoholic fermentation: on the inhibitory effect of ethanol. Biotechnol. Bioeng. 23: 201–211

Novak B & Mitchison JM (1990) Changes in the rate of oxygen consumption in synchronous cultures of the fission yeast *Schizosaccharomyces pombe*. J. Cell Science 96: 429–433

Ohnishi T (1973) Mechanism of electron transport and energy conservation in hte site I region of the respiratory chain. Biochim. Biophys. Acta 301: 105–128

Okolo B, Johnston JR & Berry DR (1987) Toxicity of ethanol, n-butanol and iso-amyl alcohol in *Saccharomyces cerevisiae* when supplied separately and in mixtures. Biotechnol. Lett. 9: 431–434

Onken U & Liefke E (1989) Effect of total and partial pressure (oxygen and carbon dioxide) on aerobic microbial processes. Adv. Biochem. Eng/Biotechnol. 40: 137–169

Ouhabi R, Rigoulet M & Guerin B (1989) Flux-yield dependence of oxidative phosphorylation at constant $\Delta\mu H^+$. FEBS Lett. 254: 199–202

Oura E (1972) Reactions leading to the formation of yeast cell material from glucose and ethanol. Alkon Keskuslaboratorio report 8078. Helsinki, Finland

Osothsilp C & Subden RE (1986) Malate transport in *Schizosaccharomyces pombe*. J. Bacteriol. 168: 1439–1443

Paca J & Gregr V (1979) Effect of pO_2 on growth and physiological characteristics of *C. utilis* in a multistage tower fermentor. Biotechnol. Bioeng. 21: 1827–1843

Pampulha ME & Loureiro V (1989) Interaction of the effects of acetic acid and ethanol on inhibition of fermentation in *Saccharomyces cerevisiae*. Biotechnol. Lett. 11: 269

Pampulha ME & Loureiro-Dias MC (1990) Activity of glycolytic enzymes of *Saccharomyces cerevisiae* in the presence of acetic acid. Appl. Microbiol. Biotechnol. 34: 375–380

Parada G & Acevedo F (1983) On the relation of temperature and RNA content to the specific growth rate in *Saccharomyces cerevisiae*. Biotechnol. Bioeng. 25: 2785–2788

Parulekar SJ, Semones GB, Rolf MJ, Lievense JC & Lim HC (1986) Induction and elimination of oscillations in continuous cultures of *Saccharomyces cerevisiae*. Biotech. Bioeng. 28: 700–710

Pascual C, Alonso A, Garcia I, Romay C & Kotyk A (1988) Effect of ethanol on glucose transport, key glycolytic enzymes, and proton extrusion in *Saccharomyces cerevisiae*. Biotechnol. Bioeng. 32: 374–378

Payne WJ (1970) Energy yields and growth of heterotrophs. Annu. Rev. Microbiol. 42: 17–52

Peinado JM, Cameira-dos-Santos PJ & Loureiro-Dias MC (1989) Regulation of glucose transport in *Candida utilis*. J. Gen. Microbiol. 135: 195–201

Perlin DS, San Francisco MJD, Slayman CW & Rosen BP (1986) H$^+$/ATP stoichiometry of proton pumps from *Neurospora crassa* and *Escherichia coli*. Arch. Biochem. Biophys. 248: 53–61

Petrov VV & Okarokov LA (1990) Increase of the anion and proton permeability of *S. carlbergensis* plasmalemma by n-alcohols as a possible cause of its de-energetization. Yeast 6: 311–318

Pinto I, Cardoso H, Leao C & van Uden N (1989) High enthalpy and low enthalpy death in *Saccharomyces cerevisiae* induced by acetic acid. Biotechnol. Bioeng. 33: 1350–1352

Pirt SJ (1965) The maintenance energy of bacteria in growing cultures. Proceedings of the Royal Society of London 163B: 224–231

Pons M-N, Rajab A & Engasser J-M (1986) Influence of acetate on growth kinetics and production control of *Saccharomyces cerevisiae* on glucose and ethanol. Appl. Microbiol. Biotechnol. 24: 193–198

Postma E, Scheffers WA & van Dijken JP (1988) Adaptation of the kinetics of glucose transport to environmental conditions in the yeast *Candida utilis* CBS 621: a continuous culture study. J. Gen. Microbiol. 134: 1109–1116

— (1989) Kinetics of growth and glucose transport in glucose-limited chemostat cultures of *Saccharomyces cerevisiae*. Yeast 5: 159–165

Reed RH, Chudek JA, Foster R & Gould CM (1987) Osmotic significance of glycerol accumulation in exponential growing yeasts. Appl. Env. Microbiol. 53: 2119–2123

Roels JA (1981) Application of macroscopic principles to microbial metabolism. Biotechnol. Bioeng. 22: 2457–2514

Romano AH (1982) Facilitated diffusion of 6-deoxy-D-glucose in bakers' yeast: evidence against phosphorylation-associated transport of glucose. J. Bacteriol. 152: 1295–1297

— (1986) Sugar transport systems of baker's yeast and filamentous fungi. In: Morgan MM (Ed) Carbohydrate Metabolism in cultured cells (pp 225–244). Plenum Publishing Corp

Rosa MF, Sá-Correia I & Novais J (1988) Improvement in ethanol tolerance of *Kluyveromyces fragilis* in jerusalem artichoke juice. Biotech. Bioeng. 31: 705–710

Rosa MF & Sá-Correia I (1991) In vivo activation by ethanol of plasma membrane ATPase of *Saccharomyces cerevisiae*. Appl. Env. Microbiol. 57: 830–835

Rouwenhorst RJ, Visser LE, van der Baan AA, Scheffers WA & van Dijken JP (1988) Production, distribution and kinetic properties of *Kluyvermyces marxianus* CBS 6556. Appl. Env. Microbiol. 54: 1131–1137

Rouwenhorst RJ, van der Baan AA, Scheffers WA & van Dijken JP (1991) Production and localization of β-fructosidase in asynchronous and synchronous chemostat cultures of yeasts. Appl. Env. Microbiol. 57: 557–562

Rutgers M, Teixeira de Mattos MJ, Postma PW & van Dam K (1987) Establishment of the steady state in glucose-limited chemostat cultures of *Klebsiella pneumonae*. J. Gen. Microbiol. 133: 445–451

Rydström J, Texeira da Cruz A & Ernster L (1970) Factors governing the kinetics and steady state of the mitochondrial nicotinamide nucelotide transhydrogenase system. Eur. J. Biochem. 17: 56–62

Sá-Correia I & van Uden N (1983) Temperature profiles of ethanol tolerance: effects of ethanol on the minimum and maximum temperatures for growth of the yeast *Saccharomyces cerevisiae* and *Kluyveromyces fragilis*. Biotechnol. Bioeng. 25: 1665–1667

Sá-Correia I (1986) Synergistic effect of ethanol, octanoic and decanoic acid on the kinetics and activation parameters of thermal death in *Saccharomyces bayanus*. Biotechnol. Bioeng. 28: 761–763

Salgueiro SP, Sá-Correia I & Novais M (1988) Ethanol-induced leakage in *Saccharomyces cerevisiae*: kinetics and relationship to yeast ethanol tolerance and alcohol fermentation productivity. Appl. Environ. Microbiol. 54: 903–909

Salmon JM (1987) L-Malic acid permeation in resting cells of anaerobically grown *Saccharomyces cerevisiae*. Biochim. Biophys. Acta 901: 30–34

Serrano R (1977) Energy requirements for maltose transport in yeast. Eur. J. Biochem. 80: 97–102

— (1988) Structure and function of proton translocating ATPase in plasma membranes of plants and fungi. Biochem. Biophys. Acta 947: 1–28

Shul'govskaya EM, Pozmogova & Rabotnova (1988) Growth of a culture of *Candida utilis* in the chemostat on a balanced medium. Microbiology 56: 496–499

Stanier RY, Ingraham JL, Wheelis ML & Painter PR (1987) Effect of the environment on microbial growth. In: General Microbiology (pp 207). Macmillan Education Ltd., London

Steinbüchel A, Frund C, Jendrossek D & Schlegel H (1987) Isolation of mutants of *Alcaligenes eutrophus* unable to derepress the fermentative alcohol dehydrogenase. Arch. Microbiol. 148: 178–186

Stouthamer AH (1973) A theoretical study on the amount of

ATP required for synthesis of microbial cell material. A. v. Leeuwenhoek 39: 545–565

Stouthamer AH & Bettenhaussen CW (1973) Utilization of energy for growth and maintenance in continuous and batch culture of microorganisms. Biochim. Biophys. Acta 301: 53–70

Stouthamer AH & van Verseveld HW (1987) Microbial energetics should be considered in manipulating metabolism for biotechnological purposes. Trends in Biotechnol. 40–46

Taylor GT & Kirsop BH (1977) The origin of medium chain length fatty acids present in beer. J. Inst. Brew. 83: 241–243

Tempest DW & Neijssel OM (1984) The status of Y_{ATP} and maintenance energy as biologically interpretable phenomena. Annual Rev. Microbiol. 38: 459–486

Vallejo CG & Serrano R (1989) Physiology of mutants with reduced expression of plasma membrane H^+-ATPase. Yeast 5: 307–319

Van Dijken JP & Harder W (1975) Growth yields of microorganisms on methanol and methane. A theoretical study. Biotechnol. Bioeng. 17: 15–30

Van Dijken JP, Otto R & Harder W (1976) Growth of *Hansenula polymorpha* in a methanol-limited chemostat. Arch. Microbiol. 111: 137–144

Van Dijken JP & Scheffers WA (1986) Redox balances in the metabolism of sugars by yeasts. FEMS Microbiol. Rev. 32: 199–224

Van Uden N & Madeiro-Lopes (1976) Yield and maintenance relation of yeast growth in the chemostat at superoptimal temperatures. Biotechnol. Bioeng. 18: 791–804

Van Uden N (1984) Temperature profiles of yeasts. Adv. Microbial Physiol. 25: 195–251

Van Uden N (1989) Effects of alcohols on membrane transport in yeasts. In: Alcohol Toxicity in Yeasts and Bacteria (pp 135–146). CRC Press Inc, Boca Raton

Van Urk H, Postma E, Scheffers WA & van Dijken JP (1989) Glucose transport in Crabtree-positive and Crabtree-negative yeasts. J. Gen. Microbiol. 135: 2399–2406

Van Urk H, Voll WSL, Scheffers WA & van Dijken JP (1990) A transition state analysis of metabolic fluxes in Crabtree-positive and Crabtree-negative yeasts. Appl. Env. Microbiol. 56: 281–287

Van Zyl PJ, Kilian SG & Prior BA (1990) The role of an active transport mechanism in glycerol accumulation during osmoregulation by *Zygosaccharomyces rouxii*. Appl. Microbiol. Biotechnol. 34: 231–235

Veenhuis M, van Dijken JP & Harder W (1983) The significance of peroxisomes in the metabolism of one-carbon compounds in yeasts. Adv. Microb. Physiol. 24: 1–82

Veenhuis M, Mateblowski M, Kunau WH & Harder W (1987) Proliferation of microbodies in *Saccharomyces cerevisiae*. Yeast 3: 77–84

Verduyn C, Giuseppin MLF, Scheffers WA & van Dijken JP (1988a) Hydrogen peroxide metabolism in yeasts. Appl. Env. Microbiol. 54: 2086–2090

Verduyn C, Breedveld GJ, Scheffers WA & van Dijken JP (1988b). Purification and properties of dihydroxyacetone re-

ductase and 2,3-butanediol dehydrogenase from *Candida utilis* CBS 621. Yeast 4: 127–133

— (1988c). Metabolism of 2,3-butanediol in yeasts. Yeast 4: 135–142

Verduyn C, Postma E, Scheffers WA & van Dijken JP (1990a) Energetics of *Saccharomyces cerevisiae* in anaerobic glucose-limited chemostat cultures. J. Gen. Microbiol. 136: 405–412

— (1990b) Physiology of *Saccharomyces cerevisiae* in anaerobic glucose-limited chemostat cultures. J. Gen. Microbiol. 136: 395–403

Verduyn C, Stouthamer AH, Scheffers WA & van Dijken JP (1991a) A theoretical evaluation of growth yields of yeasts. A. v. Leeuwenhoek 59: 49–63

Verduyn C, van Wijngaarden CJ, Scheffers WA & van Dijken JP (1991b) Hydrogen peroxide as an electron acceptor for mitochondrial respiration in the yeast *Hansenula polymorpha*. Yeast 7: 137–146

Verduyn C (1992) Energetic aspects of metabolic fluxes in yeasts. PhD. thesis, Delft, The Netherlands

Viegas CA, Sá-Correia I & Novais JM (1985) Synergistic inhibition of the growth of *Saccharomyces bayanus* by ethanol and octanoic and decanoic acids. Biotechnol. Lett. 7: 611–614

Viegas CA, Rosa MF, Sá-Correia I & Novais JM (1989) Inhibition of yeast growth by octanoic and decanoic acids produced during ethanolic fermentation. Appl. Env. Microbiol. 55: 21–28

Viegas CA & Sá-Correia I (1991) Activation of the plasma membrane ATPase of *Saccharomyces cerevisiae* by octanoic acid. J. Gen. Microbiol. 137: 645–651

Von Jagow G & Klingenberg M (1970) Pathways of hydrogen in mitochondria of *Saccharomyces carlbergensis*. Eur. J. Biochem. 12: 583–592

Von Stockar U & Marison IW (1989) The use of calorimetry in biotechnology. Adv. Biochem. Eng./Biotechnol. 40: 93–136

Vreeland RH (1987) Mechanisms of halotolerance in microorganisms. CRC Critical Reviews in Microbiology 14: 311–356

Walkercaprioglio HM, Casey WM & Parks LW (1990) *Saccharomyces cerevisiae* membrane sterol modifications in response to growth in the presence of ethanol. Appl. Env. Microbiol. 56: 2853–2857

Wallace RJ & Holms WH (1986) Maintenance coefficient and rates of turnover of cell material in *Escherichia coli* ML308 at different growth temperatures. FEMS Microbiol. Lett. 37: 317–320

Warth AD (1988) Effect of benzoic acid on growth yield of yeasts differing in their resistance to preservatives. Appl. Env. Microbiol. 54: 2091–2095

— (1989) Transport of benzoic and propanoic acids by *Zygosaccharomyces bailii*. J. Gen. Microbiol. 135: 1383–1390

Watson TG (1970) Effects of sodium chloride on steady-state growth and metabolism of *Saccharomyces cerevisiae*. J. Gen. Microbiol. 64: 91–99

Whitworth DA & Ratledge C (1977) Phosphoketolase in *Rhodotorula glutinis* and other yeasts. J. Gen. Microbiol. 102: 397–401

Winter JF, Loret MO & Uribelarrea JL (1989) Inhibition and

growth factor deficiencies in alcoholic fermentation by *Saccharomyces cerevisiae*. Current Microbiol. 18: 247–252

Zwart KB, Overmars EH & Harder W (1983) The role of peroxisomes in the metabolism of D-alanine in the yeast *Candida utilis*. FEMS Microbiol. Lett. 19: 225–231

Antonie van Leeuwenhoek **60**: 355–371, 1991.

Formation of fermentation products and extracellular protease during anaerobic growth of *Bacillus licheniformis* in chemostat and batch-culture

Ben A. Bulthuis[1], Caius Rommens, Gregory M. Koningstein, Adriaan H. Stouthamer &
Henk W. van Verseveld*
*Department of Microbiology, Biological Laboratory, Free University, de Boelelaan 1087,
1081 HV Amsterdam, The Netherlands; ([1] present adress: Genencor International Inc., 180 Kimball Way,
South San Francisco, CA 94080, USA) (* requests for offprints)*

Key words: Bacillus, chemostat, extracellular protease, fermentation, HPLC-analysis

Abstract

For a relaxed (*rel⁻*), protease producing (A-type) and a stringent (*rel⁺*), not-protease producing (B-type) variant of *Bacillus licheniformis* we determined fermentation patterns and products, growth parameters and alkaline protease-production (if any) in anaerobic, glucose-grown chemostats and batch-cultures. Glucose is dissimilated via glycolysis and oxidative pentose phosphate pathway simultaneously; the relative share of these two routes depends on growth phase (in batch) and specific growth rate (in chemostat). Predominant products are lactate, glycerol and acetaldehyde for A-type batches and acetaldehyde, ethanol, acetate and lactate for B-type batches. Both types show a considerable acetaldehyde production. In chemostat cultures, the fermentation products resemble those in batch-culture.

From the anaerobic batches and chemostats, we conclude that the A-type (with low ATP-yield) will have a Y_{ATP}^{max} of probably 12.9 g/mol and the B-type (with high ATP-yield) a Y_{ATP}^{max} of about 10.1 g/mol. For batch-cultures, both types have about the same, high $Y_{glucose}$ (12 g/mol). So, the slow-growing A-type has a relatively high efficiency of anaerobic growth (i.e. an efficient use of ATP) and the fast-growing B-type a relatively low efficiency of anaerobic growth. In aerobic batch-cultures, we found 48, respectively 41% glucose-carbon conversion into mainly glycerol and pyruvate, respectively acetate as overflow metabolites in the A- and B-type.

In both aerobic and anaerobic batch-cultures of the A-type, protease is produced predominantly in the logarithmic and early stationary phase, while a low but steady production is maintained in the stationary phase. Protease production occurs via *de novo* synthesis; up to 10% of the total protease in a culture is present in a cell-associated form. Although anaerobic protease production (expressed as protease per amount of biomass) is much higher than for aerobic conditions, specific rates of production are in the same range as for aerobic conditions while, most important, the substrate costs of anaerobic production are very much higher than for aerobic conditions.

Abbreviations and symbols: **C-rec** = %-age carbon-recovery; **C-rec-Fp** = agreement between glucose dissimilated and fermentation products (including CO_2) found; **DW** = dry weight of biomass (g/L); E_{440} = light-extinction at 440 nm; **Fp** = fermentation products; γ = reduction degree (no dimension); γ**(ferm)** = average reduction degree of all 'products' i.e. fermentation products + biomass + exocellular protein); **L** = liter; **m** = maintenance requirement (mol/g DW × h); M_b = 'molar weight' of bacteria (147.6 g/mol) with the general elementary cell composition $C_{6.0}H_{10.8}O_{3.0}N_{1.2}$); μ = specific growth-rate (h^{-1}); **O/R** = average oxidation/reduction quotient (of all fermentation products); **q** = specific rate of consumption or production (mol/g DW × h); **rel$^+$**, **rel$^-$** = stringent, relaxed genotype; **RI** = refraction index; **rpm** = rotations per minute; **UV** = ultraviolet; **X (or x)** = biomass (g/L); **Y** = molar growth yield (g DW/mol); *sub-/super-scripts:* b = biomass, corr = corrected for exocellular protein, pr = exocellular protein, s = substrate, t = time.

Introduction

Bacillus licheniformis is one of the microorganisms that are of importance in biotechnological industries for their capability of producing a variety of extracellular enzymes, among which alkaline proteases. For a relaxed (*rel$^-$*), protease producing *Bacillus licheniformis* strain and its spontaneously occurring stringent (*rel$^+$*), non-producing variant, we earlier investigated a number of morphological and physiological characteristics and their relation to the production of alkaline protease (Bulthuis et al. 1988). Moreover, Frankena et al. (1985, 1986, 1988) have published on a large series of physiological experiments with this organism, elucidating growth and production characteristics. In their work on the substrate- and energy-costs of growth and protease production, they had assumed a minimum of 14.6 g/mol for Y_{ATP}^{max}, one of the parameters that is of utmost importance in calculations of this kind (Stouthamer & van Verseveld 1985; Stouthamer 1985; Frankena et al. 1988). As its minimal value can be most reliably determined in anaerobic experiments (in the absence of complicating factors like variable composition of the electron transfer chain and assumed values for P/O and H$^+$/O), we decided to carry out these experiments. These were furthermore of interest as we had no information whatsoever on anaerobic protease production levels nor on the dissimilation route (glycolysis, pentose phosphate route *etc.*) that operates in our variants of *Bacillus licheniformis*.

Materials and methods

Bacterial strains and growth conditions

Bacillus licheniformis S1684 was provided by the company Koninklijke Gist-brocades nv at Delft, The Netherlands. This relaxed (*rel$^-$*), protease-producing strain is referred to as the A-type. Its spontaneously arising stringent (*rel$^+$*), non-producing variant is referred to as the B-type (Frankena et al. 1985; Bulthuis et al. 1988). In all experiments it was grown in the glucose minimal medium (pH = 6.8) as described by Frankena et al. (1985). For the A-type, addition of yeast-extract (0.1 g/L; Oxoid) was found to be necessary for anaerobic growth. Since both types are capable of converting to their counterparts, their identity was checked daily by Gram-stains and plating and occasionally by phase contrast microscopy. API-20 E and API-50 CH tests (API-System S.A.) were carried out each 6 to 8 months (Bulthuis et al. 1988).

Aerobic batch-cultures were carried out with 0.5 L flasks (containing no more than 150 ml of culture) placed in a rotatory shaker (New Brunswick Scientific G-25; 260 rpm). Anaerobic batch-cultures were carried out with completely filled 100 ml flasks, placed in anaerobic containers (N$_2$/CO$_2$-atmosphere). Growth in batch-cultures proceeded at 35 °C; the pH was not controlled but final values were never below 5.8.

Aerobic chemostat cultures were run as described by Frankena et al. (1985), with stirrer speed at 600–1000 rpm and air-supply at 60 L/h, bubbled through the culture. Anaerobic chemostats were run essentially the same as aerobic chemostats

(37 °C); the air-supply was replaced by N_2 that was led *over* the culture (14 L/h) and bubbled through the nutrient. Stirrer speed was reduced to 400 rpm; every day, wall-growth was loosened by wiping the inner wall of the fermentor vessel with a magnetic bar.

Analytical procedures

Growth was monitored in three ways. Dry weight was measured by filtrating culture with membrane filters (Sartorius, 40 mm $\varnothing$, pore-width 0.2 μm), as described by de Vries and Stouthamer (1968) and by measurement of carbon with a Total Organic Carbon Analyzer (TOCA 915A; Beckman). Optical density was measured at 660 nm (E_{660}) with a Zeiss PMQ3 spectrophotometer.

The %-age of lysis (if any) was established by a determination of glycolytic and/or PP-pathway enzymes (see below) in the supernate; this amount was compared to that determined in known amounts of biomass.

Protein was determined as described by Lowry et al. (1951), using Bovine Serum Albumin as a standard. For this and other colorimetric assays, light absorption was measured with a Zeiss PMQ3 spectrophotometer.

Glucose was determined with the Boehringer GOD-PERID assay as prescribed by the company, except for the use of our own glucose-standards.

Alkaline protease was determined by a modification of the method of Hanlon and Hodges (1981), as described by Frankena et al. (1985); protease is expressed as extinction units (E_{440}) per ml supernatant or mg biomass.

Acetaldehyde was determined by HPLC-analysis (see below) and enzymatically as described by Bergmeyer (1974).

The amount of total exocellular product (in mol carbon/L or mg carbon/L) was inferred from the ppm carbon in supernates (obtained by 5 minutes centrifugation at 12000 × g, 4 °C, determined with the TOCA mentioned above. The measured amounts of ppm carbon were, as a control, compared with the amounts of residual glucose (if any), the amount of biomass determined as outlined

above and the amount of fermentation products as determined by HPLC-analysis (described below).

Enzyme analysis

Samples of 100 ml culture were centrifuged (12000 × g, 10 minutes, 4 °C) and the pellets washed twice and resuspended in 10 ml Tris/HCl (100 mM, pH 7.4; 4 °C). The cells were disrupted by passage through a French Press cell. Debris was removed by centrifugation and the cell-free extracts were used in enzyme activity determinations. The *in vitro* determination of fructose-6P-kinase (F-6-PK; EC 2.7.1.11), fructose-1,6- diphosphate-aldolase (aldolase; EC 4.1.2.13), glucose-6-phosphate-dehydrogenase (G6P-DH; EC 1.1.1.49) and 6-phosphogluconatedehydrogenase (6-PGDH; EC 1.1.1.44) was carried out essentially as described by Bergmeyer (1974), by measuring changes in the E_{340} due to NAD(P)(H)-reduction (-oxidation). Presence of F-6-PK and aldolase, respectively G6P-DH and 6-PGDH, was interpreted as a qualitative indicator of a functional glycolysis respectiv. oxidative pentose-phosphate pathway (= PP-pathway).

Isocratic HPLC-analysis of fermentation products

Samples were de-proteinated prior to HPLC-analysis by the addition of concentrated perchloric acid up to a final concentration of 2% v/v (= 0.33 M). The acidified samples were left on ice for 30 minutes, centrifuged for 10 minutes at 12000 × g (4 °C) and filtrated through Millex HV$_4$ filters (Waters/ Millipore). The samples were then either directly analyzed by HPLC or stored at 4 °C for no more than 2 weeks.

HPLC-analysis was carried out with a Gilson HPLC-system (Meyvis en co.). Separation proceeded on an Aminex HPX-87H cation exchange column (300 × 7.8 mm; BioRad) with an Aminex HPX-85H guard-column (40 × 4.6 mm) at an eluens flow-rate of 0.5 ml/minute (5 mM H_2SO_4; 50 μl sample, loop-injected) at room-temperature or at 30 °C, depending on what products were pre-

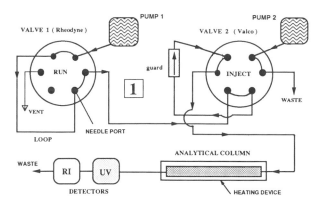

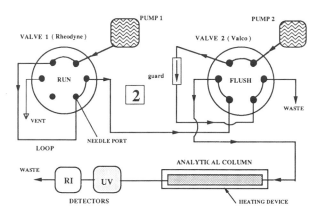

Fig. 1. Scheme showing the use of two valves (Rheodyne 7125 and Valco C6UW) in an HPLC-system for the separation of interfering substances from compounds to be analyzed (fermentation products, substrate). At zero time, the sample is injected (from loop), guard column and analytical column are on-line [1]. At t = 80 s, the guard column is switched off-line and flushed backwards while separation on the analytical column (30 min) proceeds [2].

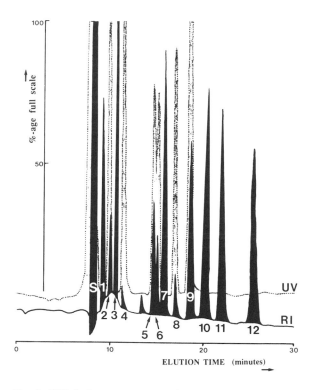

Fig. 2. HPLC-chromatogram as obtained with a 50 μl calibration mixture (at room-temperature) containing: (1) maltose, 0.15 mM; (2) citrate, 0.13 mM; (3) glucose, 0.30 mM; (4) pyruvate, 0.096 mM; (5) succinic acid, 0.5 mM; (6) lactic acid, 0.6 mM; (7) glycerol, 1.1 mM; (8) formic acid, 0.66 mM; (9) acetic acid, 2.0 mM; (10) acetaldehyde, 2.5 mM; (11) butane-2,3-diol, 1.2 mM; (12) ethanol, 3.8 mM; S = solvent. Full scale = 0.010 absorbance units (210 nm) for UV, 0.06 × 10⁻⁴ refraction units (for RI).

sent. Constant temperature was maintained by enclosing the analytical column in an alumina waterjacket (mechanics workshop, Vrije Universiteit). Analytical and guard column were coupled as shown in Fig. 1.

Sample components were identified and quantitated by simultaneous UV-absorbtion (210 nm; 0.010 AUFS; Gilson Holochrome UV/VIS) and refraction index (RI) measurement (0.06 × 10⁻⁴ RIUFS; Gilson model 131). From a calibration-mixture of glucose and fermentation products, a calibration curve was constructed from 16 data

points (4 concentrations per compound, 4 analyses per concentration) and stored in the data-collection/manipulation device of the HPLC-system (Apple IIe PC with data module 'Data Master 620' (Gilson)). Unknown samples were then automatically identified and quantitated after a run. Proper identification was checked by a control of elution time, by UV and/or RI-response and by the ratio (if any) of UV/RI-response. In case of doubt, samples were enriched with the calibration-mixture. Figure 2 shows an example of a calibration run of high concentration. Some examples of lower detection limits (50 μl sample-injection) at which reliable quantitation is still possible are 20 mM (RI) for glucose and pyruvate, 50 mM (RI) for succinate, lactic, acetic and formic acid and 250 mM for butane-diol, ethanol and glycerol (RI).

Calculations

$Y_{glucose}^{corr}$ was calculated by correcting the amount of glucose assimilated for the amount of glucose that was used for extracellular protein-synthesis. These latter costs were calculated by using 138.48 g/mol for the molar weight of protein with the general formula $C_6H_{9.48}N_{1.74}O_{2.04}$ (Babel & Müller 1985) and assuming that synthesis of 1.00 mol of extracellular protein would cost 1.03 mmol of glucose (Frankena et al. 1988). From $Y_{glucose}^{corr}$ and the ATP-yield (which follows from the fermentation balance), Y_{ATP}^{corr} was derived (in the absence of exocellular product, $Y = Y^{corr}$).

Under many of our growth conditions, considerable amounts of lactate were produced. It has been shown (Otto et al. 1980) that lactate-extrusion can be responsible for an increase in ATP-yield because of lactate/proton-symport, leading to an increase in $\Delta\mu H^-$ (and thus ATP-synthesis). We have assumed an ATP-yield of 0.5 mol per mol lactate extruded (ATP-yield and Y_{ATP}-values in which this has been taken into account are distinguished by an additional 'L'; so, ATP-L and Y_{ATP-L}). However, since we did not ourselves determine whether lactate-export indeed yields extra ATP, we have also given our 'not lactate-corrected' values for those parameters.

Fermentation balances were constructed as follows. In most cases it was impossible to reach proper redox-balances if the functioning of glycolysis only was taken into account. We therefore assumed the simultaneous functioning of glycolysis and the oxidative pentose-phosphate pathway (PP-pathway) and used the following equation:

$$2 \text{ [glucose dissimilated]} - \text{[glycerol produced]} + [A] = NADH_{Fp} \quad (1)$$

The left-hand term (all in mmoles) equals NADH produced during dissimilation of glucose; [A] is mmoles glucose dissimilated via the PP-route, the remainder is dissimilated via glycolysis. The amount of dissimilated glucose is easily inferred from : %-age dissimilation = $\{1 - Y_{glucose}/(147.6)\}100\%$ (147.6 is the 'molar weight' of bacteria with the general elementary cell composition $C_{6.0}H_{10.8}O_{3.0}N_{1.2}$). The right-hand term (in mmoles) of eq. 1 equals the amount of NADH consumed by the formation of fermentation products *in the final step* of their synthesis. These amounts are: glycerol = 1 (from glyceraldehyde-3-P); succinate = 2 (from phosphoenolpyruvate); butane-2,3-diol = 1, acetoine = 0, lactate = 1, formate ($\rightarrow H_2 + CO_2$) = 0 (from pyruvate); acetaldehyde = 1, ethanol = 2 (from acetyl-CoA) and acetate = 0 (from acetyl-P). Of course, for a proper redox-balance, the left-hand term has to equal the right-hand term. With this information, for each step in glucose dissimilation the yields or costs of NADH (as a control), ATP, H_2 and CO_2 could be calculated, using the experimental data on dissimilated glucose and fermentation products synthesized (production of H_2 and CO_2 was tested at random by measurement with respectively a gas chromatograph (Hewlett-Packard 5730A) and an infra-red gas analyzer (URAS 2T, Hartmann and Braun)). We followed the reactions given by Gottschalk (1986) for dissimilation of glucose via glycolysis and PP-route and subsequent formation of fermentation products.

We verified the validity of the assumption on %-age glycolysis and PP-route, implicit in eq. 1, by calculating the *average oxidation/reduction quotient* (O/R-value) of all fermentation products with equation 2:

$$\sum (O/R^+ \times mmol)/\sum (O/R^- \times mmol) = O/R \quad (2)$$

O/R-values follow from assigning a value of zero to compounds with the formula $(CH_2O)_x$; each surplus (or deficit) of '2H' means a value of -1 (or $+1$) for the O/R (Dawes et al. 1971). Ideally, of course, the average O/R (absolute value) should be 1.

As a final control, we calculated the *average reduction degree* (= γ_{ferm}) (the term 'reduction degree' has been discussed by Roels 1980) of all products by equation 3:

$$\gamma_{\text{ferm}} = \frac{\sum_1^n(\gamma(F_p)_n \times [\text{mmol}]_n \times \text{C-at}_n) + (\gamma_b \times [\text{mmol}]_b \times \text{C-at}_b) + (\gamma_{pr} \times [\text{mmol}]_{pr} \times \text{C-at}_pr)}{\sum_1^n([\text{mmol}]_n \times \text{C-at}_n) + ([\text{mmol}]_b \times \text{C-at}_b) + ([\text{mmol}]_{pr} \times \text{C-at}_{pr})} \quad (3)$$

(Fp = fermentation products, C-at = number of carbon-atoms, b = biomass, pr = exocellular protein).

Since glucose has a reduction degree $\gamma_s = 4.00$, the average γ of {fermentation products + biomass +(eventual) exocellular protein} should also be 4 (the reduction degree of biomass $\gamma_b = 4.20$ and of exocellular protein $\gamma_{pr} = 4.03$).

Results

Observations on growth and stability

From aerobic batch-cultures, the maximal specific growth rate μ_{max} was earlier found to be 0.6 and 0.9 h^{-1} for respectively the A- and the B-type (Bulthuis et al. 1988). For anaerobic batch-cultures we now found μ_{max} to be respectively 0.04 and 0.40 h^{-1}. In anaerobic A-type cultures, extensive flocculation occurred; the B-type hardly showed any flocculation. Like with aerobic A-type cultures, a microscopic view of the anaerobic A-type showed chain-formation (longer chains than under aerobic conditions) of Gram-positive cells; the anaerobic B-type still consisted of single cells, reacting negatively in a Gram-stain. The characteristic beige/brown to orange hue of aerobic A-type cultures, probably caused by pulcherriminic acid, was absent from anaerobic cultures. After exposure of anaerobic cultures to air, however, the colour did develop, in agreement with the proposed O_2-dependent conversion of pulcherrimine to pulcherriminic acid (Kupfer et al. 1967; Bulthuis et al. 1988).

Conversion of the A-type to the B-type occurred frequently. Unlike under aerobic conditions (Bulthuis et al. 1988), conversion was never complete. A typical example of partial conversion and 'reversion' is an anaerobic chemostat culture of the A-type, growing at $\mu = 0.03$ h^{-1} (generation time = 23.1 h). After 70 h of growth (100% A-type), the culture turned into a mixed population of 70% A-type/30% B-type (this took another 70 h) and then back into a 100% A-type culture within 20 h.

Under anaerobic conditions, some lysis of the A-type cells occurred. As mentioned in the 'Methods and materials'-section, we corrected our data for this. Lysis occurred for maximally 8% of the total biomass. For the B-type, lysis was absent or negligible.

Fermentation under anaerobic conditions

We analyzed the anaerobic production of fermentation products by both the A- and the B-type of *Bacillus licheniformis* in glucose batch-cultures and in glucose-limited chemostats. Our findings are summarized in Table 1 and 2.

The bottom lines of Tables 1 and 2 give a percentage of total substrate-carbon that is dissimilated via the glycolysis; the remainder of glucose is dissimilated via the PP-pathway. The figures for the above-mentioned calculated balances are very good (except for γ(ferm) for the B-type in chemostat at $\mu = 0.204$ h^{-1}); from those calculations, the theoretical amounts for CO_2- and H_2-production followed. For H_2-production, we experimentally verified the presence of H_2 by gaschromatography. This qualitative determination of H_2 agreed with our calculated results: H_2-production could only be detected for the B-type. We could not, however, detect the presence of H_2 in a sample of the A-type batch-culture at 28 h, although our calculations showed that some H_2 must have evolved (Table 1). In a control on calculated CO_2-production (mmol/100 mmol glucose), we found for the A-type in chemostat at $\mu = 0.03$ h^{-1} a calculated value of 46.2 (Table 1) and a measured value of 51.8; for the B-type in chemostat at $\mu = 0.071$ h^{-1} we found a calculated value of 86.0 (Table 2) and a measured value of 67.6. For the A-type, this is a reasonable agreement, for the B-type less so. The calculated data on CO_2 follow from the fermentation-patterns inferred from the experimental data. Since the data for C-recovery(-Fp), γ(ferm)- and O/R-balances (in which we used calculated CO_2 amounts) are quite good (Table 1 and 2), while calculated CO_2-

values match the measured values relatively well, we used the calculated data on CO_2-production in our further calculations.We consider our figures on NADH/NAD$^+$-balances, O/R-balances and γ(ferm)-balances as sufficient proof for the functioning of both glycolysis and PP-pathway; this is confirmed by a qualitative determination that showed the presence of the key-enzymes of glyco-

lysis (F-6-PK and aldolase) and PP-pathway (G6P-DH and 6-PGDH).

From the data in Tables 1 and 2 we inferred the experimental fermentation balances for anaerobic growth in batch-cultures; these balances are given in Table 3, which includes also eqs. 4–7. A first comparison between the A- and the B-type can be made by examining eq. 5 and eq. 7, both for cultures that had reached early stationary phase. Clearly,

Table 1. Fermentation products and growth parameters of the A-type of *Bacillus licheniformis* grown in anaerobic glucose batches and in anaerobic, glucose-limited chemostat (30 mM glucose).

	Batch (time in hours)				Chemostat	
	early log 28	log 42	log 48	stat. 66	$\mu = 0.030$	$\mu = 0.033$
lactate	33.3	40.7	43.0	48.3	89.0	59.2
glycerol	1.6	8.7	31.0	38.0	45.0	51.0
formate	–	–	0.7	4.0	–	–
acetate	0.8	3.3	2.0	5.0	5.0	–
acetaldehyde	–	3.3	34.0	53.3	26.2	44.4
acetoine	–	–	–	–	1.8	4.2
butanediol	–	0.3	9.6	18.3	5.8	12.2
pyruvate	(trace)	(trace)	0.2	(trace)	0.2	0.2
CO_2 (calc)	1.6*	(nd)	(nd)	67.9	46.2	70.1
exoprotein	0.12	0.14	0.31	0.11	0.58	0.40
glucose-residue	79.3	68.3	24.3	0.0	–	–
$Y_{glucose}$	14.8	(nd)	(nd)	11.8	8.4	6.6
$Y_{glucose}^{corr}$	17.3	(nd)	(nd)	11.9	8.8	6.8
ATP/glucose	1.88	(nd)	(nd)	0.68	0.71	0.45
Y_{ATP}^{corr}	9.2	(nd)	(nd)	17.5	12.4	15.1
ATP-L/glucose	2.05	(nd)	(nd)	0.92	1.16	0.75
Y_{ATP-L}^{corr}	8.5	(nd)	(nd)	12.9	7.9	9.1
% dissimilation	88.3	(nd)	(nd)	91.2	90.0	92.7
C-recovery Fp	98.5	89.2	83.8	94.8	100.3	100.0
C-recovery	101.2	98.2	89.4	96.4	100.6	100.0
γ (ferm)	4.02	(nd)	(nd)	4.04	4.01	4.02
O/R	1.00	(nd)	(nd)	0.96	1.00	0.99
NADH/NAD$^+$	0.99	(nd)	(nd)	1.00	1.00	1.00
% glycolysis	100.0	(nd)	(nd)	46.0	66.0	57.0

The amounts are expressed as mmol per 100 mmol of glucose (exoprotein in g/100 mmol); log and stat. = logarithmic and stationary phase in batch-culture; Y = growth yield (g biomass/mol), Y^{corr} = Y, corrected for the amount of exocellular protein; ATP-L/glucose and Y_{ATP-L} are, respectively. ATP-yield and Y_{ATP}^{corr}, recalculated under the assumption of an extra ATP-yield of 0.5 ATP per mol lactate extruded (see 'Methods and materials'); CO2 (calc) = mmol CO_2/100 mmol glucose, as inferred from the calculated material-balances; C-recovery Fp = agreement between mmol glucose dissimilated and mmol fermentation-products (including CO_2) recovered; γ(ferm) = reduction degree of {fermentation products + exocellular protein + biomass}; O/R = oxidation/reduction balance of fermentation products; (nd) = not determined; (trace) = only trace amounts (not quantitated) were found; *: from our calculations, it followed that a same amount (1.61 mmol/100 mmol glucose) of H_2 must have been produced.

the fermentation patterns differed largely, caused by the use of different dissimilation pathways (46% and 85% glycolysis for the A- and B-type, respectively (Tables 1 and 2)). Furthermore, H_2-production was present in the B-type but absent in the A-type. The ATP-yields differed very much and so, although $Y_{glucose}^{corr}$ for these cultures were nearly the same (respectively 11.9 and 12.5 g/mol; Tables 1 and 2), the Y_{ATP}^{corr} for the A- and B-type were very unlike: 17.5 for the A-type and 5.95 for the B-type.

It is common procedure to draw up fermentation balances from the results obtained with outgrown batch-cultures; we have done the same with eq. 5 and eq. 7. But this procedure, although common practice, might be a source of errors in a determination of ATP-yield (which depends on the dissimilation pathway being followed). This can be shown by examining the balances in Table 3 and the data on which these are based, shown in Tables 1 and 2. In Table 1, the measured values on fermentation-products show clearly that lactate was predominantly formed in the early phases of batch-growth

Table 2. Fermentation products and growth parameters of the B-type of *Bacillus licheniformis* grown in anaerobic glucose batches and in anaerobic, glucose-limited chemostat (30 mM).

	Batch (time in h)	Chemostats			
	24.0 (stat)	$\mu = 0.291$ h^{-1}	$\mu = 0.204$ h^{-1}	$\mu = 0.191$ h^{-1}	$\mu = 0.071$ h^{-1}
succinate	6.7	5.6	13.0	11.9	37.8
lactate	18.3	13.4	12.2	13.8	–
glycerol	3.3	4.8	–	–	–
formate	38.3	47.6	39.8	40.5	12.0
acetate	33.3	53.2	61.0	60.5	68.4
acetaldehyde	51.7	40.0	7.0	7.4	–
acetoine	(trace)	0.8	–	–	–
butanediol	10.0	0.4	–	–	–
ethanol	48.3	69.2	70.0	70.3	52.8
pyruvate	(trace)	1.0	0.3	–	0.2
CO_2 (calc)	108.3	95.3	93.7	95.5	86.0
H_2 (calc)	24.0	58.9	93.7	93.3	116.4
$Y_{glucose}$	12.5	19.8	20.1	20.2	19.4
ATP/glucose	2.10	2.04	2.56	2.55	2.70
Y_{ATP}	5.95	9.71	7.84	7.92	7.20
ATP-L/glucose	2.20	2.11	2.62	2.62	2.70
Y_{ATP-L}	5.68	9.38	7.67	7.71	7.20
% dissimilation	91.5	86.6	85.9	86.3	86.9
C-recovery-Fp	99.0	105.0	96.7	96.8	94.3
C-recovery	99.0	106.0	96.3	97.3	96.3
γ (ferm)	4.00	3.99	3.51	3.99	4.04
O/R	0.99	1.00	0.99	1.01	0.99
NADH/NAD$^+$	0.99	0.99	0.99	1.00	1.00
% glycolysis	85.1	54.0	84.6	84.9	91.5

The amounts are expressed as mmol per 100 mmol of glucose. Stat. = stationary phase in batch culture (all glucose depleted); Y = growth yield (g biomass/mol); as no exocellular protein (or protease) was found, Y is identical to Y^{corr} (cf). Table 1); ATP-L/glucose and Y_{ATP-L} are respectively. ATP-yield and Y_{ATP}, recalculated under the assumption of an extra ATP-yield of 0.5 ATP per mol lactate extruded (see 'Materials and Methods') CO_2 (calc) and H_2 (calc) = mmol gas/100 mmol glucose, as inferred from the calculated material-balances; C-recovery Fp = agreement between mmol glucose dissimilated and mmol fermentation-products (including CO_2) recovered; γ (ferm) = reduction degree of {fermentation products + exocellular protein + biomass}; O/R = oxidation/reduction balance of fermentation products; (trace) = only trace amounts (that could not be quantitated) were found.

(69% of the final [lactate] was already present after 28 hours) while glycerol and acetaldehyde were synthesized predominantly in the intermediate and late stages of batch-growth. From this it will be apparent, even without calculations on percentage glycolysis and PP-pathway, that shifts in the use of dissimilation pathways had occurred. A comparison of eqs. 4 and 5 will be illustrative. In the early logarithmic phase of the A-type (eq. 4), lactate-production accounted for 91% of the glucose-carbon dissimilation; the final balance for batch-growth (eq. 5) shows an only 27% -conversion into lactate. The ATP-yield in the early logarithmic phase (eq. 4) was three times higher than that found with cumulative product-data (eq. 5).

Clearly, a differential approach to growth in a batch-culture would give the most reliable information on fermentation-pattern and ATP-yield in a certain stage of that culture. We have calculated the changes occurring in the A-type batch-culture between 28 and 42 hours of growth (Table 1); this resulted in the fermentation balance of eq. 6 (Table 3). Equation 4 (representative for the first stage of growth, the early logarithmic phase) and eq. 6 (representing the fermentation pattern during mid-logarithmic phase) can now be compared. Obviously, a remarkable shift in fermentation-pattern occurred: from 100% to 61% glycolysis, a large decrease in lactate production, a very large increase

in glycerol production, a commencement of acetaldehyde production and an almost four-fold reduction in ATP-yield. For the period of 28 to 42 hours, a $Y_{glucose}^{corr}$ of 4.9 g/mol was calculated (for 0–28 hours: 17.3 g/mol) which, due to the large differences in ATP-yield between 0–28 and 28–42 hours, leads to a comparable Y_{ATP}^{corr} of 9.9 g/mol (0–28 hours: 9.2 g/mol). Unfortunately, we did not succeed in a differential approach to the data from 42 to 48 and 48 to 66 hours. Now, comparing these growth-parameters to the ones found with the cumulative product-data (eq. 5), it will be clear that a fermentation-balance based on products measured in stationary phase gives insufficient insight into what physiological processes have led to those products. The ATP-yield following from eq. 5, and thus Y_{ATP}, can be no more than a good approximation. It must be close, though, for otherwise the carbon-balances and $NADH/NAD^+$-, O/R- and γ(ferm)-balances would have been far from perfect.

In order to support and extend our findings on ATP-yield and Y_{ATP}, we carried out chemostat experiments with both types. For practical reasons, the A-type was cultured at a μ-value close to μ_{max} (= 0.04 h^{-1}). Since the anaerobic μ_{max} of the B-type is much higher (0.40 h^{-1}), this variant could be cultured at several μ-values. Like with the data on batch-cultures, we inferred fermentation balances

Table 3. Experimental fermentation balances (inferred from the data in Tables 1 and 2) for anaerobic growth in glucose batch-cultures of *Bacillus licheniformis.*

	Equations
A-type, 28 hours:	
1 glucose → 1.82 lactate + 0.09 glycerol + 0.09 CO_2 + 0.04 acetate + 0.09 H_2 + 1.88 ATP	(4)
A-type, 66 hours:	
1 glucose → 0.53 lactate + 0.42 glycerol + 0.58 acetaldehyde + 0.20 butanediol + 0.74 CO_2 + 0.05 acetate + 0.04 formate + 0.68 ATP	(5)
A-type, between 28 and 42 hours:	
1 glucose → 0.69 lactate + 0.66 glycerol + 0.31 acetaldehyde + 0.61 CO_2 + 0.23 acetate + 0.03 butanediol + 0.17 H_2 + 0.50 ATP	(6)
B-type, 24 hours:	
1 glucose → 1.18 CO_2 + 0.57 acetaldehyde + 0.53 ethanol + 0.36 acetate + 0.20 lactate + 0.11 butanediol + 0.42 formate + 0.07 succinate + 0.04 glycerol + 0.89 H_2 + 2.1 ATP	(7)

Products are arranged (left to right) in decreasing order of carbon-amount.

from the experimental data in Tables 1 and 2. These balances are given in Table 4, which includes also eqs. 8–13.

Predominant products in A-type chemostats were lactate, glycerol and acetaldehyde, in agreement with the findings in batch-cultures. With the B-type, ethanol and acetate were always found while, depending on μ, succinate (at low μ) or acetaldehyde (at high μ) was an important product. As already apparent from the batch-cultures, a large difference between the A- and B-type was the production of H_2 by the B-type (and absence or low production of H_2 by the A-type) and the higher ATP-yield for the B-type. Some growth-parameters of the chemostat cultures are summarized in Table 5.

Acetaldehyde production

Since the production of acetaldehyde might be interesting for its commercial application (see 'Discussion'), we confirmed its presence not only by HPLC-analysis but also by enzymatic analysis. Furthermore, we examined the influence of acetalde-

hyde on growth in an aerobic batch-culture of the A-type. To a culture in the logarithmic growth-phase, acetaldehyde was added up to a final concentration of 20 mM (highest [acetaldehyde] were found in the early stationary phase (Table 1, batch, 66 hours), to wit $0.3 \times 53.3 = 16$ mM)). Upon addition of acetaldehyde, μ dropped from 0.63 h^{-1} to about 0.3 h^{-1}; after 3.5 hours, growth was resumed at approximately the initial rate (0.57 h^{-1}).

Fermentation under aerobic conditions

As the production of overflow-metabolites during glucose-sufficient growth in batch-cultures is a well-known phenomenon, we decided to examine the occurrence of overflow products for our types of *Bacillus licheniformis*. The results are shown in Fig. 3. For the A-type, glycerol was the predominant product in the early logarithmic phase; overall, the most predominant products were acetate, glycerol and pyruvate. The average reduction degree of overflow products, γF_p, was about 3.8. Finally, about 48% of glucose had been turned into overflow products.

Table 4. Experimental fermentation balances (inferred from the data in Tables 1 and 2) for anaerobic growth in glucose-limited chemostat cultures of *Bacillus licheniformis*.

A-type	Equations
$\mu = 0.030$ h^{-1}:	
1 glucose → 0.99 lactate + 0.50 glycerol + 0.29 acetaldehyde + 0.51 CO$_2$ + 0.06 butanediol + 0.06 acetate + 0.02 acetoine + 0.71 ATP	(8)
$\mu = 0.033$ h^{-1}	
1 glucose → 0.64 lactate + 0.55 glycerol + 0.48 acetaldehyde + 0.76 CO$_2$ + 0.13 butanediol + 0.05 acetoine + 0.45 ATP	(9)
B-type	
$\mu = 0.071$ h^{-1}:	
1 glucose → 0.43 succinate + 0.79 acetate + 0.61 ethanol + 0.99 CO$_2$ + 0.14 formate + 1.34 H$_2$ + 2.70 ATP	(10)
$\mu = 0.191$ h^{-1}:	
1 glucose → 0.81 ethanol + 0.70 acetate + 1.11 CO$_2$ + 0.14 succinate + 0.16 lactate + 0.47 formate + 0.17 acetaldehyde + 1.08 H$_2$ + 2.55 ATP	(11)
$\mu = 0.204$ h^{-1}:	
1 glucose → 0.81 ethanol + 0.71 acetate + 1.09 CO$_2$ + 0.15 succinate + 0.46 formate + 0.14 lactate + 0.08 acetaldehyde + 1.09 H$_2$ + 2.56 ATP	(12)
$\mu = 0.291$ h^{-1}:	
1 glucose → 0.80 ethanol + 0.61 acetate + 1.10 CO$_2$ + 0.46 acetaldehyde + 0.55 formate + 0.15 lactate + 0.06 succinate + 0.06 glycerol + 0.01 acetoine + 0.68 H$_2$ + 2.04 ATP	(13)

Products are arranged (left to right) in decreasing order of carbon-amount.

For the B-type, acetate was the most predominant product (with CO_2 in a stoichiometric ratio of 1:1). The average reduction degree of overflow products, γF_p, was about 3.4. Finally, 41% of glucose had been turned into overflow products that were apparently being dissimilated during the stationary phase.

Protease production in aerobic and anaerobic batch-cultures

We compared the protease production (by the A-type) of aerobic and anaerobic batch-cultures. Our results are shown in Fig. 4. From panel A it is clear that the final protease-production in these batches was approximately equal ($E_{440} = 0.46$/ml supernate) and that protease was produced during the logarithmic and early stationary phase. Due to the large differences between anaerobic and aerobic growth yield ($Y_{glucose} = 83$, respectively 12 g/mol for aerobic and anaerobic growth; panel B), the specific production (E_{440}/mg biomass) for the anaerobic condition is about 9 times higher than for the aerobic condition (respectively 2.21 and 0.25). The specific production rates (E_{440}/mg biomass $\times$ hour) relate as 0.07 (aerobic) and 0.09 (anaerobic), while the substrate costs are about equal: 7.0 and 7.7 E_{440}/mol glucose for aerobic, respectively anaerobic production. For the anaerobic chemostat (Table 1), an average specific production rate of 0.03 E_{440}/(mg biomass $\times$ h) was found, with substrate costs of 8.5 E_{440}/mol glucose. We have summarized these data in Table 6 and added data on protease production in aerobic, glucose-limited chemostat cultures for comparison.

The amount of protease during stationary phase was quite stable (Fig. 4). We tested whether this was due to *de novo* synthesis or to a 'delayed' export of already synthesized protease. In an aerobic glucose-sufficient batch-culture, growth was arrested immediately and completely by the addition chloramphenicol (CA) in mid-logarithmic phase (170 mg CA/ml = 125 mg CA/mg biomass). The amount of protease in the supernate began to decline immediately; a 50% reduction in initial amount was reached within 5 hours.

This result most probably means *de novo* synthesis of protease. Nevertheless, another explanation for the stable amounts of protease in stationary phase was still possible, to wit: the occurrence of a steady release of cell-associated protease (synthesized in logarithmic phase) into the culture fluid, more or less compensating for the decrease due to extracellular turnover (as determined in the experiments with chloramphenicol). We therefore repeated our experiments with chloramphenicol addition. We measured the amounts of protease in the supernates and in the cells (centrifuged and washed twice) in both a control and a chloramphenicol-treated batch-culture. During the early stages of logarithmic growth, the amount of cell-associated protease increased rapidly to a constant value of about 10% of total protease (= cell-associated + soluble protease). This percentage remained constant during the stationary phase (control) and during arrested growth (chloramphenicol addition); the decrease in absolute amounts of cell-associated protease was proportional to the decrease in absolute amounts of soluble protease.

Discussion

For both the relaxed, protease-producing type (A) and the stringent, protease-negative type (B) of *Bacillus licheniformis* we have analyzed anaerobic growth characteristics and product formation. In anaerobic batch-cultures, we found μ_{max} to be 0.04 and 0.40 h^{-1} for the A- and B-type respectively, an even larger difference than for aerobic growth (respectively 0.6 and 0.9 h^{-1} (Bulthuis et al. 1988)). The observed flocculation (predominantly in batch-, but also in chemostat-culture) in A-type cultures might be explained by a combination of low carbon- and energy-availability (anaerobiosis, low growth rate ($\mu < 0.04$ h^{-1})), low shear forces (as compared to aerobic conditions) and the occurrence of relatively (compared to aerobic conditions) long cell-chains (Kwok & Prince 1988). The competitive character of the B-type under anaerobic conditions is apparently less pronounced than under aerobic conditions (Bulthuis et al. 1988) as we never observed a 100% reversion of the A- to

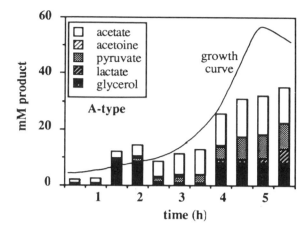

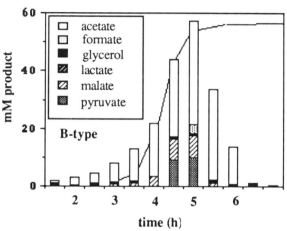

Fig. 3. Formation of overflow metabolites during aerobic, glucose-sufficient (15 mM) batch-growth of *Bacillus licheniformis*. Cumulative amounts are shown.

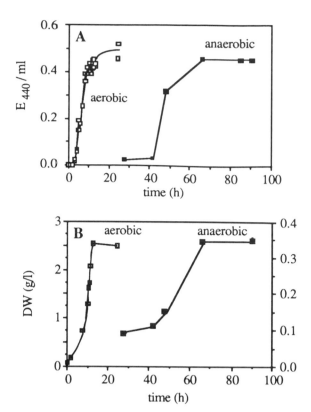

Fig. 4. Aerobic and anaerobic protease-production in glucose (30 mM) batch-cultures of the A-type of *Bacillus licheniformis*. Panel B shows the growth curves of the batch-cultures (NB: vertical axes – left for aerobic, right for anaerobic).

the B-type. This is somewhat surprising, considering the competitive advantage the B-type would have due to its much higher μ_{max}. Apparently, other factors (such as affinity constants for substrate uptake) must play a role. The large differences in μ_{max} (respectively, 0.04 and 0.40 h^{-1}) may be related to the production of protease by the A-type and the absence of production in the B-type. Considering, however, that protease accounts for only 3 to 4.3% of the total carbon-conversion (Table 5), this cannot be all of the explanation (this is further dealt with in the discussion on the efficiency of growth, below).

In Table 1 and 2 we summarized our experimental data on fermentation products and growth parameters. As mentioned in the 'Results'-section, proper NADH/NAD$^+$-, O/R- and $\gamma_{(ferm)}$-balances (and carbon-balances; calculated CO_2-production) could only be obtained when simultaneous functioning of glycolysis and PP-pathway was assumed. Since any description of microbial growth should satisfy these balances, the correct balances themselves are the best proof for the functioning of these two dissimilation routes simultaneously. The confirmation of this by our qualitative demonstration of the occurrence of key-enzymes of these dissimilation routes is therefore not surprising.

In general, not much has been published on the *simultaneous* use of catabolic pathways in *Bacillus* species. For *B. thuringiensis* it has been found that glucose is dissimilated via 'concurrent operation of the EMP-route and the PP-pathway' (Bulla et al. 1980, 1970). But for *B. subtilis*, Dawes et al. (1971)

mention data from Blackwood et al. (1947) that can be arranged in the following balance:

$$1 \text{ glucose} \rightarrow 0.55 \text{ butanediol} + 0.57 \text{ glycerol} +$$
$$1.18 \text{ } CO_2 + 0.18 \text{ lactate} +$$
$$0.08 \text{ ethanol} + 0.02$$
$$\text{acetoine} + 0.01 \text{ succinate} +$$
$$0.01 \text{ formate} + 0.002 \text{ } H_2 \qquad (14)$$

The carbon- and O/R-balance for this equation are 0.98 and 0.99, while we calculated an average γ_{Fp} of 4.05 for the fermentation-products. These values hardly changed when we recalculated them on the assumption of a 3.3% carbon-flow via the PP-route (that percentage follows from eq. 1).

Earlier data from Frankena et al. (1985) on anaerobic batch-cultures of *Bacillus licheniformis* (A-type) led to the following balance:

$$1 \text{ glucose} \rightarrow 0.46 \text{ butanediol} + 0.51$$
$$\text{glycerol} + 1.0 \text{ } CO_2 + 0.20$$
$$\text{lactate} + 0.06 \text{ acetate} +$$
$$0.04 \text{ ethanol} + 0.01$$
$$\text{formate} + 0.05 \text{ } H_2 \qquad (15)$$

Table 5. Growth parameters of anaerobic glucose-limited chemostat cultures of *Bacillus licheniformis*.

μ (h^{-1})	Y_{glu}	Y_{glu}^{corr}	ATP-yield	Y_{ATP}^{corr}	Glycol	Exopr
A-type						
0.030	8.4	8.8	0.71–1.16	12.4–7.9	66.0	4.3
0.033	6.6	6.8	0.45–0.75	15.1–9.1	57.0	3.0
B-type						
0.071	19.4	19.4	2.70–2.70	7.2–7.2	91.5	(none)
0.191	20.2	20.2	2.55–2.62	7.9–7.7	84.9	(none)
0.204	20.1	20.1	2.56–2.62	7.8–7.7	84.6	(none)
0.291	19.8	19.8	2.04–2.11	9.7–9.4	54.0	(none)

$Y_{glu}^{corr} = Y_{glu}$, corrected for the amount of glucose used in exoprotein synthesis (for the B-type, $Y_{glu} = Y_{glu}^{corr}$). Y_{ATP} (g/mol) and ATP-yield (mol ATP/mol glucose): data on the left in these columns are those in which any ATP from lactate-export is ignored; the data on the right were calculated assuming an extra ATP-yield of 0.5 mol ATP/mol lactate (so, ATP-L and Y_{ATP}-L^{corr}; see 'Materials and methods'); glycol(ysis) and exopr(otein) in % of total glucose catabolized respectively metabolized.

(carbon-and O/R-balance: 0.94 and 0.93; we calculated an average γ_{Fp} of 3.98), from which we calculated a 100% dissimilation via the glycolysis.

For organisms other than *Bacillus* species, results comparable to ours (as far as the simultaneous functioning of two dissimilation routes is concerned) have been found. For instance, Meyer & Papoutsakis (1988) showed that for glucose fermentation of *Enterobacter cloacae*, near-perfect balances could be obtained by assuming the functioning of glycolysis only; but for glucose-fermentation of *Escherichia coli*, proper balances were only obtained when the simultaneous functioning of glycolysis and PP-route was assumed. Similar findings were reported for aerobic, glucose-limited cultures of *Klebsiella aerogenes* (Neijssel & Tempest 1975), for the cyanobacterium *Cyanothece* PCC 7822 by van der Oost et al. (1989) and, as early as 1958, by Wang et al. for aerobic *E. coli*.

It is needless to say that the simultaneous functioning of two dissimilation routes, each with its own characteristic redox-balance and ATP-yield, constitutes a metabolic system capable of meeting environmental changes in a most flexible way. This simultaneous functioning of two dissimilation routes cannot be as special as one might be inclined to think from the relatively scarce amount of reports on this subject. Since under anaerobic conditions no NADPH can be obtained from the tricarboxylic acid cycle, the functioning of glycolysis (or Entner-Doudoroff route) alone might be insufficient to meet the demand for NADPH for biosynthetic purposes, a partial dissimilation via the PP-route would seem indispensable (e.g. Schlegel 1986). The functioning of two dissimilation routes is a major reason for the large variety of fermentation patterns that we have found. In the A-type batch-cultures, the early logarithmic phase (Table 3: eq. 4) is characterized by lactate production (91% C-conv; here, and below, '% C-conv' represents the percentage of dissimilated substrate-carbon that is converted into the fermentation product(s) mentioned). Mid-logarithmic phase and stationary phase cultures (Table 3: eqs. 6 and 5) show predominant formation of lactate, glycerol and acetaldehyde (together 78 and 67% C-conv, respectively).

B-type batch-cultures (in stationary phase; Table 3: eq. 7) show predominant formation of acetaldehyde, ethanol, acetate and lactate (65% C-conv). So, characteristic differences in fermentation patterns between the A- and the B-type of *B. licheniformis* in batch-culture are a large lactate production for the A-type (B-type: lactate is only one of several important products), a relatively low ATP-yield (B-type: relatively high) and minor or no production of H_2 by the A-type (B-type: substantial H_2-production).

In A-type chemostat cultures, the fermentation patterns and ATP-yield are in good agreement with those found in batch-culture (Table 4: eqs. 8 and 9); the main fermentation products are lactate, glycerol and acetaldehyde (84 and 76% C-conv for $\mu = 0.030$ and 0.033 h^{-1}, respectively).

The B-type chemostat cultures are also comparable to the batch-cultures; the product-pattern shifts from succinate, acetate, ethanol (at $\mu = 0.071$ h^{-1}) via ethanol, acetate and acetaldehyde (at $\mu = 0.191$ and 0.204 h^{-1}) to ethanol, acetate and acetaldehyde (at $\mu = 0.291$ h^{-1}). Carbon conversions for these fermentation products range from 61 to 75% (Table 4: eqs. 10–13). Table 5 shows how this shift comes about; as μ increases, the relative importance of glycolysis decreases.

As mentioned above (eq. 15), Frankena et al. (1985) found butanediol, glycerol and lactate (66% C-conv) as predominant products in a stationary phase batch-culture of the *B. licheniformis* A-type, nearly identical to the balance drawn up for *B. subtilis* by Dawes et al. (1971; cf. eq. 14). This differs from our results in the *predominance* of products (Table 3: eq. 5), but not very much in the *identity* of products. From the discussion above it will be clear that small differences in growth conditions and sampled growth phase can easily lead to such differences.

The ability of *Bacillus* species to produce a large variety of fermentation products is clear from our results as well as from literature data. For instance, Shibai et al. (1974) found 2,3-butylene-glycol, lactate and acetate as predominant products in anaerobic, glucose-grown chemostat cultures of *B. subtilis*; these products were replaced by inosine as the main product upon increasing the O_2-supply. In the early logarithmic phase of anaerobic batch-cultures of *B. thuringiensis* (growing on glucose-tryptone), lactate and (somewhat later) pyruvate were produced while in mid-logarithmic phase and early stationary phase acetoin and acetate appeared. Acetate was almost completely reclaimed during the stationary phase (Benoit et al. 1990). With nitrate as electron-acceptor, anaerobic glucose-grown chemostat cultures of *B. stearothermophilus* showed lactate and acetate as the sole products (Reiling & Zuber 1983). Comparing glucose batch-cultures of *B. polymyxa* under anaerobic and aerobic conditions, a shift in fermentation pattern of butanediol, lactate and ethanol to acetate and acetoin was found by de Mas et al. (1988) (the presence of these products under aerobic conditions is the result of overflow metabolism, as has been elaborately discussed by Neijssel & Tempest (1975, 1976, 1979)). Our aerobic batch-cultures (Fig. 3) show similar shifts in predominant products, as compared to the anaerobic condition. For the A-type, main products are glycerol and pyruvate (all products: 48% C-conv), for the B-type acetate (all products: 41% C-conv). Like in the experiments of Reiling and Zuber (1983) mentioned above, acetate (in the B-type) was apparently being metabolized in the stationary phase. The percentages of C-conversion into overflow-metabolites in our aerobic batches are comparable to the results of others. For instance, aerobic, K-limited and glucose-grown chemostat cultures of *B. stearothermophilus* showed acetate production up to 34% of total carbon metabolized (Pennock & Tempest 1988) while, for another strain of this species, Coultate & Sundaram (1975) found a C-conversion of over 18% for an aerobic, glucose-grown batch-culture. In aerobic, sulphate-limited cultures of *Klebsiella aerogenes*, Teixiera de Mattos et al. (1982) found up to 31% of metabolized glucose in acetate and pyruvate.

The production of acetaldehyde might be of commercial interest because of the application of acetaldehyde in food industry and its use as a 'raw material' in the synthesis of a variety of simple compounds. Wecker & Zall (1987) reported on the selection of allyl alcohol mutants of *Zymomonas mobilis* for increased acetaldehyde production. In glucose batch-cultures, their wild type strain produced up to 17.7 mM acetaldehyde, a 2.7% conversion of glucose-carbon into product; the highest acetaldehyde production-level reached was 56.4 mM (a 16.3% C-conv) with their most suitable mutant.

Our A-type batch-cultures produced up to 16 mM acetaldehyde (17.8% C-conv), chemostats up to 13.3 mM (14.8% C-conv). In B-type batches (stationary phase) we found 15.5 mM (17.2% C-conv), in chemostats 12 mM ($\mu = 0.291$ h^{-1}; 13.3% C-conv). These concentrations and carbon-conversions compare favourably with the figures from Wecker & Zall (1987) (although these concentrations are quite high and may have influenced the growth rates in batches, toxification has apparently not occurred since all batches were glucose-depleted and all chemostats glucose-limited). Moreover, those authors reached maximum productivity (26.5% C-conversion) under conditions of continuous air-stripping of acetaldehyde from the batch-fermentor, a doubling compared to the same culture that was not continuously being air-stripped. Since our B-type chemostat cultures were *not* selected for acetaldehyde-production, *not* continuously air-stripped, nor growing at the maximum growth rate (which, quite likely, will show the highest acetaldehyde production rates), the B-type of our *B. licheniformis* might be induced to produce even higher amounts of acetaldehyde. Wecker & Zahl (1987) mention the commercial attractiveness of acetaldehyde; if a market-analysis for this product would turn out favourable for today's situation,

our B-type *B. licheniformis* would probably be a suitable organism for acetaldehyde fermentation.

We have shown (Fig. 4) that the production of alkaline protease by A-type batch-cultures of *B. licheniformis* under both aerobic and anaerobic conditions predominantly occurs in the logarithmic and early stationary phase while a low but steady production level is maintained during the stationary phase (given our results with chloramphenicol treated cultures). So, protease production in our *B. licheniformis* clearly occurs (at least in stationary phase) *de novo*, possibly by translation of a pool of preformed mRNA's. This is akin to the results of Semets et al. (1973), who found chloramphenicol-sensitivity of the production of a neutral protease of *B. subtilis* and concluded on the presence of a large pool of mRNA's, maintained by a high rate of transcription. Probably as a consequence of the process of cotranslational export, up to 10% of the total protease in our cultures will be present in a cell-associated form.

The production of protease in aerobic chemostat cultures has elaborately been investigated by Frankena et al. (1986). A very clear increase in protease production was found in glucose-grown, O_2-limited cultures (as compared to O_2-sufficient cultures). Our results show that an even further decreased O_2-supply (i.e. anaerobic growth) is not an attractive alternative to O_2-limited growth. For although the anaerobic batch-cultures showed the highest levels of protease per mg of biomass (Table 6), in accordance with Frankena's observations (1985, 1986) on increased production under conditions of

Table 6. Comparison of protease production data obtained with aerobic and anaerobic, glucose-grown batch-cultures and chemostat cultures of the A-type of *Bacillus licheniformis*.

	E_{440}/mg	E_{440}/mg × h	E_{440}/ mol glucose
batch, anaerobic, $\mu = 0.04$ h^{-1}:	2.21	0.09	7.7
batch, aerobic, $\mu = 0.6$ h^{-1}:	0.25	0.07	7.0
chemostat, anaerobic, $\mu = 0.03$ h^{-1}:	1.04	0.03	8.5
chemostat, aerobic, $\mu = 0.03$ h^{-1}:	1.89[1]	0.06[1]	94.5[1]
	1.29[2]	0.04[2]	56.9[2]

For the batch-cultures and anaerobic chemostat, the figures on production costs (E_{440}/mol glucose) were inferred from Table 1 and Fig. 4. For aerobic chemostat cultures, the equations on glucose-consumption and protease-production published by (1) Frankena et al. (1985) and (2) Bulthuis et al. (1989) were used. Specific rates of production (E_{440}/mg biomass × h) in chemostat were obtained by multiplication with μ; for the batch-cultures, the hours during which production actually took place were used.

energy-limitation the, specific rates of protease production for both anaerobic batches and chemostats are in the same range as that of aerobic chemostats while, most important, the substrate costs of anaerobic production are very much higher than those of aerobic production (Table 6).

Part of the rationale for doing anaerobic experiments was our interest in the values of Y_{ATP}^{max} and $Y_{glucose}^{max}$ under anaerobic conditions. Reliable indications for the value of notably Y_{ATP}^{max} are of utmost importance for theoretical calculations on the efficiencies of growth and protease-production (Stouthamer 1985; Stouthamer & van Verseveld 1985). Frankena et al. (1985, 1986, 1988) have published experimental and theoretical work on this. Aerobic experiments resulted in the conclusion that the *B. licheniformis* A-type might have a high efficiency of growth; the aerobic $Y_{glucose}^{max}$ was high, presumably caused by a high Y_{ATP}^{max}. Anaerobic batch-culture experiments resulted in a value for Y_{ATP} of 14.6 g biomass/mol ATP. We corrected Frankena's value for lactate-extrusion and found 13.0 g/mol for Y_{ATP-L}, which then would be the probable minimal value for Y_{ATP}^{max}. Our experimental results point in the same direction. From our chemostat experiments (Table 5), a Y_{ATP-L}^{corr} of 9.1 g/mol follows (at $\mu = 0.033$ h^{-1}); for the batch-culture in stationary phase we found $Y_{ATP-L}^{corr} = 12.9$ g/mol (Table 1). We therefore propose a probable (minimal) value of $> 9.1–12.9$ g/mol for Y_{ATP}^{max}. This value is confirmed by experiments on the proton motive force, phosphorylation potential and P/O-ratios (Bulthuis et al., submitted).

The differences with regard to the growth-parameters of the A- and the B-type are interesting. Under the anaerobic condition there is, first, the large difference in μmax, which was mentioned above. Second, the ATP-yield (ATP/glucose) is lower for the A-type than for the B-type, while the Y_{ATP} for the A-type is much higher than Y_{ATP} for the B-type. Linear regression on the data for the B-type in Table 5 results in a Y_{ATP-L}^{max} of 10.1 g/mol at $\mu = 0.030$ h^{-1}; at that μ-value, the A-type has a Y_{ATP-L}^{corr} of 7.9 g/mol. The values for $Y_{glucose}$ in batch-culture are alike (Tables 1, 2) but due to the changes in metabolism (variable %-ages PP-route/glycolysis) $Y_{glucose}^{max}$ cannot be determined. Summarizing, the A-type shows a more efficient use of ATP; at the μ-values of 0.030 and 0.033 h^{-1} the Y_{ATP-L}^{max} values are comparable to those of the B-type at much higher μ-values (Table 5). Also, Y_{ATP-L}^{max} in batch-cultures is higher for the A- than for the B-type (Tables 1, 2). So, if Y_{ATP} is taken as a measure of growth-efficiency, there is a relatively high efficiency of anaerobic growth in the slow-growing A-type and a relatively low efficiency of anaerobic growth in the fast-growing B-type. The efficiency of aerobic growth (and aerobic Y_{ATP}^{max}) is subject of another publication (Bulthuis et al., submitted).

References

Babel W & Müller RH (1985) Correlation between cell composition and carbon conversion efficiency in microbial growth: a theoretical study. Appl. Microbiol. Biotechnol. 22: 201-207

Benoit TG, Wilson GR & Baugh CL (1990) Fermentation during growth and sporulation of *Bacillus thuringiensis* HD-1. Lett. Appl. Microbiol. 10: 15–18

Bergmeyer HU (1974) Methoden der enzymatischen Analyse, 2nd edition. Verlag Chemie GmbH, Weinheim (Germany)

Bulla LA, St Julian G, Rhodes RA & Hesseltine CW (1970) Physiology of spore forming bacteria associated with insects. Can. J. Microbiol. 16: 243–248

Bulla LA, Bechtel DB, Kramer KJ, Shetna YI, Aronson AI & Fitz-James PC (1980) Ultrastructure, physiology, and biochemistry of *Bacillus thuringiensis*. CRC Critical Revs. Microbiol. 8: 147–204

Bulthuis BA, Frankena J, Koningstein GM, van Verseveld HW & Stouthamer AH (1988) Instability of protease production in a *rel/rel⁻*-pair of *Bacillus licheniformis* and associated morphological and physiological characteristics. A. van Leeuwenhoek 54: 95–111

Bulthuis BA, Koningstein GM, van Verseveld HW & Stouthamer AH (1989) A comparison between aerobic growth of *Bacillus licheniformis* in continuous culture and partial-recycling fermentor, with contributions to the discussion on maintenance energy demand. Arch. Microbiol. 152: 499–507

Dawes EA, McGill DJ & Midgley M (1971) Analysis of fermentation products. In: Norris JR & Ribbons DW (Ed) Methods in Microbiology (pp 53–217). Academic Press Inc., London, UK

Frankena J, van Verseveld HW & Stouthamer AH (1985) A continuous culture study of the bioenergetic aspects of growth and production of exocellular protease in *Bacillus licheniformis*. Appl. Microbiol. Biotechnol. 22: 169–176

Frankena J, Koningstein GM, van Verseveld HW & Stouthamer AH (1986) Effect of different limitations in chemostat cultures on growth and production of exocellular protease by

Bacillus licheniformis. Appl. Microbiol. Biotechnol. 24: 106–112

Frankena J, van Verseveld HW & Stouthamer AH (1988) Substrate and energy costs of the production of exocellular enzymes by *Bacillus licheniformis*. Biotechnol. Bioeng. 32: 803–812

Hanlon GW & Hodges NA (1981) Bacitracin and protease production in relation to sporulation during exponential growth of *Bacillus licheniformis* on poorly utilized carbon and nitrogen sources. J. Bacteriol. 147: 427–431

Ingraham JL, Maaløe O & Neidhardt FC (1983) Chemical synthesis of the bacterial cell: polymerization, biosynthesis, fuelling reactions, and transport. In: Growth of the Bacterial Cell (pp 87–173). Sinauer Associates, Inc., Sunderland (USA)

Kwok KH & Prince IG (1988) Flocculation of *Bacillus* species for use in high-productivity fermentation. Enzyme. Microb. Technol. 11: 597–603

Kupfer DG, Uffen RL & Canale-Parola (1967) The role of iron and molecular oxygen in pulcherrimin synthesis by bacteria. Arch. Mikrobiol. 56: 9–21

Logan BE & Hunt JR (1988) Bioflocculation as a microbial response to substrate limitations. Biotechnol. Bioeng. 31: 91–101

Lowry OH, Rosebrough NJ, Farr AL & Randall RJ (1951) Protein measurement with the Folin phenol reagent. J. Biol. Chem. 193: 265–275

deMas C, Jansen NB & Tsao GT (1988) Production of optically active 2,3-butanediol by *Bacilus polymyxa*. Biotechnol. Bioeng. 31: 365–377

Meyer CL & Papoutsakis ET (1988) Detailed stoichiometry and process analysis. In: Erickson LE & Fung DYC (Eds) McGregor WC (series Ed) Handbook on anaerobic fermentations, Vol 3, Bioprocess Technology (pp 83–118). Marcel Dekker, Inc., New York, USA

Neijssel OM & Tempest DW (1975) The regulation of carbohydrate metabolism in *Klebsiella aerogenes* NCTC 418 organisms, growing in chemostat culture. Arch. Microbiol. 106: 251–258

— (1976) The role of energy-spilling reactions in the growth of *Klebsiella aerogenes* NCTC418 in aerobic chemostat culture. Arch. Microbiol. 110: 305–311

— (1979) The physiology of metabolic overproduction. In: Bull AT, Ellwood DC & Ratledge C (Eds) Microbial Technology: Current State and Future Prospects. Symp Soc Gen Microbiol 29, pp 53–82. University Press, Cambridge, UK

van der Oost J, Bulthuis BA, Feitz S, Krab K & Kraayenhof R (1989) Fermentation metabolism of the unicellular cyanobacterium *Cyanothece* PCC 7822. Arch. Microbiol. 152: 415–419

Otto R, Sonnenberg ASM, Veldkamp H & Konings WN (1980) Generation of an electrochemical proton gradient in *Streptococcus cremoris* by lactate efflux. Proc. Natl. Acad. Sci. USA 77: 5502–5506

Pennock J & Tempest DW (1988) Metabolic and energetic aspects of the growth of *Bacillus stearothermophilus* in glucose-limited and glucose-sufficient chemostat culture. Arch. Microbiol. 150: 452–459

Reiling HE & Zuber H (1983) Heat production and energy balance during growth of a prototrophic denitrifying strain of *Bacillus stearothermophilus*. Arch. Microbiol. 136: 243–253

Roels JA (1980) Application of macroscopic principles to microbial metabolism. Biotechnol. Bioeng. 22: 2457–2514

Schlegel HG (1986) General microbiology 6th edition (pp 213–235). Press Syndicate of the University of Cambridge, UK

Semets EV, Glenn AR, May BK & Elliott WH (1973) Accumulation of messenger ribonucleic acid specific for extracellular protease in *Bacillus subtilis* 168. J. Bacteriol. 116: 531–534

Shibai H, Ishizaki A, Kobayashi K & Hirose Y (1974) Simultaneous measurement of dissolved oxygen and oxidation-reduction potentials in the aerobic culture. Agric. Biol. Chem. 38: 2407–2411

Stouthamer AH (1985) Towards an integration of various aspects of microbial metabolism: energy generation, protein synthesis and regulation. In: Proceedings third European Congress on Biotechnology 4 (pp 223–239). Verlag Chemie-Dechema, Weinheim (Germany)

Stouthamer AH & van Verseveld (1985) Stoichiometry of microbial growth. In: Cooney CL & Humphrey AE (Eds) Comprehensive Biotechnology: The Principles, Applications and Regulations of Biotechnology in Industry, Agriculture and Medicine (pp 215–238). Pergamon press, Oxford, UK

Teixeira de Mattos MJ, Neijssel OM & Tempest DW (1982) Influence of aerobic and anaerobic nutrient-limited environments on metabolite over-production by *Klebsiella aerogenes* NCTC 418. In: Krumphanzl V, Sikyta B & Vanek Z (Eds) Overproduction of microbial products (pp 581–592). Academic Press, London UK

de Vries W & Stouthamer AH (1968) Fermentation of glucose, lactose, galactose, mannitol and xylose by *Bifidobacteria*. J. Bacteriol. 96: 472–478

Wang CH, Stern I, Gilmour CM, Klungsoyr S, Reed DJ, Bialy JJ, Christensen BE & Cheldelin VH (1958) Comparitive study of glucose metabolism by the radiorespirometric method. J. Bacteriol. 76: 207–216

Wecker MSA & Zall R (1987) Production of acetaldehyde by *Zymomonas mobilis*. Appl. Environ. Microbiol. 53: 2815–2820

Antonie van Leeuwenhoek **60**: 373–382, 1991.

Quantitative aspects of glucose metabolism by *Escherichia coli* B/r, grown in the presence of pyrroloquinoline quinone

R.W.J. Hommes,[1,4] J.A. Simons,[1] J.L. Snoep,[1] P.W. Postma,[2] D.W. Tempest[3] & O.M. Neijssel[1,*]

[1] *Department of Microbiology, Biotechnology Centre, University of Amsterdam, P.O. Box 20245, 1000 HE Amsterdam, The Netherlands;* [2] *Department of Biochemistry, Biotechnology Centre, University of Amsterdam, Amsterdam, The Netherlands;* [3] *Department of Microbiology, University of Sheffield, Sheffield S10 2TN, England; (*[4] *present address: Gist-Brocades NV, P.O. Box 1, 2600 MA Delft, The Netherlands) (*[*] *requests for offprints)*

Key words: glucose metabolism, *Escherichia coli*, pyrroloquinoline quinone, glucose dehydrogenase

Abstract

Escherichia coli B/r was grown in chemostat cultures under various limitations with glucose as carbon source. Since *E. coli* only synthesized the glucose dehydrogenase (GDH) apo-enzyme and not the appropriate cofactor, pyrroloquinoline quinone (PQQ), no gluconate production could be observed. However, when cell-saturating amounts of PQQ (nmol to μmol range) were pulsed into steady state glucose-excess cultures of *E. coli*, the organisms responded with an instantaneous formation of gluconate and an increased oxygen consumption rate. This showed that reconstitution of GDH *in situ* was possible.

Hence, in order to examine the influence on glucose metabolism of an active GDH, *E. coli* was grown aerobically in chemostat cultures under various limitations in the presence of PQQ. It was found that the presence of PQQ indeed had a sizable effect: at pH 5.5 under phosphate- or sulphate- limited conditions more than 60% of the glucose consumed was converted to gluconate, which resulted in steady state gluconate concentrations up to 80 mmol/l. The specific rate of gluconate production (0.3–7.6 mmol·h^{-1}·(g dry wt cells)$^{-1}$) was dependent on the growth rate and the nature of the limitation. The production rate of other overflow metabolites such as acetate, pyruvate, and 2-oxoglutarate, was only slightly altered in the presence of PQQ. The fact that the cells were now able to use an active GDH apparently did not affect apo-enzyme synthesis.

Abbreviations: HEPES: N-2-hydroxy-ethylpiperazine-N'-2-ethane sulphonic acid, MES: 2-morpholino-ethane sulphonic acid, PQQ: pyrroloquinoline quinone (systematic name: 2,7,9-tricarboxy-1*H*-pyrrolo-(2,3-*f*)-quinoline-4,5-dione), WB: Wurster's Blue (systematic name: 1,4-bis-(dimethylamino)-benzene perchlorate

Introduction

The initial step in the catabolism of glucose by members of the Enterobacteriaceae, like *Escherichia coli* and *Klebsiella pneumoniae*, is often thought to be the uptake, with concomitant phos- phorylation, of this compound by the phosphoe- nolpyruvate-dependent glucose phosphotransfe- rase system (PTS) (Postma & Lengeler 1985). *K. pneumoniae* (previously called *K. aerogenes*), how- ever, metabolizes glucose also by means of a PTS-independent, direct glucose oxidative pathway, similar to that described for *Pseudomonas* species (Dawes 1981; Wood & Schwerdt 1953). This system consists of two membrane-bound enzymes, the pyrroloquinoline quinone(PQQ)-linked glucose dehydrogenase (GDH) (Neijssel et al. 1983) and the flavoprotein gluconate dehydrogenase (Matsushita et al. 1982), and it is via this pathway that gluconate and 2-ketogluconate are produced under some glucose-excess conditions (Hommes et al. 1985; Neijssel & Tempest 1975a,b).

In contrast to *K. pneumoniae*, *E. coli* strains never were observed to produce gluconate and/or 2-ketogluconate under glucose-sufficient growth conditions (e.g. see: Leegwater 1983). Hence, not surprisingly, no GDH or gluconate dehydrogenase activity could be detected in these cells. Nevertheless, several strains of *E. coli*, including the type strain, were found to synthesize the (inactive) GDH apo-enzyme (Hommes et al. 1984) and one must conclude that this organism is unable to synthesize the cofactor PQQ, at least under the culture conditions that have been tested until now. The apparent inability to make the appropriate cofactor for the GDH is not restricted to *E. coli*, but seems to be rather widespread in Nature (van Schie et al. 1987).

It was subsequently shown that a PTS-negative mutant of *E. coli* that was unable to grow on glucose, could regain its capacity to grow on glucose when 1 μM PQQ was present in the medium. Presumably, the glucose was oxidized to gluconate, which could be metabolized further (Hommes et al. 1984). Similarly, *Acinetobacter lwoffi* could convert glucose into gluconate when 0.2 μM PQQ was added (van Schie et al. 1984); thus, reconstitution of the GDH apo-enzyme *in vivo* was possible by adding small concentrations of PQQ to the growth medium. These results suggested that organisms that only possess GDH apo-enzyme might be auxotrophic for PQQ.

In view of these observations, it will be clear that the effect of the growth environment on glucose metabolism by *E. coli* might substantially change when PQQ is present in the medium, because it could enable the cells to metabolize part of the glucose via GDH. Since studies done so far with *E. coli* strains were always carried out without addi-

tion of PQQ, we therefore thought it worthwhile to examine the effect of PQQ on the pattern of overflow metabolism shown by *E. coli*, when grown in chemostat cultures under different nutrient limitations. In this contribution the effect of an active (holo)GDH on the metabolism of glucose by *E. coli* will be described.

Materials and methods

Organism

Escherichia coli B/r H266 was kindly provided by Dr. C.L. Woldringh (Department of Electron Microscopy and Molecular Cytology, University of Amsterdam) and was maintained by monthly subculture on tryptic meat-digest agar slopes.

Growth conditions

Organisms were cultured in a Porton-type chemostat (Herbert et al. 1965). In order to obtain glucose-, ammonia-, sulphate-, phosphate- or potassium-limited growth conditions, simple salts media were used as specified by Evans et al. (1970) with glucose added as the carbon source. Carbon-limited media contained 5 g/l glucose, whereas all other media contained 12–40 g/l glucose to ensure carbon excess under potassium (1 mM KCl input), sulphate (0.2 mM sulphate input), ammonia (15 mM ammonia input), or phosphate limitation (1 mM phosphate input). The pH value of the culture was maintained automatically at pH 6.9 ± 0.2 or at pH 5.5 ± 0.2, using sterile 4N or 2N NaOH as titrant, and the temperature was set at 35 °C. Fully aerobic conditions were maintained throughout by injecting air, at a rate of 600–700 ml/min, into the region of the impeller which was rotating at about 1400 rpm. When the cells were to be grown in the presence of PQQ, then this compound was supplied to the culture vessel by means of an extra medium pump. A solution of PQQ (3.4 μM PQQ input) in the above mentioned medium of Evans et al. (1970), without the growth limiting substrate and without glucose, was added to the culture vessel at a rate of 1/17 of the rate of medium addition to provide a steady state concentration of 0.2 μM PQQ.

PQQ-pulse procedure

In order to determine the maximal glucose dehydrogenase activity *in situ*, in chemostat cultures of *E. coli* growing in the absence of PQQ, 2 ml of a sterilized solution of PQQ was injected into the culture by means of a syringe, providing a final concentration of 0.2 μM, while all other culture conditions remained unchanged. Samples were taken at regular intervals, the cells were sedimented in an Eppendorf centrifuge and the supernatants were frozen to be analyzed later.

Incubations with washed cell suspensions

Incubations with washed suspensions were performed essentially according to Hommes et al. (1985). Cell suspensions were prepared either in a 100 mM HEPES/NaOH buffer (pH 7.0) containing 1 mM $MgCl_2$ or in a 100 mM MES/NaOH buffer (pH 5.5) containing 1 mM $MgCl_2$. The same buffer was used for the incubations. Before addition of glucose (40 mM) the cells were preincubated with PQQ for 15 min at 35 °C.

Preparation of cell-free extracts and enzyme assay

Cell-free extracts were prepared as reported previously (Hommes et al. 1984). The buffer used for the cell-free extracts was a 10 mM phosphate buffer (pH 6.0) containing 5 mM $MgCl_2$.

The spectrophotometric assay of glucose dehydrogenase using Wurster's Blue (WB) as the electron acceptor was carried out at 35 °C with freshly-prepared cell-free extracts using a Beckman DU40 spectrophotometer (Beckman Instruments Inc., Irvine, USA) according to Hommes et al. (1985). To determine the maximal glucose dehydrogenase activity (as a measure for the apo-enzyme level),

the cell-free extracts were incubated for 15 min. with 100 μM PQQ at 35 °C prior to the assay.

Analyses

Bacterial dry weight was measured by the procedure of Herbert et al. (1971). Glucose was assayed with glucose oxidase using the diagnostic kit of Sigma (St. Louis, MO, USA). Gluconate, pyruvate, 2-oxoglutarate, and acetate were assayed by the methods of Möllering & Bergmeyer (1974), Czok & Lamprecht (1974), Bergmeyer & Bernt (1974), Holz & Bergmeyer (1974) respectively. The above-mentioned compounds were also determined by HPLC (LKB, Bromma, Sweden) with either an Aminex HPX 87H Organic Acids Column (Biorad, Richmond, USA) with 5 mM H_2SO_4 as eluent or an Aminex A28 column (Biorad, Richmond, USA) with 0.3 M formic acid adjusted to pH 5.5 with ammonia as eluent, using a 2142 Refractive Index Detector (LKB, Bromma, Sweden), an SP 4270 Integrator (Spectra Physics, San Jose, USA), at 55 °C. Protein was determined according to Gornall et al. (1949) using bovine serum albumin as a standard. Oxygen consumed and carbon dioxide produced by the cultures were determined by passing the gas from the fermenter through an oxygen analyzer (Taylor Servomex Type OA 272, Crowborough, Sussex, England) and a carbon dioxide analyzer (Servomex IR Gas Analyzer PA404, Crowborough, Sussex, England).

Reagents

All reagents used were of the best analytical grade commercially available. Wurster's Blue was prepared according to Michaelis & Granick (1943). PQQ was kindly provided by Prof. Dr. Ir. J. A. Duine (Dept. of Microbiology, Delft University of Technology, The Netherlands).

Results

It has been shown that in mutants of *Escherichia*

coli, lacking the intact glucose uptake system (PTS), growth on glucose could be restored by adding PQQ to the growth medium (Hommes et al. 1984). Whereas an active GDH in those mutants was a necessity for growth to proceed on glucose, wild-type *E. coli* is perfectly able to grow on this carbon source without having this alternative glucose-metabolizing system available. Therefore, it was interesting to investigate whether reconstitution of an active GDH in growing wild-type cells would have an effect on the pattern of glucose metabolism.

Hence, *E. coli* B/r was grown in a phosphate-limited chemostat culture, conditions where a considerable GDH-apo-enzyme level in cell-free extracts could be detected (Hommes et al. 1984) and where *Klebsiella pneumoniae* showed a high rate of gluconate production (Neijssel & Tempest 1975b). When steady state conditions prevailed, a sterilized solution of PQQ was added to the culture to provide an end concentration of 10 μM, while all other culture parameters remained constant. Immediately, without any lag, the culture started to produce gluconate at a rate of 4.4 mmol·h^{-1}·(g dry wt cells)$^{-1}$ (Fig. 1). Moreover, whereas the carbon dioxide production rate virtually did not change, the specific oxygen consumption rate increased by 2.4 mmol·h^{-1}· (g dry wt cells)$^{-1}$, which was in good agreement with the gluconate production rate. When less PQQ (2 μM final concentration) was injected into the culture a lower gluconate production was measured (1 mmol·h^{-1}· (g dry wt cells)$^{-1}$).

Although these results clearly showed that one could also reconstitute GDH in growing wild-type cells of *E. coli*, the production of gluconate stopped after some time, which suggested that it might only be a transient effect. However, repeated addition of PQQ resulted again in gluconate production (results not shown). To study a possible influence of PQQ on the metabolism of *E. coli* it would be preferable to grow the organism under precisely defined conditions in chemostat cultures in the presence of PQQ. Therefore, we tried to determine the optimal conditions for reconstitution of active GDH *in situ*.

Since *K. pneumoniae* showed a maximal GDH activity at low pH values (around pH 5.5,

(Hommes et al. 1989b)) it seemed logical to grow *E. coli* at a low pH value. The affinity for PQQ of the apo-enzyme was found to be much higher at pH 5.5 when compared with the affinity determined at pH 7.0. When washed cell suspensions of cells grown at pH 6.9 were incubated at pH 5.5, maximal GDH activity was already found with PQQ concentrations of 0.1 μM. whereas at pH 7.0 10 μM of PQQ was needed (Table 1). Furthermore, when *E. coli* was grown under phosphate-limited conditions at pH 5.5 (D = 0.14 h^{-1}), a pulse of 0.2 μM PQQ was more than sufficient for maximal GDH activity (7.7 mmol·h^{-1}·(g dry wt cells)$^{-1}$; Fig. 2), which was significantly higher than the activity observed at pH 7.0. Unlike at pH 7.0, a lag of about 10 min was observed before maximal gluconate production was measured at pH 5.5. However, more important, gluconate production was found to proceed at an undiminished rate for at least two hours. Taking these results together we considered it best to examine the effect of an active GDH on glucose metabolism at pH 5.5.

The organism was grown in continuous cultures under phosphate-limited glucose-excess conditions at pH 5.5 and at different dilution rates either without PQQ or in the presence of 0.2 μM PQQ. When steady state conditions had been attained, culture fluids were quantitatively analyzed and carbon balances constructed (Table 2). No gluconate production could be observed in cultures without added PQQ, whereas minute amounts of PQQ in the growth medium resulted in a huge accumulation of gluconate (up to 80 mM of gluconate in steady state). As might be expected 0.2 μM PQQ was sufficient for almost maximal reconstitution of GDH activity *in situ*. Increasing the concentration of PQQ to 0.8 μM resulted in only marginal changes of gluconate production (results not shown), whereas in every supernatant of cultures grown in the presence of PQQ still some free PQQ could be detected. Although the apparent affinity for glucose of the glucose dehydrogenase, as measured *in vitro* with right-side out vesicles or cell-free extracts using the Wurster's Blue assay, ranged from 0.78 to 1.08 mM, the steady state glucose concentration in the chemostat cultures had to be higher than about 20 mM to give maximal GDH activity.

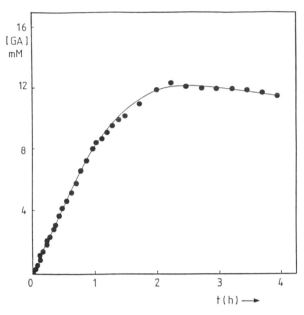

Fig. 1. The effect of a pulse of PQQ (end concentration 10 μM) to a phosphate-limited chemostat culture of *Escherichia coli* B/r, grown at pH 6.9 (D = 0.10 h^{-1}; 35 °C). GA = gluconate concentration in mmol/l.

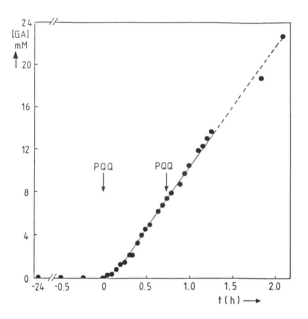

Fig. 2. The effect of a pulse of PQQ (end concentration 0.2 μM; t =0 min) to a phosphate-limited chemostat culture of *Escherichia coli* B/r, grown at pH 5.5 (D = 0.14 h^{-1}; 35 °C). At t = 45 min extra PQQ was added to an end concentration of 0.8 μM. GA = gluconate concentration in mmol/l.

At every growth rate, the addition of PQQ had a drastic effect on the metabolism of glucose by *E. coli* (Table 2). As expected no effect of PQQ could be observed on the amount of biomass that was synthesized under these conditions, because these cultures were phosphate-limited. But in the presence of PQQ the yield on glucose (g cells synthesized/mol glucose consumed) decreased signif-

Table 1. Rates of gluconate formation of washed suspensions of cells of *Escherichia coli* B/r, grown phosphate-limited at pH 6.9 (35 °C; D =0.1 h^{-1}).

PQQ μM	Gluconate production nmol·min^{-1}·mg drw^{-1}	
	pH 5.5	pH 7.0
0	0	0
0.1	360	nd
1.0	nd	160
10.0	310	320

Buffers used: 100 mM HEPES (pH 7.0) + 1 mM MgCl$_2$ or 100 mM MES (pH 5.5) + 1 mM MgCl$_2$. Before addition of glucose (40 mM) the cells were preincubated with the desired concentration of PQQ for 15 min at 35 °C. nd = not determined.

icantly due to the increased rate of glucose consumption. The alternative glucose metabolizing pathway was very active at low dilution rates: more than 60% of the glucose consumed was converted in gluconate at D = 0.15 h^{-1}.

The question arose as to whether the effect of PQQ on glucose metabolism by *E. coli* would be the same under different nutrient limitations. Hence, the organism was grown in aerobic chemostat cultures at a pH value of 5.5 under, respectively, potassium-, sulphate-, ammonia-, and carbon-limited conditions in the presence and absence of 0.2 μM PQQ. The PQQ-dependent production of gluconate was indeed not restricted to phosphate-limited chemostat cultures of *E. coli*. As is shown in Table 3 all glucose-excess cultures responded to the presence of PQQ by producing gluconate, although the specific rates of gluconate formation varied with the different limitations and specific growth rates. Naturally the presence of PQQ did not influence glucose-limited growth of *E. coli* due to the apparent high K_m for glucose of the GDH (about 1 mM as compared with an apparent K_m of 15 μM of the PTS).

Finally we examined whether *E. coli* had adjusted the GDH activity *in situ* and/or the synthesis of the apo-enzyme as a consequence of the availability of PQQ. Therefore we determined the maximal rates of gluconate production in chemostat cultures, grown without PQQ, after a cell-saturating pulse of PQQ (Table 4). When comparing these results with the steady state production rates of cultures grown in the presence of PQQ (Table 2, Table 3), it is evident that usually the activity of the direct glucose oxidative pathway in cultures grown in the absence of PQQ and subsequently pulsed with PQQ was at least equal to, or even higher than the activity of the pathway in cultures that were grown in the presence of this cofactor. An indication for the amount of apo-enzyme synthesis can be obtained by measuring the maximal GDH activity in cell-free extracts after preincubation with saturating concentrations of PQQ. Thus, from a number of cultures grown with and without PQQ samples were taken, cell-free extracts prepared and their maximal glucose dehydrogenase activities measured using Wurster's Blue as electron acceptor (Table 5). Again the results showed that the presence of PQQ in the culture medium exerted no great influence on the synthesis of GDH apo-enzyme.

Discussion

Since some micro-organisms synthesize only the inactive glucose dehydrogenase (GDH) apo-enzyme, whereas other bacteria excrete the appropriate cofactor PQQ into their environment, it has been suggested that PQQ ought to be regarded as a vitamin (for review see: Duine et al. 1986). Indeed, as has been shown now, in some bacteria an active GDH could be easily demonstrated when small concentrations of PQQ (nmolar to μmolar range) were present in the growth medium (Hommes et al. 1985; van Schie et al. 1984; van Schie et al. 1987).

From the results presented in this study it becomes clear that reconstitution of GDH-holoenzyme *in situ*, as already described for mutants of *Escherichia coli* PC 1000 (Hommes et al. 1984), was also possible in actively growing wild-type cells of *E. coli* B/r when minute amounts of PQQ were added to the medium. Under glucose-excess conditions, cultures of *E. coli* responded to the presence of PQQ by producing gluconate via the active GDH, although they were perfectly able to grow without this direct glucose oxidative pathway.

Despite reports of *E. coli* possessing holoenzyme activity (Ameyama et al. 1985) and even excreting PQQ (0.5 nmol/l; (Ameyama et al. 1984)), in our experiments gluconate production was strictly dependent upon the addition of PQQ to the medium.

Table 2. Rates of glucose utilization, oxygen utilization, and product formation expressed in a phosphate-limited chemostat culture of *Escherichia coli* B/r growing at various dilution rates in the presence and absence of 0.2 μM PQQ (pH 5.5; 35 °C).

Dilution rate (h^{-1})	0.15	0.15	0.30	0.29	0.58	0.58
PQQ (μM)	0	0.2	0	0.2	0	0.2
Glucose used	3.2	10.0	4.8	6.9	6.9	8.7
Gluconate	0	6.3	0	2.5	0	1.3
Acetate	0.1	0.5	0.3	0.9	0.5	1.0
Carbon dioxide	12.9	13.1	16.1	14.8	17.4	18.7
Carbon recovery (%)	99	97	100	104	98	99
Oxygen used	12.5	15.2	14.8	15.4	16.4	18.8
Respiratory quotient (q_{CO_2}/q_{O_2})	1.03	0.86	1.09	0.96	1.06	0.99
Gluconate/glucose ratio ($\times$ 100%)	0	63	0	36	0	15
$Y_{glucose}$	46.3	15.3	62.5	42.0	84.1	67.7
Y_O	6.0	5.0	10.1	9.4	17.7	15.4

Specific rates are expressed in mmol·h^{-1}·(g dry wt cells)$^{-1}$. $Y_{glucose}$ is expressed as g dry weight cells formed per mol glucose consumed and Y_O is expressed as g dry weight cells formed per 0.5 mol oxygen consumed.

It has been reported that the low concentrations of PQQ, detected in *E. coli* by Ameyama and co-workers (Ameyama et al. 1984; Ameyama et al. 1985), might have been obtained from the medium and/or glassware rather than from biosynthesis (Duine et al. 1987; van Schie et al. 1987).

The physiological role of GDH *apo*-enzyme synthesis in bacteria is not well understood. For a number of micro-organisms able to synthesize both the apo-enzyme and the cofactor PQQ it has been suggested that the active GDH may function as an auxiliary energy system (de Bont et al. 1984; Hommes et al. 1985; Müller & Babel 1986). Moreover, it has been shown that membrane vesicles of *E. coli* were able to energize the uptake of amino

acids in the presence of PQQ, using glucose as energy donor (van Schie et al. 1985). Therefore it is reasonable to assume that the physiological role of the glucose dehydrogenase in *E. coli*, when PQQ is present in its growth environment, could also be that of an additional energy-generating system.

At pH 5.5, in the presence of PQQ, the cultures of *E. coli* showed almost no overflow metabolism, only small amounts of acetate, pyruvate (Table 2; Table 3), and sometimes 2-oxoglutarate (Table 3, ammonia limitation) could be observed. These results do not differ much from those obtained with cultures of *E. coli*, grown at pH 6.8, where a similar pattern of overflow metabolism was observed (Leegwater 1983).

Table 3. Rates of glucose utilization, oxygen utilization, and production formation expressed in variously-limited chemostat cultures of *Escherichia coli* B/r growing at various dilution rates in the presence and absence of 0.2 μM PQQ (pH 5.5; 35 °C).

D	PQQ	q_{Glc}	q_{GA}	q_{Og}	q_{Pyr}	q_{Ac}	q_{CO_2}	C-rec	q_{O_2}	RQ	GA/Glc	Y_{Glc}	Y_O
Potassium limitation													
0.16	0	4.6	0	0	0.6	0.9	16.3	95	15.4	1.06	0	34.8	5.5
0.19	0.2	6.0	1.1	0	0.3	1.0	16.2	92	16.2	1.00	18	31.7	5.9
0.30	0	6.3	0	0	0.7	1.0	19.6	94	17.9	1.09	0	47.6	8.7
0.29	0.2	8.9	2.6	0	0.3	1.6	18.0	93	18.5	0.97	29	32.6	7.8
0.60	0	9.4	0	0	0.2	2.0	22.4	90	19.6	1.14	0	63.5	15.6
0.61	0.2	12.8	3.3	0	0.1	2.0	20.7	90	22.3	0.93	26	47.7	13.7
Sulphate limitation													
0.15	0	2.7	0	0	0.5	0.7	7.7	101	8.1	0.95	0	54.9	9.3
0.16	0.2	7.3	4.4	0	0.7	0.6	7.9	100	10.5	0.75	60	21.9	7.6
0.30	0	4.9	0	0	0.1	0.6	12.8	90	12.1	1.06	0	61.7	12.4
0.30	0.2	12.2	7.6	0	0.1	0.5	10.5	95	14.1	0.74	62	25.4	10.9
0.58	0	7.3	0	0	0.1	0.7	17.3	97	16.4	1.05	0	80.6	17.9
0.58	0.2	10.6	3.3	0	0.1	0.4	13.5	92	14.4	0.94	31	53.4	19.5
Ammonia limitation													
0.14	0	2.9	0	0.5	0	0.1	8.8	98	9.3	0.95	0	48.3	7.5
0.15	0.2	3.6	0.3	0.8	0	0.2	9.6	101	10.7	0.90	8	41.7	7.0
0.48	0	6.2	0	0	0	0.5	14.5	97	14.5	1.00	0	77.4	16.6
0.49	0.2	7.1	0.3	0	0	0.7	16.5	92	16.3	1.01	4	69.0	15.0
Glucose limitation													
0.15	0	2.0	0	0	0	0	5.2	96	5.2	1.00	0	75.8	14.4
0.15	0.2	1.9	0	0	0	0	5.6	102	5.6	1.00	0	81.2	13.9

Abbreviations used: Glc = glucose; GA = gluconic acid; Og = 2-oxo-glutarate; Pyr = pyruvate; Ac = acetate; C-rec = carbon recovery; RQ = respiratory quotient = q_{CO_2}/q_{O_2}. The dilution rate is given in h^{-1}, and the PQQ-concentration in μM. Specific rates are expressed in mmol·h^{-1}·(g dry wt cells)$^{-1}$. Carbon recovery is given as percentage of the glucose consumed that is retrieved in carbon products and the ratio GA/Glc is given as the percentage of the glucose consumed that is retrieved in gluconic acid. Y_{Glc} is expressed as g dry weight cells formed per mol glucose consumed and Y_O is expressed as g dry weight cells formed per 0.5 mol oxygen consumed.

When comparing cultures of *E. coli* growing with and without PQQ (Table 2; Table 3), it becomes clear that an active GDH also caused the respiratory quotient (rate of carbon dioxide production/rate of oxygen consumption) to decrease. This is clearly what one would expect when some of the glucose is directly oxidized to gluconate. The increased oxygen consumption rate that was usually observed after the addition of PQQ, indicated that the respiratory chain was not working at its maximal velocity before PQQ addition, and that it must have possessed some overcapacity to cope with the extra electron flow caused by the oxidation of glucose to gluconate. Experiments of Matsushita et al. (1987) with reconstituted proteoliposomes containing respiratory chain components and GDH from *E. coli* indicated that the electrons were transported via ubiquinone 8 and cytochrome *o*. Thus, biologically available energy might have been obtained from the conversion of glucose into gluconate via the GDH. Sometimes, active GDH resulted also in the lowering of the CO_2 production by the culture (Table 2; Table 3). This might be indeed an indication for energy being produced by means of the GDH, thereby causing the tricarboxylic acid cycle to operate at a reduced rate. Although it is reasonable to assume that the formation of a cell will cost as much ATP when growing in the presence as in the absence of PQQ, the rates of ATP synthesis during growth under these conditions may vary widely. Therefore it is impossible to calculate from these data of glucose-excess cultures, how many ATP equivalents were obtained per mol of gluconate formed.

The presence of PQQ hardly affected the production of the other overflow metabolites. This is in good agreement with the observation that inclusion of PQQ in the growth medium had no effect on the production of extracellular polysaccharide in ammonia-limited cultures of *Agrobacterium radiobacter*, an organism that also possesses only the apo-enzyme of glucose dehydrogenase (Linton et al. 1987).

It is interesting to note that the type of overflow metabolism exhibited by cultures of *E. coli* growing in the presence of PQQ resembled the overflow metabolism of *K. pneumoniae* (Neijssel & Tempest 1975a; Neijssel et al. 1983). Thus, under potassium- or phosphate-limited growth conditions gluconate was now a major overflow product and,

Table 4. The maximal gluconate production rates in variously-limited chemostat cultures *Escherichia coli* B/r growing at various dilution rates (pH 5.5; 35 °C) after a cell saturating pulse of PQQ (0.2 μM end concentration).

Limitation	Dilution rate h^{-1}	q_{max} gluconate mmol/g.h
Phosphate	0.15	7.6
	0.30	5.8
	0.57	4.1
Sulphate	0.15	4.3
	0.30	3.9
	0.58	2.6
Potassium	0.17	2.8
	0.30	5.3
Glucose[a]	0.15	1.7
Glucose + PQQ[b]	0.15	1.5

[a] Simultaneously with PQQ glucose was injected into the glucose-limited culture to an end concentration of 20 mM. [b] Since this glucose-limited culture, although grown in the presence of 0.2 μM PQQ, could not produce gluconate (due to the apparent high K_m for glucose of the glucose dehydrogenase), maximal gluconate production under these conditions was also determined by a simultaneous pulse of glucose (20 mM end concentration) and PQQ (end concentration 0.4 μM).

Table 5. Glucose dehydrogenase activities in cell-free extracts of *Escherichia coli* B/r, grown in chemostat culture (pH 5.5; 35 °C; with or without 0.2 μM PQQ), after preincubation with 100 μM PQQ (15 min at 35 °C).

Limitation	Dilution rate h^{-1}	Glucose dehydrogenase nmol WB reduced·min^{-1}· mg protein^{-1}
Glucose	0.15	15
Glucose + PQQ	0.15	12
Phosphate	0.15	350
Phosphate + PQQ	0.15	290
Sulphate	0.15	310
Sulphate + PQQ	0.15	270
Sulphate	0.30	190
Sulphate + PQQ	0.30	250
Sulphate	0.58	90
Sulphate + PQQ	0.58	200